Fundamentals of fluid power

Fundamentals of fluid power

William D. Wolansky
Iowa State University of Science and Technology

John Nagohosian
Henry Ford Community College

Russell W. Henke

WAVELAND PRESS, INC.
Prospect Heights, Illinois

For information about this book, write or call:
    Waveland Press, Inc.
    P.O. Box 400
    Prospect Heights, Illinois 60070
    (708) 634-0081

Copyright © 1977 by Houghton Mifflin Company. Reprinted by arrangement with Houghton Mifflin Company, Boston. 1986 reissued by Waveland Press, Inc.

ISBN 0-88133-194-5

All rights reserved. No part of this book may be reproduced, stored in a retrieval system, or transmitted in any form or by any means without permission in writing from the publisher.

Printed in the United States of America

9 8 7 6 5 4

Contents

Preface xii

Safety practices and precautions xiv

Chapter 1 Fluid power technology 1
- **Historical development of fluid power** 4
- **Advantages of fluid power systems** 7
- **Comparison of three power-transmission methods** 11
  - Electrical 11
  - Mechanical 12
  - Fluid 12
- **How a fluid power system works** 15
  - Pascal's law 16
  - Components of a hydraulic system 18
- **Fluid power applications** 19
- **Summary** 20
- **Review questions and problems** 21
- **Selected readings** 21

Chapter 2 Fundamental laws of fluid mechanics 23
- **Physical properties of fluids** 24
  - Density 25
  - Temperature 26
  - Pressure 28
- **Compression and expansion of fluids** 29
  - Gas laws 30
  - Effect of heat on fluids 33
- **Fluids in motion** 33
- **Conservation of energy** 36
  - Bernoulli's theorem 36
  - Torricelli's theorem 39
  - Reynolds number 40
- **Force-pressure-area relationships** 43
- **Energy-work-power relationships** 45
- **Applications of fluid power theory** 50
- **Summary** 54

Review questions and problems 55
Selected readings 56

Chapter 3 Fluids and auxiliaries 57

Functions of fluids within a system 58
Properties of fluids 58
Viscosity 59
Oxidation stability 62
Foaming 63
Pour point 64
Antiwear 64
Water separation 64
Types of hydraulic oils and fluids 65
Fire-resistant 66
Fluids of the future 68
Fluid conditioning 71
Filtration 73
Lubrication and heat transfer 77
Heat exchangers 78
Fluid reservoirs 79
Sizing a reservoir 80
Accessories 81
Fluid conductors 83
Types of conductors 83
Selection of conductors 83
Sizing pipes 84
Semirigid conductors 85
Flexible hose 88
Fittings 90
Threaded 90
Tube 91
Flared 91
Flareless 92
Welded 92
Quick-disconnect couplings 92
Future trends in conductor development 93
Summary 93
Review questions and problems 94
Selected readings 95

Chapter 4 Energy input devices 97

Pump performance and rating 98
Nonpositive-displacement pumps 100
Positive-displacement pumps 101

Variable- and fixed-delivery pumps 101
Volumetric output and pressure capabilities 101
Pump slippage 102
Pump ratings for continuous and intermittent use 103
Classification of pumps by design 104
Centrifugal pumps 104
Reciprocating pumps 105
Rotary pumps 107
Gear 107
Vane 112
Piston 117
Air compressors 122
Rotary 123
Reciprocating 125
Efficiency of energy input devices 127
Volumetric 127
Mechanical 128
Overall 130
Compressor performance 130
Comparing efficiency 131
Determining capacity 132
Maintenance 137
Summary 138
Review questions and problems 138
Selected readings 139

Chapter 5 Energy modulation devices 141

Pressure-control valves 143
Simple-relief 143
Compound-relief 145
Unloading 147
Sequence 148
Counterbalance 149
Pressure-reducing 150
Directional-control valves 151
Internal control elements 152
Ways 154
Position 157
Center 158
Actuation 158
Spring condition 160
Servo 162
Feedback loop 162

Flow-control valves 164
Noncompensated 165
Compensated 166
Summary 168
Review questions and problems 169
Selected readings 170

Chapter 6 Energy output devices 171

Principles of actuator operation 172
Actuating cylinder 175
Velocity 177
Construction 181
Body style 185
Mounting configurations 186
Air cylinders 190
Construction 191
Selection 192
Fluid motors 193
Operation 193
Fixed- and variable-displacement motors 194
Speed 195
Torque 196
Gear motors 199
Vane motors 200
Piston motors 202
Air motors 204
Vane motors 204
Piston motors 206
Speed 207
Limited-rotation motors 209
Actuator designs 209
Construction 211
Summary 211
Review questions and problems 212
Selected readings 213

Chapter 7 Fluid power circuits and systems 214

Open- and closed-loop circuits 215
Open-loop circuit classification 216
Function 216
Control 217
System 217
Open-loop circuit design 218
Criteria for selecting circuits 218

Circuit load requirements 218
Cycle time requirements 220
Pump pressure requirements 222
Actuator sizing 224
Output speed control 224
Closed-center circuits 225
Comparison of open-center and closed-center circuits 230
Circuits for effecting pressure control 231
Maximum pressure-limiting 231
Unloading 232
High-low 233
Controlling cylinder pressure 234
Accumulator circuits 236
Flow-control circuits 238
Meter-in 238
Meter-out 239
Bleed-off 240
Regenerative 240
Intermittent-feed control 241
Deceleration control 242
Flow divider 243
Sequencing 243
Synchronizing actuators and motors 244
Fail-safe circuits 248
Pneumatic power circuits 249
Effect of fluid characteristics on motor performance 249
Examples of pneumatic power circuits 252
Summary 258
Review questions and problems 259
Selected readings 263

Chapter 8 Instrumentation 264

Instrument output 266
Accuracy 270
Measurement standards 271
Instrument classification 272
Signal-generating instruments 272
Measuring pressure 273
Mechanical 274
Electric 275
Measuring force 276
Measuring torque 278
Measuring displacement 282
Electric signal output 282

Electric instruments for measuring displacement 284
Potentiometric 284
Variable transformers 284
Digital transducers 285
Optical transducers 285
Synchros 286
Mechanical instruments for measuring displacement 287
Measuring velocity 288
Measuring acceleration 289
Measuring viscosity 290
Measuring flow of fluids 290
Volume 290
Mass 293
Summary 295
Review questions and problems 296
Selected readings 297

Chapter 9 Fluidics 298

Historical development of fluidics 301
Advantages of fluidics 302
Disadvantages of fluidics 303
Basic fluidic principles 303
Coanda effect 304
Entrainment 305
Jet interaction 306
Momentum exchange 307
Pressure differential 308
Monostable jet-interaction amplifier 309
Construction of fluidic devices 309
Active devices 314
Flip-Flop 314
OR-NOR 315
AND-NAND 316
Schmitt trigger 317
Binary counter 318
Memory logic 320
Fluid amplifiers 321
Vortex 322
Turbulence 323
Focused-jet 324
Diaphragm 325
Impact modulator 326
Induction 327
Double-leg elbow 327

Passive circuit elements 328
Capacitors 328
Fluid resistors 329
Fluidic diode 330
Fluid oscillators 330
Digital logic functions 331
Addition with binary numbers 334
Symbolic logic notation 335
Applications of fluidic devices 336
Basic fluidic circuit applications 336
Moving-parts fluid controls 340
Pneumatic moving-parts systems 341
Summary 343
Review questions and problems 344
Selected readings 344

Appendix A Conversion of English gravitational unit system to international 347

Supplementary conversion units 347
Prefixes for decimal multiples and submultiples 349

Appendix B The nature of fluids 350

Appendix C Important fluid power facts and formulas 351

Theory 351
Fluids 351
Area 352
Volume 352
Cylinders 352
Conductors 353
Power 353
Motors 354

Appendix D ANSI graphic symbols for fluid power diagramming 355

Appendix E Symbols and abbreviations used in this text 385

Index 390

Preface

Fundamentals of Fluid Power is an introductory textbook designed to provide the technical information needed as a foundation for work with fluid power components and systems. The text will help students understand the "why" and the "how" of various operating principles of fluid power. It covers those subjects considered essential to understanding the supporting theory: design, function, and operation of components and fabrication and analysis of circuits.

This book is prepared for industrial education students and apprentices, engineering technicians, and industrial technicians. Scientific principles and mathematical problems are explained in such a way that students with high school algebra and physics can solve the problems in the examples and review exercises.

A systems approach facilitates understanding important concepts and relationships within a dynamic operational system. Chapter 1 helps students understand why fluid power systems are used to transfer and control energy. The potential of fluid power in diverse applications is thoroughly discussed and illustrated. Basic theory related to fluids, covered in Chapter 2, enables students to explore and understand the applications of theory to practical problems encountered in industry and elsewhere. Chapter 3 contains technical and environmental information on fluid selection and contamination control.

The next three chapters, 4 through 6, discuss hardware from the systems approach of energy input devices, energy modulation devices, and energy output devices; these chapters also provide for close examination of a functioning system. Chapter 7 introduces problems in practical circuit applications. The topic "instrumentation" in Chapter 8 is included because of its importance in controlling physical variables with precision in today's fluid power systems. The last chapter introduces the student to fluidics, a rapidly emerging technology within fluid power systems.

We would like to take this opportunity to express our appreciation for the efforts of all those who helped to bring this project to completion. We are indebted to many organizations and fluid power component manufacturers who submitted photographs and other illustrative materials for inclusion in this text.

We also extend our appreciation to our former students to whom much credit belongs for their evaluation and suggestions during the early stages of the preparation of this manuscript. Steve Madsen, an Industrial Education student at Iowa State University, did the initial artwork for the illustrations and prepared many captions as perceived by a student.

Special acknowledgment is due the Houghton Mifflin staff and the invaluable suggestions by their reviewers—John U. Gonzales, Milwaukee Area Technical College; Alfred Penko, Cuyahoga Community College, Ohio; John Bellman, Erie Community College, New York; and Victor Bridges, Umpqua Community College, Oregon—which greatly added to many improvements over the original draft. The typing of the manuscript by Claire Wolansky and Julie Brinkmeyer, who were students of a new vocabulary and emerging technology, is much appreciated. To all others who helped in any way, we express our appreciation.

Safety practices and precautions

Safety practices in an educational fluid power laboratory and field service have become essential to effective learning and efficient service. The introduction of higher-pressure systems and of expensive test and instrumentation equipment has necessitated greater emphasis on intelligent and safe use of test equipment when servicing fluid power equipment. Fluid power laboratory learning systems, or trainers, have been designed to minimize potential hazards to learners. Pressure-relief valves at pump and reservoir units should be adjusted only with the instructor's immediate supervision. Hoses and all other components should not exceed the working pressure limits to provide an ample safety margin for operational circuits.

A list of general safety practices follows to help students become aware of possible hazards and how to avoid or cope with them:

- Wear industrially approved eye-protection devices.
- Wear proper clothing to avoid the possibility of getting entangled in rotating or moving parts.
- Remember that hydraulic oil can be under extremely high pressure. An escaping stream of it can easily penetrate the skin and cause blood poisoning. All lines and connections must be in good condition and properly fastened.
- Understand the function and operation of the control valves.
- Know how to shut down the laboratory learning unit and to prepare it for safe inspection and maintenance.
- When disassembling a fluid power component, be careful not to unload a spring force that may cause parts to fly.
- Always discharge an accumulator before working on a circuit or the accumulator.
- Be sure to relieve all hydraulic pressure before disconnecting any lines or pipes in a fluid power system.
- Do not service components on the hydraulic panel while the motor is running.
- Fluid power components should be kept clean, especially when being repaired or overhauled.
- Do not overtorque housing bolts or other fasteners if critical adjustments may affect the internal moving elements.

- Be certain to secure all hose connections after repairs or any changes are made.
- Report to the instructor any change in operating characteristics such as erratic gage pressures, unusual sounds, and leakage from components.
- Maintain the proper level of fluid in the reservoir and keep all working areas safe and clean. Oily floors can be very hazardous.
- Report any occurrence or observation that might improve preventative maintenance practices.
- Observe all precautions when starting up the fluid power laboratory learning system for the first time as well as after any repair or change in circuit hookup.
- Keep all guards and shields in place.

Chapter 1
Fluid power technology

In this chapter, the development of fluid power technology is viewed as our increasing ability to convert, transmit with control, and apply fluid energy to perform useful work. An understanding of the fundamental laws and principles of fluid mechanics has enabled engineers to design components and systems of increasing complexity and compatibility with other mechanical and electrical systems. Diversified applications of fluid power systems are illustrated to provide the reader with an idea of the expanding role of fluid power in many industries.

Key terms

- **Energy** The capacity of doing work and of overcoming inertia, as by heat, mechanical, pressurized-fluid, or chemical forces. The two basic energy states are *potential* and *kinetic*. Potential energy is that due to the position of one body relative to another. It is regarded as stored energy such as that in a compressed spring or pulled bow. Kinetic energy is manifested by bodies in motion such as a wind-driven boat.
- **Fluid** A substance that has definite mass and volume, at a constant temperature and pressure, but no definite shape. There are two classes of fluids—liquids and gases.
- **Fluid power** The transmission and control of energy by means of a pressurized fluid.
- **Hydraulics** Engineering science pertaining to liquid pressure and flow.
- **Pneumatics** Engineering science pertaining to gaseous pressure and flow.
- **Power** The time rate at which energy is transferred or converted into work.
- **Pressure** Force per unit area. The distributed reaction (pressure) on a confined fluid is measured in pounds per square inch (psi).
- **Work** The product of the applied force times the distance through which it moves.

The significance of any power technology will be seen clearly if it is examined within the framework of a continuing search to relieve people of manual labor and enable them to gain greater control of their environment.

The human capacity to think and invent has set people apart from the rest of the animal kingdom. In prehistoric times people created stone tools to search for food and protect themselves. The bow and arrow is an example of human inventiveness in harnessing muscle power to pull a bowstring, which in turn with its pent-up energy (stored or potential) mechanically multiplied the effectiveness of a person's muscles enough to fling the arrow at the target at immense speed. Using muscle power derived from the conversion of food into energy and applying it to tools and machines extended and multiplied human manual capabilities.

Civilized people have continually sought substitutes for muscle power (Figure 1.1). Indeed, the history of civilization might be seen as a reflection of the search for ways to exploit different and greater sources of energy. People domesticated animals and invented machines to harness and control energy from the sun, water, wind, fossil fuel, and the atom. Sails were made to capture the potential

Figure 1.1 Technological progress involves the use of hydraulically equipped machines. (a) Fork lift. (b) Heavy road-construction equipment. (Courtesy Caterpillar Tractor Co.)

(a)   (b)

energy of the wind, the water wheel to capture the fluid energy of moving water (Figure 1.2), the engine and turbine to harness the heat energy of fossil fuel, electric generators to harness the tidal energy of the sea, fuel cells to harness the chemical energy of hydrogen and oxygen, and the cyclotron to extract atomic energy from atomic fission. Today research is devoted to developing exotic ways of extracting energy from almost inexhaustible sources such as geo-

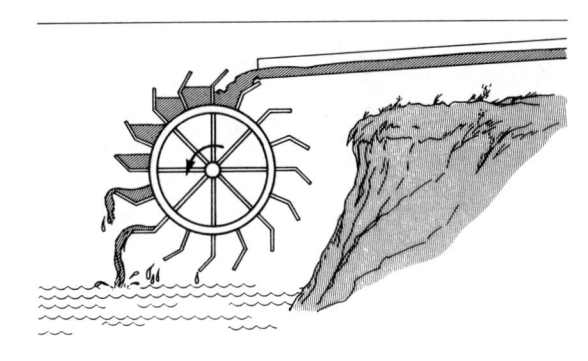

Figure 1.2 The water wheel changes the kinetic energy of moving water into mechanical energy.

thermal energy from the earth, tidal energy from the sea, nuclear fusion, and even cosmic rays from other galaxies (Figure 1.3).

Figure 1.3 The ever-expanding array of power sources.

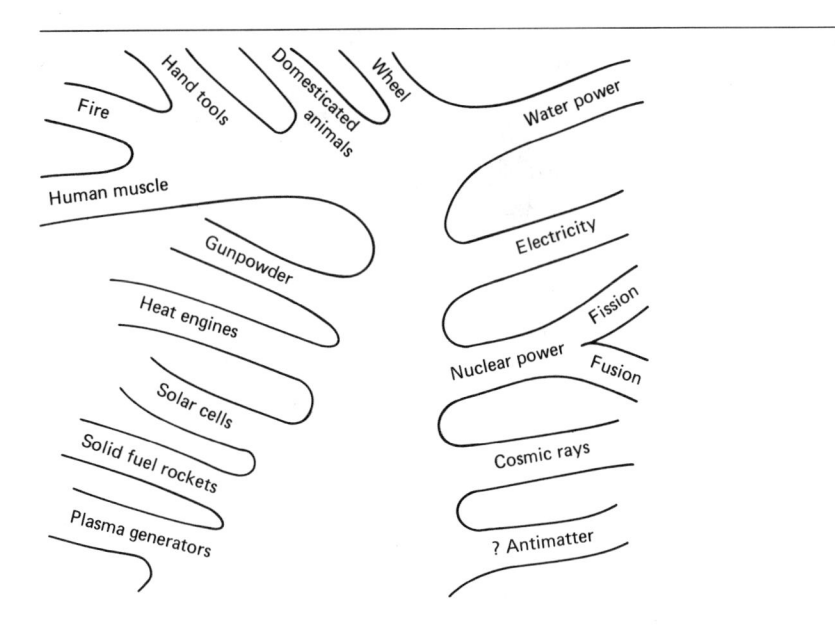

In the area of scientific and technological knowledge and skills, people are even more concerned with acquiring large, more diverse sources of energy. The increase in world population, the quantity of energy consumed per capita, and the large amounts of energy used for single purposes such as launching a rocket—all make this concern a relevant one. Technologists and scientists are also concerned with converting energy to useful forms, transmitting it to where it is needed, storing it, and controlling it to perform the ever-greater number of services required (Figure 1.4).

Figure 1.4 This equatorially mounted 140-ft radio telescope is controlled by hydraulic systems, electrical equipment, and instrumentation housed in a concrete base. (Courtesy National Radio Astronomy Observatory, Green Bank, West Virginia)

Even though we have applied our ingenuity to search for and create many ways to acquire new sources of energy, we have succeeded in developing only three methods of transmitting power that are of any major importance. These are mechanical, electrical, and fluid power.

Historical development of fluid power

The history of all these developments sets a background for the importance of studying fluid power, an emerging and rapidly expanding technology (Figure 1.5).

Fluid power is the technology that deals with the transmission of energy by means of a pressurized fluid. It includes *hydraulics,* which is the engineering science pertaining to liquid pressure and flow, and *pneumatics,* which is the engineering science pertaining to gaseous pressure and flow. There is no record of the first use of wind to do work in moving sailing ships or of moving water to turn water wheels (Figure 1.6), but millenniums ago, probably five thousand years,

5 Historical development of fluid power

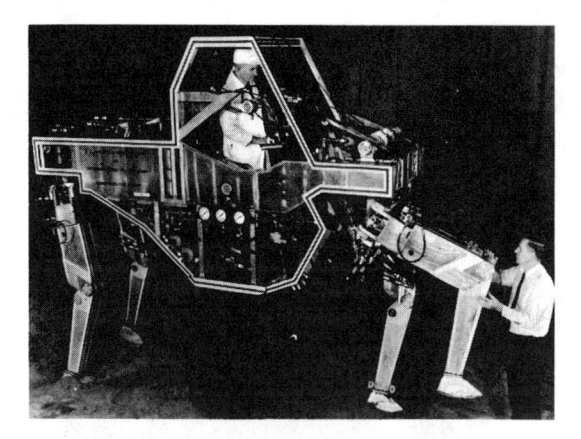

Figure 1.5 Hydraulic, electronic, and mechanical devices are used in the walking machine designed for military use. (Courtesy General Electric Research and Development Center)

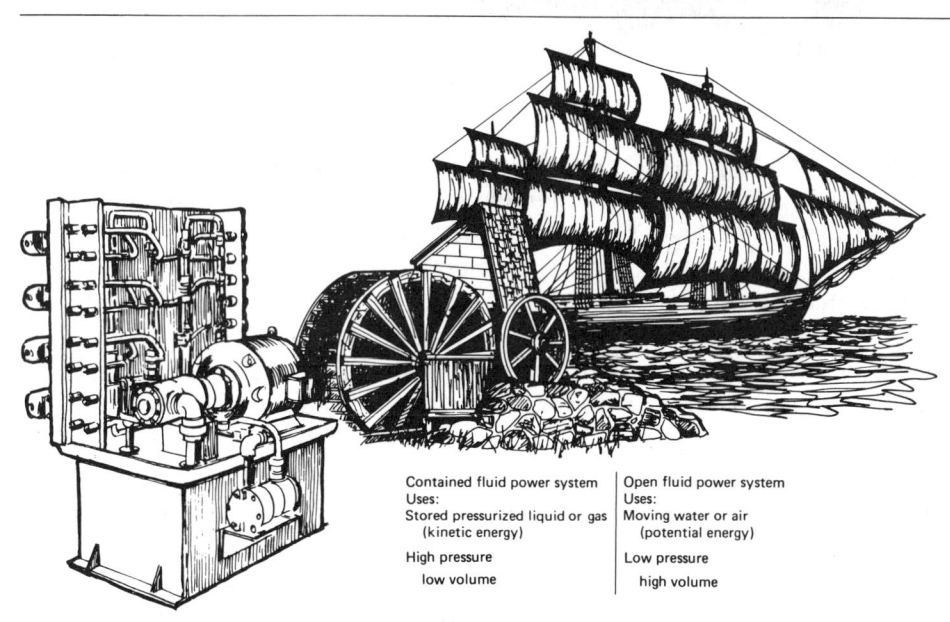

Figure 1.6 Forerunners of fluid power technology applications.

Contained fluid power system
Uses:
Stored pressurized liquid or gas (kinetic energy)
High pressure
 low volume

Open fluid power system
Uses:
Moving water or air (potential energy)
Low pressure
 high volume

China and Egypt made use of these two labor-saving devices. Early Chinese records show the use of wood valves to control water through bamboo pipes in about 4000 B.C. Ancient pharaohs built a masonry dam across the Nile, 14 miles south of present-day Cairo, for control of irrigation water by canals, sluices, brick conduits, and ceramic pipes. The Roman Empire also had extensive water systems using aqueducts, reservoirs, and valves to carry water to their cities. The use of fluids to perform useful work is nearly as old as civilization itself.

Historically, the use of water or air to produce power depended on the movement of vast quantities of fluid at relatively low pressures.

Also, people relied on nature to supply the pressure, as in the case of wind or waterfalls. But fluid power as it is defined and understood today is made possible through the use of pumps or compressors that create high pressures within a confined system and use relatively small quantities of fluids (Figure 1.7).

Figure 1.7 Industrial hydraulic unit provides a compact, high-capacity power system. (Courtesy Vickers Division, Sperry Rand Corporation)

In spite of the early uses of water and air for doing work, the principles of fluid flow were not understood until comparatively recently. In 1650 Blaise Pascal, a French scientist, discovered the fundamental law of physics on which fluid systems are based. Simply stated, this law says that *pressure of a confined fluid at rest is transmitted equally in all directions.* A century later Daniel Bernoulli developed the law concerning the conservation of energy in a flowing fluid. Also simply stated, this law says that *the mass flow rate of fluid past all cross sections of a tube is equal.* As was so often the case with other technologies, another century was to pass before these laws were applied and mechanical devices were developed to utilize the scientific principles for the benefit of humanity.

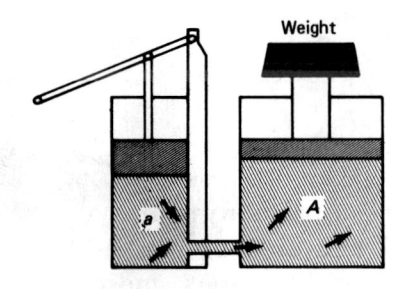

Figure 1.8 Bramah's hydraulic press (1795) had two pistons of different sizes. Force applied to Piston *a* was transmitted through pressurized fluid to Piston *A*.

In 1795 Joseph Bramah, an Englishman, built the first hydraulic press (Figure 1.8). It was not until 1850 that the demands of the Industrial Revolution in Great Britain led to the fuller development of Bramah's water press and other industrial machines. Steam-driven water pumps were used to power the presses. The brief development of fluid power technology during the Industrial Revolution soon subsided as the development of heat engines and mechanical controls and the emergence of electric power in the latter part of the nineteenth century diverted attention from the necessary developments and refinements of fluid power devices and systems. By the turn of the century Westinghouse was developing the air brake that was later used on trains, buses, and trucks. In 1906 a hydraulic system replaced the electric system for elevating and controlling guns on battleships. The first installation was made on the U.S.S. *Virginia*. By then, petroleum oil had been discovered and was replacing water as a medium for transmitting energy; precision-machined parts became available; and the troublesome seal problems were partially solved. These factors all contributed to the greater acceptance and reliability of fluid power systems, but even so it was not until after World War I that the potential for applications of fluid power was again recognized.

By 1926 the United States government developed the self-contained packaged system for a vehicle, with a reservoir, pump, controls, and actuator (Figure 1.9). Since this time and particularly near the middle of the twentieth century, the acceptance and utilization of fluid power has been so remarkable that one can scarcely visit a manufacturing or service industry, drive a car, observe a rocket being launched, or visit a dentist without coming in contact with a fluid power application.

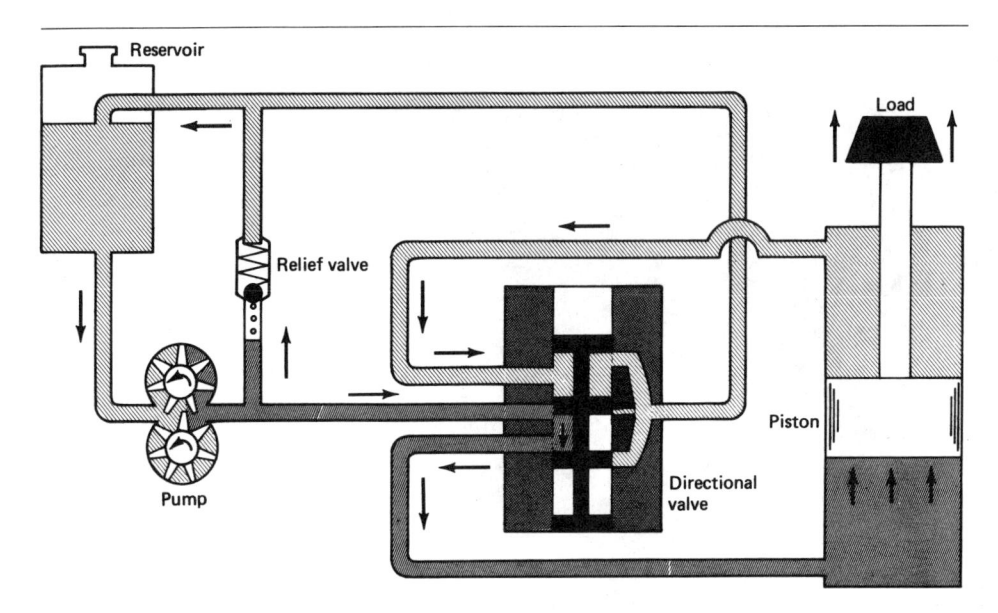

Figure 1.9 Components of a hydraulic system.

Advantages of fluid power systems

The design engineer has the option of choosing a mechanical, electric, electronic, pneumatic, or hydraulic method of power control and transmission. In many applications combinations of methods are used to produce efficient, economical, and safe operation of the equipment being designed. With the exploding technological advances, many more sophisticated tools and machines are produced, which are capable of accurate control, reliable and consistent performance, and increased safety. Whether examining the controls and means of transmitting energy in a rocket's stabilization system, a dentist's pneumatic drill, or a large Caterpillar tractor bulldozing a roadbed for a new expressway, it becomes evident that industry is designing systems that provide greater control and transmission of energy with speeds, quantity, and precision that were impossible only a decade ago.

The rapid acceptance and utilization of fluid power have been due to a number of favorable characteristics. For example, in the mechanical extraction of an impacted wisdom tooth, the mallet and chisel have given way to the air forceps capable of rocking a tooth 1000 times per minute, thus enabling a dentist to lift it gently instead of yanking it out as was done with the former tools. The crawler-transporter used to move the *Saturn V*, its mobile launcher, and its mobile service tower are a complex of hydraulic and electrical power-control systems (Figure 1.10). This vehicle uses hydraulic jacks to raise its

Figure 1.10 (a) Tracked transporter is an example of power equipment needed in the space program. Stability and steering is maintained by hydraulic power. (Courtesy NASA) (b) A tracking and communications antenna system is used for data acquisition, communication, and space position finding. Hydraulic power is used to position the antenna. (Courtesy NASA)

load of $11\frac{1}{2}$ million pounds and keep it level at all times, while it transports the load $3\frac{1}{2}$ miles to the launch pad. The mobile launcher and the service tower are leveled and stabilized when parked by use

9 Advantages of fluid power systems

of hydraulic jacks. Hydraulic applications are particularly suitable for this purpose—to control accurately immense forces (Figure 1.11).

Figure 1.11 (a) Many tons of rock are loaded hydraulically by road-building equipment. (Courtesy Caterpillar Tractor Co.) (b) This tunnel-boring machine makes use of hydraulics. (Courtesy Calweld)

(a)

(b)

Robert M. Hall's invention, the Craniotome, which is a surgical tool powered by an air turbine, has the capability of slowing down from 100,000 rpm to a dead stop in a fraction of a second and is another example of the flexibility of fluid power systems (Figure 1.12). Jumbo jets and supersonic transports are equipped with three or four independent systems for comfort, safety, and flight-control functions (Figure 1.13). Forces on these aircraft are too large to be controlled mechanically with the necessary precision (Figure 1.14).

Figure 1.12 The Craniotome is a surgical tool that uses an air turbine to power the cutter at speeds of 100,000 rpm. The Craniotome can stop in a fraction of a second.

Figure 1.13 The first Boeing 747 to fly utilized four independent 3000-psi primary hydraulic systems. Each system is associated with a different main engine. (Courtesy Boeing)

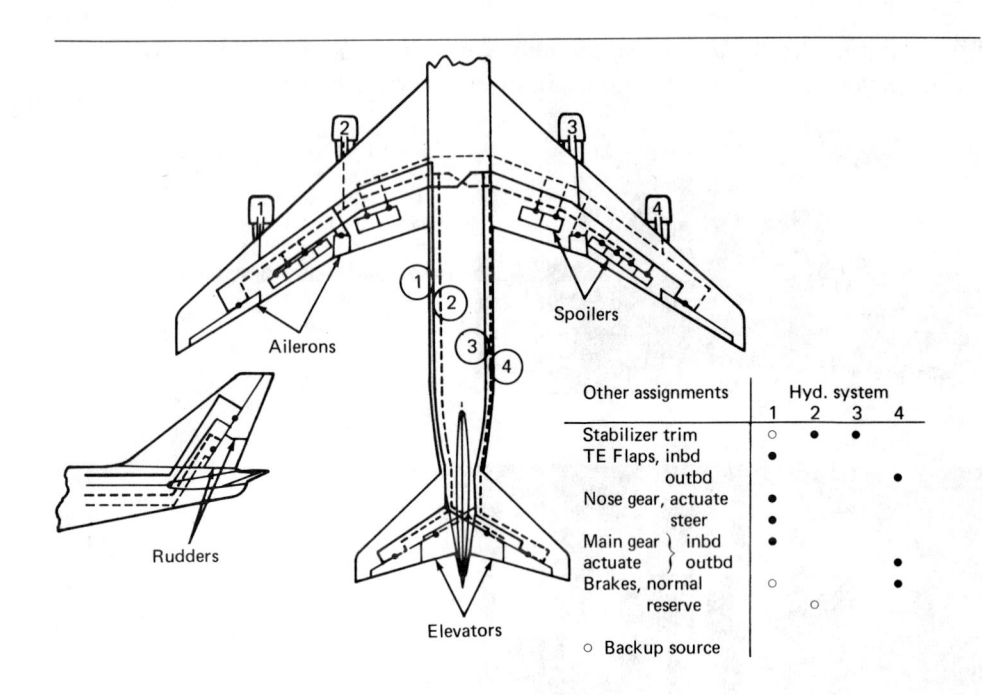

Figure 1.14 Hydraulic systems are assigned to various functions of a supersonic transport.

Other assignments	Hyd. system			
	1	2	3	4
Stabilizer trim	○	●	●	
TE Flaps, inbd	●			
outbd				●
Nose gear, actuate	●			
steer	●			
Main gear inbd actuate	●			
outbd				●
Brakes, normal	○			●
reserve			○	

○ Backup source

There are other numerous and exotic uses of fluid power in aerospace, oceanographic explorations, military devices, and the automotive industry that could illustrate the flexibility, safety, and economy of fluid power applications (Figures 1.15 and 1.16). Within limited space, it is possible to list these advantages only in general terms.

Figure 1.15 The *F. V. Hunt* is a 185-ft research vessel equipped with hydraulic pumps, motors, and controls. (Courtesy Denison Division, Abex Corporation)

- **Simplicity of design** Most components and systems in common use are simple in construction. These systems have relatively few moving parts as compared with mechanical systems and thereby fewer gears, clutches, cams, levers, etc.

- **Multiplication of forces** Fluid power systems are an efficient method of multiplying forces almost infinitely with few moving parts.

- **Ease and accuracy of control** Large forces can be controlled by much smaller ones. Fluid power systems provide constant torque at infinitely variable speeds in either direction with smooth reversals.

- **Flexibility of location** Components and systems can be located with considerable flexibility. Pipes and hoses in place of mechanical elements virtually eliminate location problems.

- **Efficiency and economy** Fluid power systems have a high efficiency with minimum friction loss, keeping the cost of power transmission to a minimum. Their compatibility with other controls such as electrical, electronic, and mechanical facilitates economic efficiency. The system's high reliability also increases its production returns and reduces maintenance costs.

- **Safety** Overload protection through use of automatic valves guards the system against breakdown from overloading. There are fewer hazards as a result of fewer exposed moving parts.

Figure 1.16 This 105-ton-capacity rear dump truck uses hydraulic cylinders. (Courtesy Euclid)

Comparison of three power-transmission methods

There are three primary methods of transmitting the power of the prime mover to a machine designed to do work, namely, (1) electric, (2) mechanical, and (3) fluid. Fluid systems use both liquid and gas media. To design successful equipment, an engineer must have sufficient knowledge about the advantages, disadvantages, and limitations of the available systems. Such knowledge enables the engineer to decide on the most suitable one or to choose some combination of systems that best fulfills the specified requirements of the equipment.

Electric

An electric control or transmission system is the most effective means of operating over long distances and is efficient when low power levels are involved. (The electric control system in a car is an example of low power level.) The large number of commercially compatible components available for electric controls has contributed to the use of electric controls where other systems such as air might be superior. Instruments for troubleshooting electric components and circuits have also contributed to the extensive use of electric systems.

Electric systems do have limitations. Contact points in relays may arc or corrode to the extent that a system malfunctions. The response time of electromechanical solenoids in some control systems is too sluggish for today's control needs. Dirt, grease, combustible vapors, and other atmospheric contamination particles create undesirable environments for electric controls. Electric arcing is a fire hazard in explosive environments, and the risk of electric shock is ever present in high-voltage transmission. Electric systems need to provide for heat dissipation as do most systems that generate heat by pressure or friction.

Mechanical

Mechanical systems are most adaptable to applications that require positive action in situations where motion is transmitted and the load is exerted over short distances. The cam on a camshaft of a car transmits the force to the valve positively through a pushrod.

Mechanical systems made up of cams and cam followers, gears, pulleys, belts, sprockets, chains, as well as inertia-responding devices such as ball governors or thermal-responding devices such as bimetallic controls, provide the designer with many choices in the selection of the most desirable system. These systems also have disadvantages that should be recognized: the difficulty of creating safety protection within the system, lubrication problems, many moving parts (creating both direct wear and abrasion particles that accelerate wear), and the limited ease of speed variation. Their performance capabilities are also generally limited to short distances.

Fluid

A fluid power system converts, transmits, controls, and applies energy through a pressurized fluid within a closed circuit. It may be simple or complex. The use of fluid power for transmitting forces and controlling motions has several advantages that make it desirable for certain applications. To summarize the advantages mentioned earlier, hydraulics and pneumatics provide a power system that offers safety for the machine and operator. Fluid power systems also permit accurate position control for both linear and rotary actuators. Such a system can transmit enormous forces at an infinitely variable speed and with rapid reversing capabilities. With the use of tubing, hoses, and pipes, fluid power systems can be placed at some distance from a machine (Figure 1.17). Also, with the use of panels, manifolds, and subplates, the system can be compactly built as an integral part of a piece of equipment.

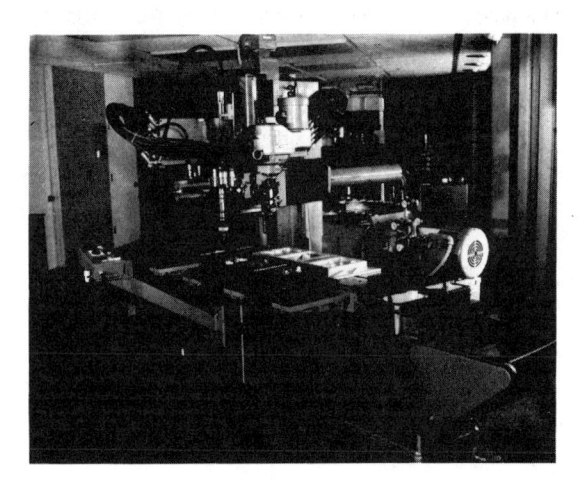

Figure 1.17 This high-precision industrial planer uses hydraulic actuators and controls to maintain precision reproduction of parts. (Courtesy True-Trace)

There are limitations to fluid power systems such as contamination, which is perhaps the most important problem in a fluid system. These systems require preventive maintenance, which can be established by the constant monitoring of filtration and regulation, and by recording the operational hours of fluids, seals, and hoses.

In many instances the equipment designer has an option as to the type of transmission and control system to install. A designer's decision is influenced by such factors as performance requirements, size and availability of commercial components for a system, and cost. It is necessary to decide which system or combination of systems will provide the most efficient, economic, safe, and dependable machine. Frequently design engineers are not sufficiently familiar with all three types of systems, which limits their design capabilities. Until such time as engineers are adequately prepared in the area of fluid power systems, these systems will not be applied to the extent that their advantages warrant.

A fluid power system is an arrangement of several components: primary mover, pump, fluid energy controls, and actuator. The primary mover supplies the electrical or mechanical energy input, which the pump converts into fluid energy. The fluid energy is transmitted to the controls, which modulate it, and the actuator, that is, a cylinder or fluid motor, converts the fluid energy to mechanical energy output.

Basically a fluid power system performs four major functions: (1) converting mechanical energy to fluid energy, (2) transmitting the fluid energy through the system to the actuator, (3) modulating the fluid, and (4) applying it to perform work (Figure 1.18).

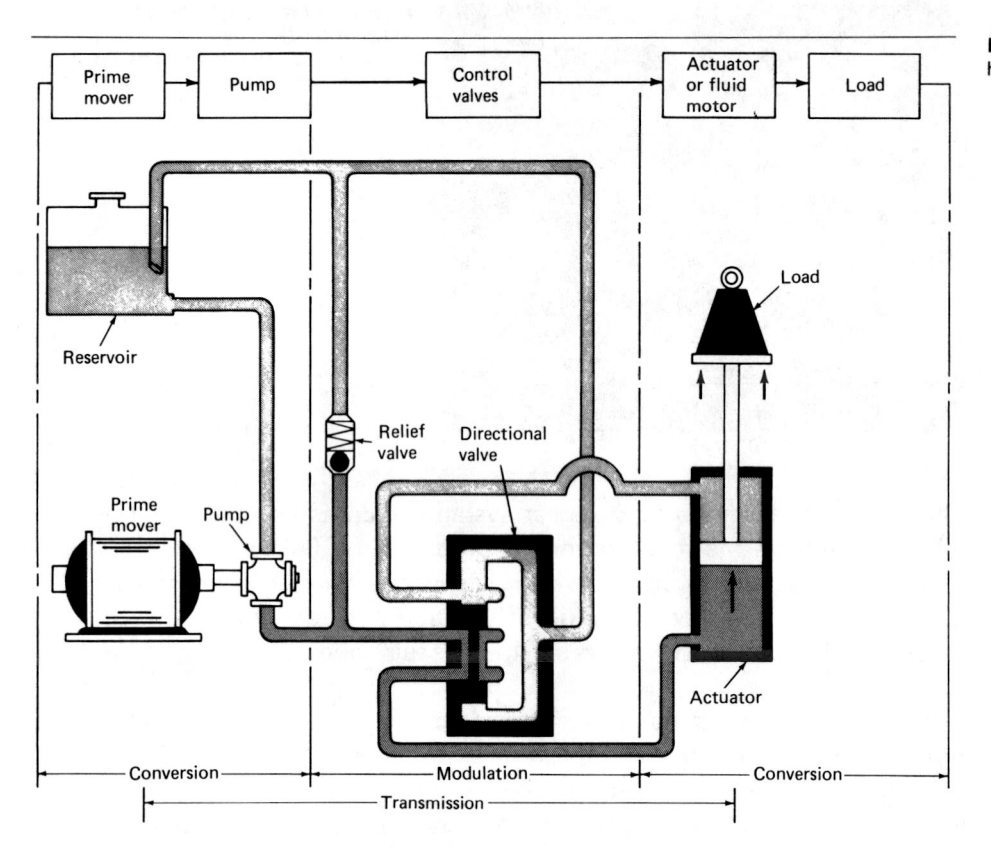

Figure 1.18 Functions of a hydraulic system.

A complete assembly of components within a system makes up a *circuit*. The fundamental hydraulic or pneumatic circuit is similar in some respects to an electric circuit in the functions performed (Figure 1.19). To be at all practicable, even the simplest circuit must include components that are compatible and will provide the degree of control desired in the transmission of forces to get a job done.

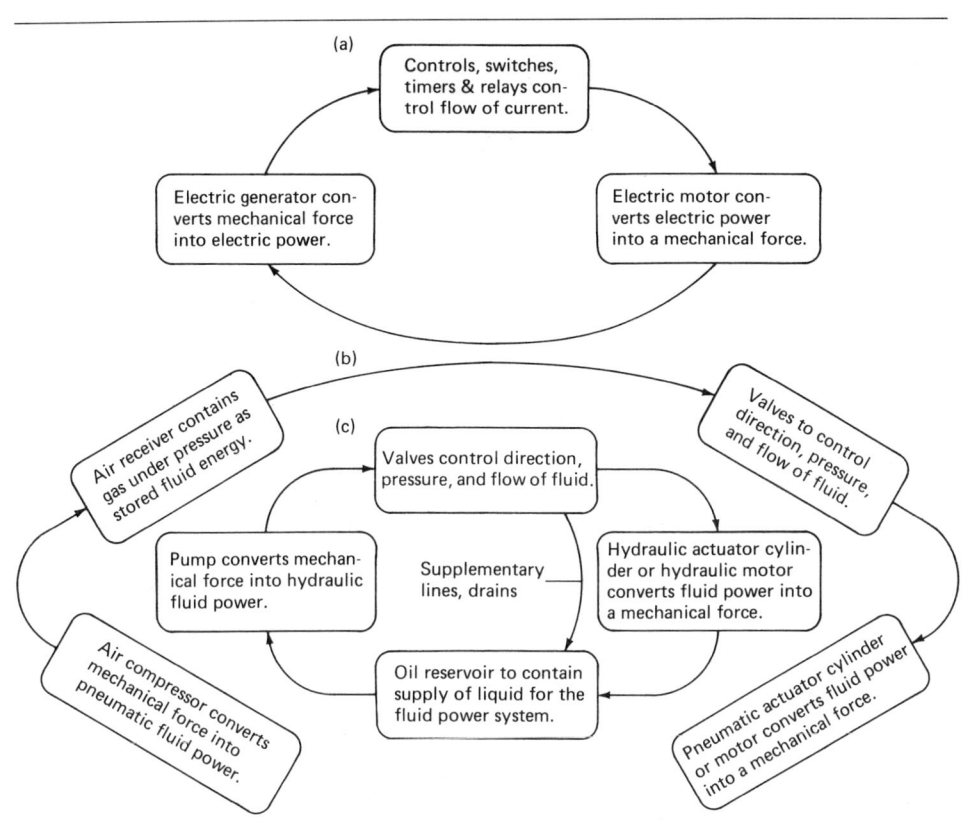

Figure 1.19 Functions performed by components in fundamental circuits are the same in (a) electric, (b) hydraulic, and (c) pneumatic circuits.

(a)
- Controls, switches, timers & relays control flow of current.
- Electric generator converts mechanical force into electric power.
- Electric motor converts electric power into a mechanical force.

(b)
- Valves control direction, pressure, and flow of fluid.
- Pump converts mechanical force into hydraulic fluid power.
- Hydraulic actuator cylinder or hydraulic motor converts fluid power into a mechanical force.
- Supplementary lines, drains
- Oil reservoir to contain supply of liquid for the fluid power system.

(c)
- Air receiver contains gas under pressure as stored fluid energy.
- Valves to control direction, pressure, and flow of fluid.
- Air compressor converts mechanical force into pneumatic fluid power.
- Pneumatic actuator cylinder or motor converts fluid power into a mechanical force.

In designing and fabricating a system, it is necessary to determine what the job requirements are and then create the most simple, efficient, reliable, and safe circuit. The proper functioning of most fluid power systems depends on accurate control of fluids. As more precise controls are developed, the applications of fluid power systems will greatly increase, and they will become more competitive with other systems.

How a fluid power system works

A hydraulic system can exert forces ranging from a delicate touch of several ounces to an enormous force of 50,000 tons or more because, first of all, liquids are essentially incompressible (Figure 1.20); and, second, a confined liquid can multiply forces (Figure 1.21). The first point—that liquids can be only slightly compressed— means the reduction of the volume that they occupy, even under immense pressure, is extremely small. For example, if a pressure of

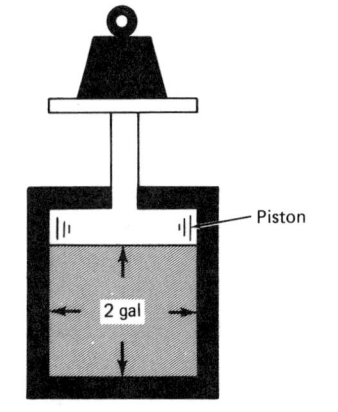

Figure 1.20 Liquids can be only slightly compressed.

Figure 1.21 A confined liquid can be used to multiply force.

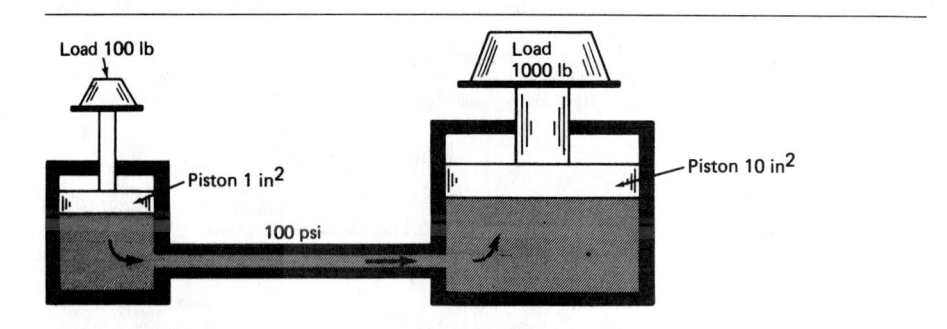

100 pounds per square inch (psi) is applied to the surface of water, the volume will decrease by only 3/10,000 of its original volume. Oil at 3000 psi will decrease in volume less than 1.2%. Therefore, because of the very small changes that occur even at high pressures, liquids are considered incompressible.

Pascal's law

Pascal's law explains the second point, the ability of confined fluids to multiply forces: *Pressure exerted on a confined liquid is transmitted undiminished in all directions and acts with equal force on all equal areas.*

As shown in Figure 1.22, a 100-lb force (F_1) acting on a 1-sq in. piston develops a pressure of 100 psi, which when transmitted through a liquid to a 10-sq in. piston enables it to support a tenfold force (100 × 10 = 1000 lb). To simplify this process of multiplication of forces, it will be useful to restate Pascal's law in words and then translate it into a mathematical expression: *Pressure equals force divided by area.*

$$p = \frac{F}{A}$$

In other words, a force of 100 lb over an area of 1 sq in. creates a pressure of

$$p = \frac{100}{1} = 100 \text{ psi}$$

Note in the illustration that 100 psi is transmitted easily around the corners and exerts 100 psi (equal pressure on equal areas) on Piston B. To calculate how much the forces are multiplied by the 100 psi

Figure 1.22 Force-pressure-area relationship in the multiplication of hydraulic force.

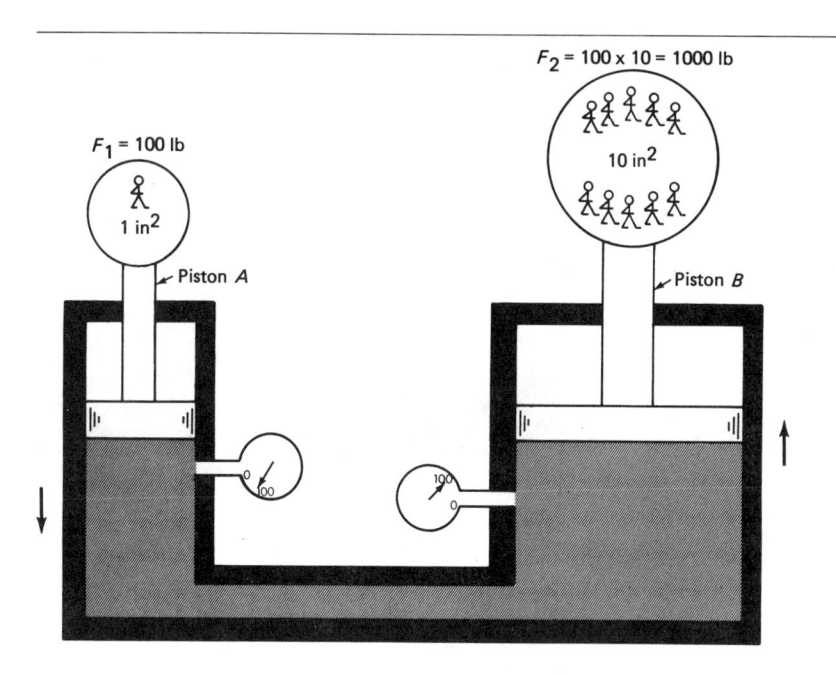

acting on a cross-sectional area of 10 sq in., the formula can be stated in terms of the force. Transposing the formula

$$p = \frac{F}{A} \quad \text{gives} \quad F = PA$$

or, referring to Figure 1.22, $F_2 = 100 \times 10 = 1000$ lb. A designer normally makes calculated allowances for losses due to friction in pipes, hoses, and fittings.

As with a mechanical lever, the multiplication of force is gained at the expense of distance moved. To move the 10-sq in. piston $\frac{1}{10}$th of an inch, the 1-sq in. piston must be moved 1 in. (Figure 1.23). When a force is applied to the end of a column of confined fluid, as in the case of a cylinder, it is transmitted straight through to the other end and also equally undiminished in every direction throughout the column. The free-moving piston within the cylinder is moved by the transmitted energy.

If gas is used instead of liquid, the force is transmitted in the same manner. The essential difference is that gas, being more compressible, provides a much less rigid force than the liquid. However, pneumatic actuators can operate at much higher speeds than hydraulic ones. These are the main response differences between the action of liquids and gases in a fluid power system.

Figure 1.23 Force-distance relationship. Piston A displaces less fluid than Piston B. Hydraulic force advantage gained (Piston B) creates a sacrifice in distance (Piston B).

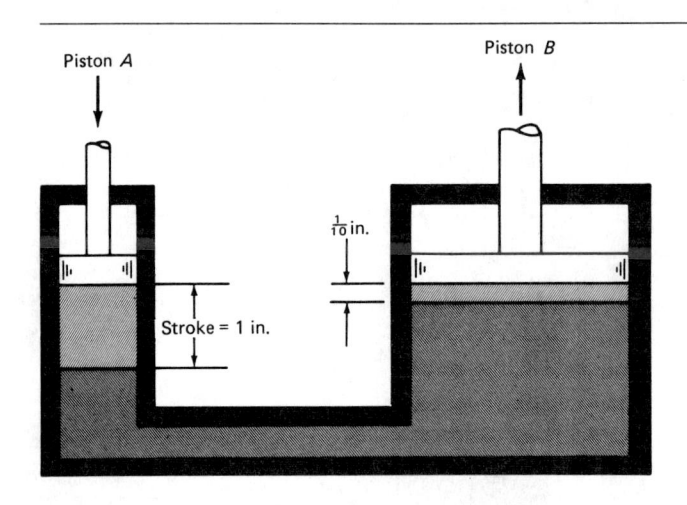

Components of a hydraulic system

A simple hydraulic circuit, regardless of its applications—machine tools, mobile equipment, aircraft, medical tools, or any other—contains four basic mechanical elements common to all hydraulic systems: a reservoir to hold the fluid; a pump to force the fluid through the system (the pump is driven by an electric motor or other primary power source); valves to control fluid pressure and flow; and an actuator (a cylinder for linear or a motor for rotary motion) to convert the energy of the compressed fluid into mechanical force to do work (Figure 1.24).

Figure 1.24 Basic hydraulic circuit.

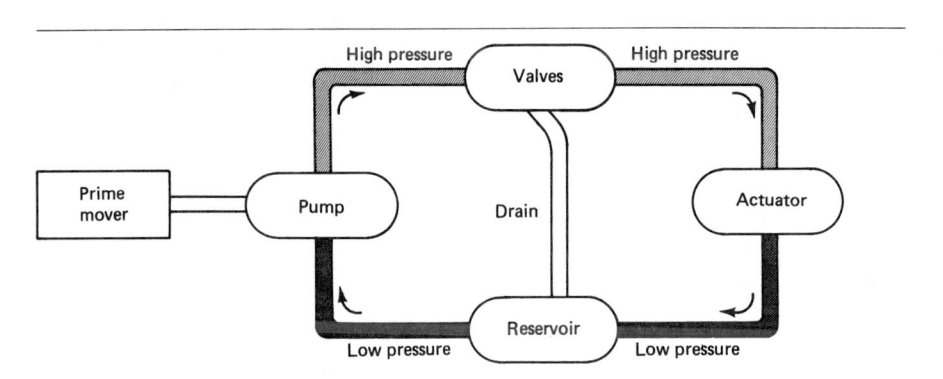

The complexity of a hydraulic power system varies according to the specific application. Some complex systems include two or more

Figure 1.25 Industrial hydraulic components including prime mover (electric motor), reservoir, pump, valves, and conductors. (Courtesy Nicholson)

pumps, numerous valves serving different functions, and several cylinders or motors (Figure 1.25).

Another essential difference between a hydraulic and a pneumatic system is that the gas is compressed and remains under high pressure (usually about 80–120 psi) throughout the entire system (Figure 1.26). All fluid power systems use the principle of multiplying, transmitting, and controlling force through a pressurized fluid.

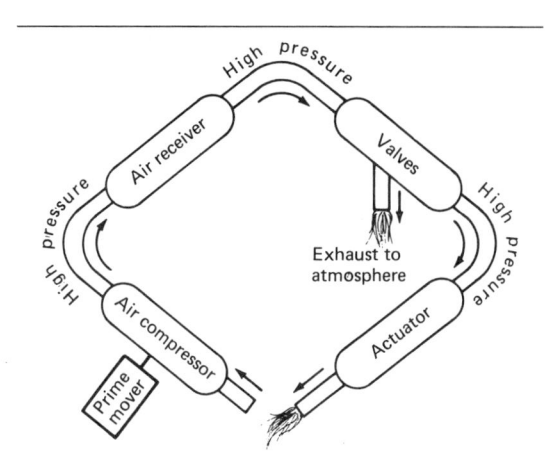

Figure 1.26 Basic pneumatic circuit.

Fluid power applications

The rapid expansion of fluid power applications in manufacturing, transportation, construction, agriculture, lumbering, material handling, mining, printing, and space exploration has made this technology noticeable in all parts of the world. As engineers design new vehicles for land, sea, air, and space, increasingly larger, heavier, and more complex fluid power systems will be used to propel, guide, and control such vehicles. Fluid power is no longer limited in its application to large loads where high pressures and flows are necessary. It is now possible to make small compact systems that can be used in surgical tools, toothbrushes, railroad switching systems, hospital beds, cars, and space vehicles.

The ever-increasing number of fluid power applications is due to research investments by fluid power component manufacturers and the efforts of creative people who design, develop, and achieve more flexible and efficient fluid power systems. Better seals, more precision in manufactured components, safer fluids, greater standardization of components, and more better-educated personnel who are involved in the design, manufacture, sales, operation, and maintenance of systems have all contributed to the reliability and acceptance of fluid power systems.

The demand for fluid power systems—ranging from the simple circuit found in a hydraulic jack to the complex fluid power-controlled, multiple-station transfer machines that perform all operations on a single casting such as an automotive transmission housing—has produced the need for numerous components. Today there are literally hundreds of fluid power systems manufacturers, some of whom build single highly specialized components while others build whole systems to meet customers' requirements. Still others manufacture components for complete systems ranging from heavy industrial and specialized aircraft systems to exotic systems for space travel and exploration.

Perhaps the single most important factor that contributed to the rapid acceleration in fluid power applications was the development of symbols and standards that have become widely accepted and used throughout the fluid power industry. The National Fluid Power Association created the standards and the Fluid Power Society developed the symbols so that the fluid power designer and user would have a common language. The American National Standards Institute (ANSI) is continuing the work in developing standards for the fluid power industry.

Summary

Several important technological developments have contributed to the growth and expansion of the fluid power industry. Understanding fluid mechanics is essential to energy transmission and control by use of pressurized fluid. The ability to design components and systems that operate predictably in many applications has contributed to wide acceptance of fluid power systems. The discovery of petroleum oil, more precision machining, and improved seals have made it possible to expand the use of hydraulic and pneumatic systems.

As we become more conscious of the need to conserve energy, we will seek more efficient systems for doing work. Fluid power systems are an efficient method of multiplying forces almost infinitely with just a few moving mechanical parts. Fluid systems can use either liquid or gas media for transmitting fluid energy, but all require a prime mover such as a gasoline or diesel engine, an electric motor, or a turbine to drive the hydraulic pump or pneumatic compressor. Fluid power systems can adapt to the full range of force requirements and can be used in a wide range of environmental conditions.

Review questions and problems

1 Why is hydraulic power especially suitable in situations where large forces have to be applied or controlled?
2 In a fluid power system, the multiplication of force is gained at the expense of what factor?
3 What are the three common methods of transmitting power?
4 What are the essential differences between the contained pressurized fluid power systems that we know today and an early hydraulic system such as the water wheel?
5 Which two scientists discovered the fundamental laws of physics on which fluid systems are based?
6 List six advantages of hydraulically operated systems or machines.
7 List 10 applications of hydraulics or pneumatics with which you are familiar.
8 Name the four basic functions performed in a fluid power system.
9 Express Pascal's law in equation form.
10 What were some of the factors that contributed to the development of more reliable and efficient hydraulic systems?

Selected readings

- Fitch, Ernest C., Jr. *Fluid Power and Control Systems.* McGraw-Hill, New York, 1966.
- *Fluid Power.* Bureau of Naval Personnel Navy Training Course, NAVPERS 16193-A, U.S. Government Printing Office, Washington, D.C., 1966.
- Halacy, D. S., Jr. *Energy and Engines.* World Publishing, Cleveland, 1967.
- Henke, Russell W. *Introduction to Fluid Mechanics.* Addison-Wesley, Reading, Massachusetts, 1966.
- *Hydraulic Fundamentals and Industrial Hydraulic Oils.* Sun Oil Company, Philadelphia, 1963.

- McNickle, L. S., Jr. *Simplified Hydraulics.* McGraw-Hill, New York, 1966.
- Newton, Donald G. *Fluid Power for Technicians.* Prentice-Hall, Englewood Cliffs, New Jersey, 1971.
- Pease, Dudley A. *Basic Fluid Power.* Prentice-Hall, Englewood Cliffs, New Jersey, 1967.
- Pippenger, John, and Tyler G. Hicks. *Industrial Hydraulics.* McGraw-Hill, New York, 1962.
- Randall, Ralph D. *Fundamentals of Pneumatic Control.* St. Regis Paper Co., Providence, Rhode Island, 1961.
- Robinson, Lister J. *Basic Fluid Mechanics.* McGraw-Hill, New York, 1963.
- Stewart, Harry L., and John M. Storer. *ABC's of Fluid Power.* Howard W. Sams, Indianapolis, Indiana, 1966.
- ———. *Fluid Power.* Howard W. Sams, Indianapolis, Indiana, 1968.

Chapter 2
Fundamental laws of fluid mechanics

This chapter introduces the basic laws and principles of fluid mechanics. It describes, in theoretical and applied terms, the physical properties and behavior of fluids used in hydraulic and pneumatic systems and also energy forms, force, work, and torque. Understanding the laws and principles will permit the reader to use basic theoretical concepts in practical applications.

Key terms

- **Absolute pressure** Is measured from absolute zero pressure, that is, no pressure at all, rather than from atmospheric pressure (approximately 14.7 psi). Absolute pressure (psia) is equal to atmospheric pressure plus gage pressure (psig). Psia = 14.7 for calculation purposes.
- **Absolute temperature** Can be expressed in terms of a quantity by converting the Fahrenheit scale to the Rankine scale. To change from the temperature T_F on the Fahrenheit scale to the temperature on the Rankine scale, use this relationship: $T_R = T_F + 460$.[1]
- **Absolute viscosity** The ratio of the shearing stress to the shear rate of a fluid. It is expressed in units of centipoise.
- **Acceleration** The change in the velocity of a body divided by the time during which the change takes place.
- **Force** The action of one body on another tending to change the state of motion of the body acted on. The unit of force associated with the work being done by a cylinder or motor may be measured in any unit of weight but is generally expressed in pounds. In fluid power, the force is determined by the relationship: force = pressure × area.
- **Horsepower** A unit for measuring power: 1 hp = 550 ft-lb/sec. Hydraulic horsepower is computed from flow rate and pressure differential. Theoretical horsepower may be determined by these equations:

$$hp = \frac{gpm \times psi}{1714}$$

$$= 0.000583 \times psi \times gpm$$

[1] A complete list of the symbols and abbreviations used in this book appears in the appendixes.

- **Kinematic viscosity** The absolute viscosity divided by the mass density of the fluid. It is usually expressed in units of centistokes.
- **Laminar flow** Occurs when the liquid particles flow smoothly in even layers, and no cross-flow takes place.
- **Mass** A derived dimension, expressed by Newton's equation: $F = m \cdot a$. The kilogram is the standard unit of mass. Using the unit of pound mass, we can state Newton's law: *One pound force will accelerate one pound mass at the rate of* 32.1739 *ft/sec² at sea level and 45°N. latitude.*
- **Mass density** The ratio of the mass of a sample of the substance to its volume.

$$\rho = \frac{m}{V}$$

- **Turbulent flow** Occurs when liquid particles flow in a random or erratic pattern, creating high energy losses due to friction.
- **Velocity** The rate of motion in a particular direction. The velocity of fluid flow is usually measured in feet per second.

$$v = \frac{L}{t}$$

- **Viscosity** A measure of the internal friction or the resistance of a fluid to flow.

Fluid mechanics is the study of the physical behavior of fluids and fluid systems and the laws describing their behavior. Several scientists studied fluid behavior, formulated laws related to liquids and gases, and thus made it possible to calculate and predict the behavior of fluids under given conditions. When fluid power systems lift a plow, power-brake an automotive vehicle, lower the landing gear on an aircraft, control automation equipment remotely, stabilize a rocket, and assist the astronaut or aquanaut to direct a vehicle with precision to its destination, the performances of such systems are based on predictable and tested control of the physical variables of fluids.

Physical properties of fluids

In general, the behavior of fluids follows laws similar to those that govern the behavior of solid bodies. Previous study, in other science courses, of states of matter and the physical characteristics of liquids will be helpful in understanding the practical applications of pressurized fluids.

Earlier, it was indicated that fluids include both liquids and gases. A *fluid* is a substance that has a definite mass and volume but no definite shape. It changes its shape easily and takes the shape of a container. *Gases* are fluids that are highly compressible; they will fill the vessel containing them but will vary in volume. Both liquids and gases readily transmit pressure. Because a liquid conforms to the shape of its container, liquids can be pumped and transmitted through pipes, tubing, hoses, and ports (Figure 2.1).

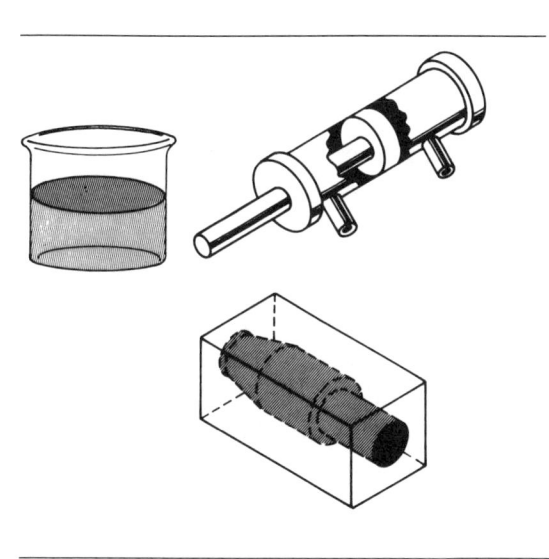

Figure 2.1 Liquids conform to the shape of the container.

Density

Liquids have weight. The *mass density* of a substance is defined as the ratio of the mass of a sample of the substance to its volume. Mathematically, this relationship can be expressed as

$$\rho = \frac{m}{V}$$

where ρ (Greek letter rho) is the density of the substance, m is the mass of the substance, and V is its volume. The densities of liquids vary slightly with changes of temperature and pressure. Most often in engineering practice and calculations, *weight density* refers to the weight per unit volume of a material. *Weight density* (w) is determined by the formula

$$w = \frac{W}{V}$$

The density of water is 62.4 pounds per cubic foot (lb/ft³) and the standard density of air is 0.0752 lb/ft³. This means that if a cubic foot of water at 39 °F (4 °C) were weighed accurately (not including the container), it would weigh 62.4 lb. The mass density of water may also be expressed as 1 gram per cubic centimeter (1 g/cm³). To find the density of a substance, its weight and volume must be known.

▶ **Example** Determine the density of hydraulic oil weighing 6 lb and occupying a volume of 231 in³ (1 gallon).

$$w = \frac{W}{V}$$

Changing cubic inches to cubic feet

$$V = \frac{231 \text{ in}^3}{12 \times 12 \times 12} = \frac{231 \text{ in}^3}{1728} = 0.1337 \text{ ft}^3$$

$$w = \frac{6 \text{ lb}}{0.1337 \text{ ft}^3} = 44.9 \frac{\text{lb}}{\text{ft}^3}$$

Oil is lighter than water, with a specific gravity in the neighborhood of 0.75 to 0.8. In other words, oil weighs about 0.8 × 62.4 lb/ft³ = 50 lb/ft³. The specific gravity of liquids or solids is the ratio of the weight of an equal volume of water, or the ratio of the densities of the material and water. Thus

$$\text{Specific gravity} = \frac{\text{weight of substance}}{\text{weight of equal volume of water}}$$

or

$$Sg = \frac{Ws}{Ww}$$

Within a fluid power system, as a liquid is subjected to high pressures, the volume will decrease slightly (1½% for 3000 psi). The density therefore will be increased. As the pressure is dissipated, the volume will return to its original value instantly provided that the temperature is also maintained at a constant value.

Kelvin to Celsius and Fahrenheit
$°K = C + 273$
$= \frac{5}{9}(F - 32) + 273$
$= \frac{5}{9}F + (273 - 17.77)$
$= \frac{5}{9}F + 255.23$

Celsius to Kelvin and Fahrenheit
$°C = K - 273$
$= \frac{5}{9}(F - 32)$

Fahrenheit to Kelvin and Celsius
$°F = \frac{9}{5}(K - 273) + 32$
$= \frac{9}{5}K - 491 + 32$
$= \frac{9}{5}K - 459.4$
$= \frac{9}{5}C + 32$

Table 2.1 Temperature conversion

Temperature

Another physical variable that must be considered in fluid mechanics is temperature. The four temperature scales used are the absolute or

27 Physical properties of fluids

Kelvin, the Celsius, the Fahrenheit, and the Rankine scales (Figure 2.2).

Figure 2.2 The four temperature scales.

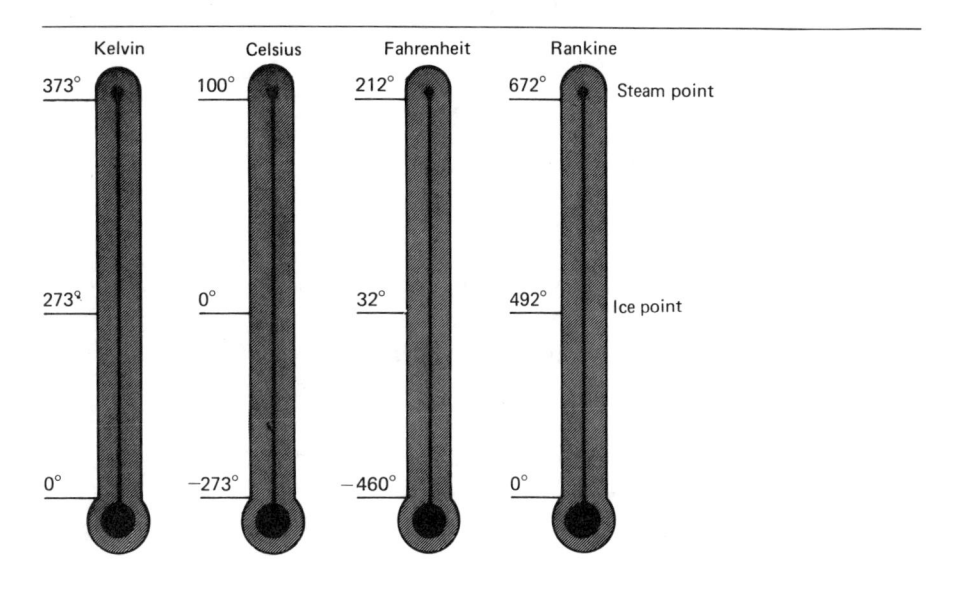

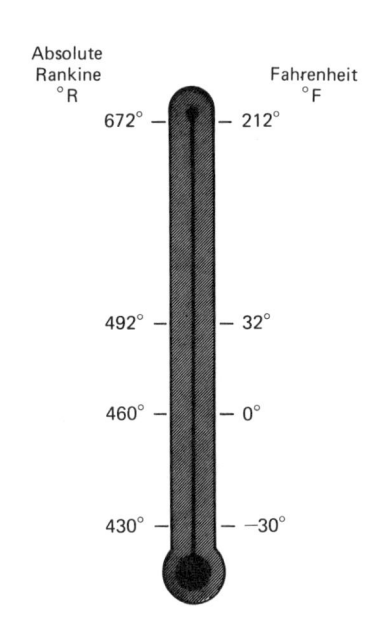

Figure 2.3 Absolute Rankine scale and Fahrenheit scale. For calculating purposes, readings have been rounded off to the nearest whole number.

The zero point of the Kelvin scale is established at -273 °C, or 459.7 °F below the freezing point of water. This absolute scale is used in the study of gases. Experiments with hydrogen indicate that if a gas is cooled to -273.16 °C ($-273°$ is used for most theoretical calculations), all molecular motion ceases and no additional heat can be extracted. Absolute zero may be expressed as 0 °K, as -273 °C, or as -459.7 °F ($-460°$ is used for most calculations) (see Table 2.1). The Celsius scale is constructed by using a freezing point of 0 °C and a boiling point of 100 °C for water, with 100 equal divisions between these points. The Fahrenheit scale uses a freezing point of 32 °F and a boiling point of 212 °F for water and has 180 divisions between these points.

For calculation purposes, the Rankine scale illustrated in Figure 2.3 is commonly used to convert Fahrenheit to absolute temperatures. For Fahrenheit readings above zero, 460° is added. Thus 72 °F converted to an absolute scale reading would be 72° + 460° = 532° absolute. If the Fahrenheit reading is below zero, that reading is subtracted from 460°. Thus for a below-zero temperature of -30 °F, the absolute scale reading would be 460° − 30° = 430° absolute. It should be noted that the Rankine scale (Figure 2.3) does not indicate absolute temperature readings in accordance with the Kelvin scale, but these conversions can be used to calculate changes in the state of gases. Because the temperature of fluids is recorded in degrees Fahrenheit (°F), the Rankine absolute equivalent is used for calculation purposes. The relationship between the temperature on the Rankine scale, T_R, to the same temperature T_K on the Kelvin scale is $T_R = \frac{9}{5} T_K$. Thus the freezing point on the Rankine scale is $T_R = \frac{9}{5}(273) = 492$ °R.

Pressure

By definition, *pressure* is force per unit area. For our purpose, this will be expressed in pounds per square inch (psi). A force can be applied to a single point of a solid, but it can be applied only over a surface of a fluid. Pressure in all hydraulic and pneumatic systems operates according to Pascal's law that states: *Pressure exerted on a confined liquid is transmitted undiminished in all directions and acts with equal force on all equal areas.* It is easier to illustrate the significance of pressure by referring to Figure 2.4. The weight, or

Figure 2.4 Pressure is expressed as pounds per square inch (psi) of surface area.

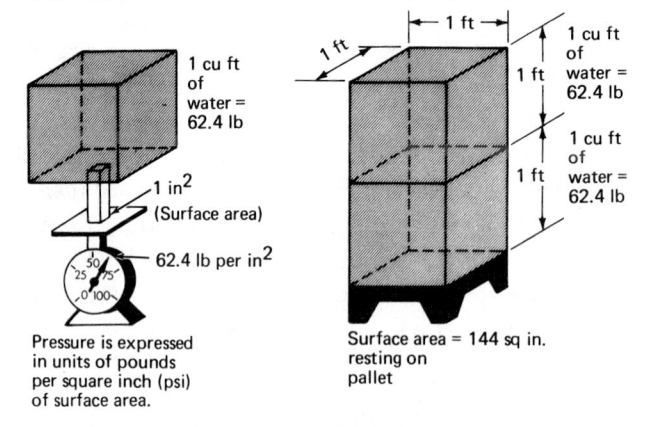

downward force, of 1 ft³ of water is 62.4 lb. The pressure on the bottom surface of this cube filled with water is the force (weight in this case) divided by the area supporting it (144 in²):

$$p = \frac{F}{A}$$

$$p = \frac{62.4 \text{ lb}}{144 \text{ in}^2} = 0.433 \text{ psi}$$

A column 2 ft high containing 2 ft³ of water would result in a pressure of 0.866 psi. Thus there is a direct relationship between the the height of a column of fluid and the pressure at its base.

Absolute temperature is used to calculate changes in the state of gases or air; it is also necessary to use *absolute pressure* for these calculations. Absolute pressure (psia) is measured from absolute zero pressure rather than from the normal or atmospheric pressure (Figure 2.5) (approximately 14.7 psi at sea level). Gage pressure (psig), as

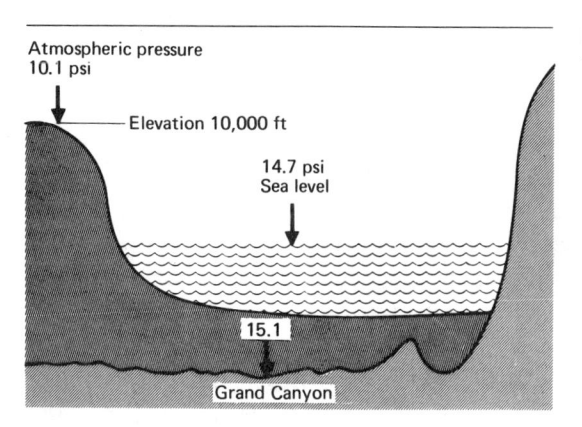

Figure 2.5 Atmospheric pressure varies with altitude.

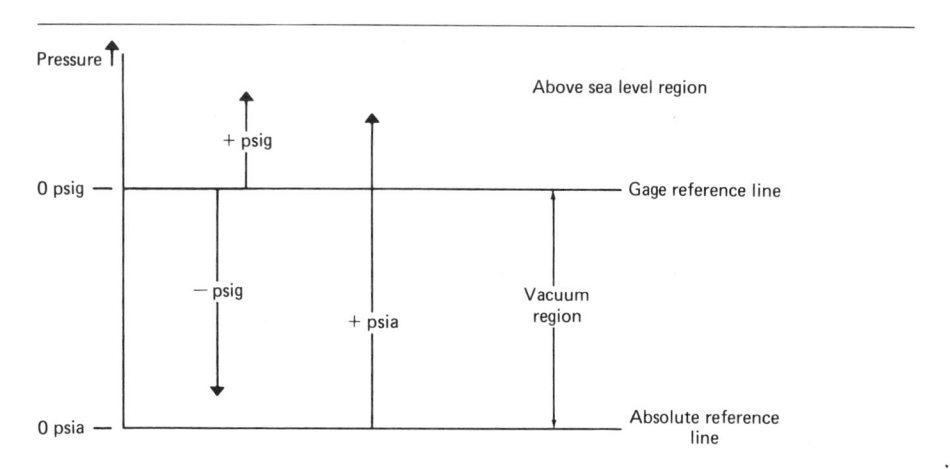

calibrated on all ordinary gages, indicates pressure in excess of atmospheric pressure. Absolute pressure is equal to atmospheric pressure plus gage pressure. For example, a gage reading of 100 psig is equivalent to 100 + 14.7 = 114.7 psia (Figure 2.6).

Figure 2.6 Relationship of gage pressure (psig) and absolute pressure (psia).

Compression and expansion of fluids

The two essential differences between gases and liquids are in their compressibility and expansibility. While liquids are incompressible for all practical purposes, gases are highly compressible. On the other hand, through expansion, gases tend to completely fill any enclosed container while liquids merely seek a level to the extent of their normal volume.

Gas laws

It is important to understand the three variables of pressure, temperature, and volume and their relationships within the operation of a pneumatic system. Air is a mixture of gases and follows the laws of a perfect gas with regard to its behavior in volumetric expansion, contraction, and absorbing and releasing heat. A given volume of gas will occupy any size of vessel whereas the same volume of liquid may not fill the given size of vessel.

Boyle's law Robert Boyle, an English scientist of the seventeenth century, studied the structure of gases in relation to their compressibility. By experimentation and direct measurement, he discovered that when the temperature of an enclosed sample of gas was kept constant and the pressure doubled, the volume was reduced to half the former value (Figure 2.7); as the applied pressure decreased, the

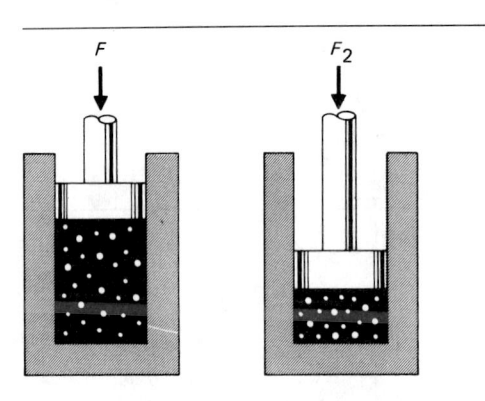

Figure 2.7 Boyle's law states, essentially, that gas is compressed to half its volume by doubling the force.

volume increased again. He concluded that for a constant temperature the product of the volume and pressure of an enclosed gas remains the same. The law that he derived (Boyle's law) states: *The volume of an enclosed gas varies inversely with its pressure provided the temperature remains constant.* Mathematically, this relationship may be expressed as

$$V_1 p_1 = V_2 p_2 \quad \text{or} \quad \frac{V_1}{V_2} = \frac{p_2}{p_1}$$

31 Physical properties of fluids

▶ **Example** In Figure 2.7, if 4 ft³ (V_1) of air are under pressure of 100 psig (p_1) and the air is compressed to a volume (V_2) of 2 ft³, what will the new gage pressure (p_2) be?

Remember that for calculation purposes, it is necessary to use absolute pressure.

$$V_1 p_1 = V_2 p_2$$

Substituting given quantities

$$4 \text{ ft}^3 \times (100 \text{ psig} + 14.7) = 2 \text{ ft}^3 \times p_2$$

$$p_2 = \frac{4 \text{ ft}^3 \times 114.7 \text{ psia}}{2 \text{ ft}^3}$$

$$= 229.4 \text{ psia}$$

Converting 229.4 psia to gage reading

$$= 229.4 \text{ psia} - 14.7 \text{ psi} = 214.7 \text{ psig}$$

Theoretically, this is what would happen with an ideal gas. However, because density varies with temperature (that is, the gases expand when heated and contract when cooled) (Figure 2.8), there will be, in practice, some slight fluctuations in the relationship of pressure and volume.

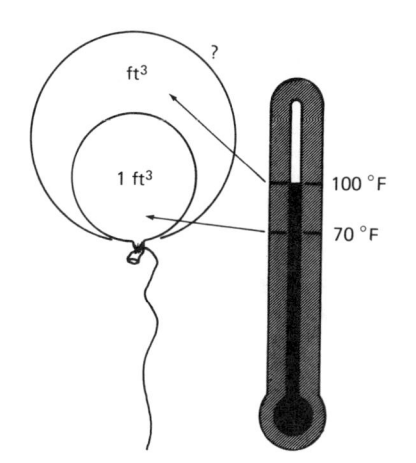

Figure 2.8 Gases expand when heated and contract when cooled.

Charles' law The laws of Robert Boyle and J. A. C. Charles (1746–1823) describe the changes of state in which one condition remains constant. Charles' law states: *At constant pressure, the volume of a gas varies in direct proportion to a change in temperature.* For solving problems, absolute values of pressure and temperature must be used; that is, gage or thermometer readings must be converted to absolute values. Mathematically, this law can be expressed as:

$$\frac{\text{Initial volume}}{\text{Final volume}} = \frac{\text{initial absolute temperature}}{\text{final absolute temperature}}$$

or using the formula $\dfrac{V_1}{V_2} = \dfrac{T_1}{T_2}$

Transposing the formula $V_2 = \dfrac{V_1 \times T_2}{T_1}$

▶ **Example** In Figure 2.9 if a tank containing 100 ft³ of air at atmospheric pressure and 50 °F is placed in direct sunlight where it is heated to 120 °F, what volume of air will escape from the tank?

$T_1 = 50\ °F + 460° = 510\ °R$ (Initial absolute temperature in degrees Rankine)

$T_2 = 120\ °F + 460° = 580\ °R$ (Final absolute temperature in degrees Rankine)

$$V_2 = \frac{100\ ft^3 \times 580\ °R}{510\ °R}$$

$V_2 = 113.7\ ft^3$

Volume escaping $= V_2 - V_1$
$= 113.7\ ft^3 - 100\ ft^3$
$= 13.7\ ft^3$

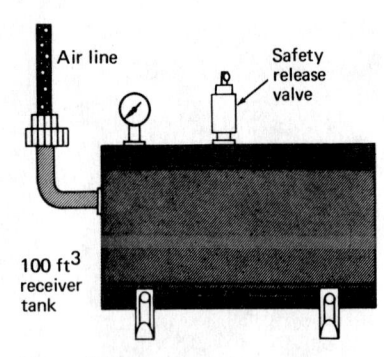

Figure 2.9 An air-receiver tank with a safety valve.

Gay-Lussac's law Supplementing the gas laws of Charles and Boyle, J. L. Gay-Lussac observed: *If the volume of a gas remains constant, the pressure exerted by the confined gas is directly proportional to the absolute temperature.* This law can be expressed as follows:

$$\frac{\text{Initial pressure}}{\text{Final pressure}} = \frac{\text{initial absolute temperature}}{\text{final absolute temperature}}$$

or, using the mathematical formula, $\dfrac{p_1}{p_2} = \dfrac{T_1}{T_2}$

Transposing the formula $p_2 = \dfrac{p_1 T_2}{T_1}$

▶ **Example** In Figure 2.9, if a tank containing 100 ft³ of air at 14.7 psig and 50 °F is heated to 120 °F, what will be the final pressure?

$T_1 = 50\ °F + 460° = 510\ °R$ (Rankine)

$T_2 = 120\ °F + 460° = 580\ °R$

$p_1 = 14.7\ \text{psig} + 14.7 = 29.4\ \text{psia}$

$p_2 = \dfrac{p_1 T_2}{T_1}$

$= \dfrac{29.4\ lb/in^2 \times 580\ °R}{510\ °R}$

$= 33.4\ \text{psia}$

Change in gage pressure: p_2 = absolute pressure − 14.7
                              = 33.4 psia − 14.7
                              = 18.7 psig

Effect of heat on fluids

It is a known phenomenon that liquids and gases expand when heated. The amount of gas expansion can be calculated by applying Charles' law, which describes the relationship of volume and temperature, as in the case of compressed air. For example, leaving a filled aerosol container or a cylinder filled with liquid exposed to intense sunlight with no room for expansion can create extensive pressure build-up within the vessel. Any excessive pressure within a system creates undue strain on seals and packing.

Fluids in motion

When liquid flows through a pipe or any other conductor in such a way that it completely fills the pipe and as much liquid enters one end of the pipe as leaves the other end in the same time, then the liquid is considered to be flowing at a *steady rate* (Figure 2.10).

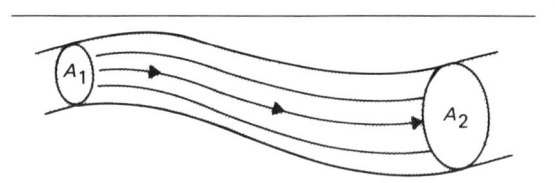

Figure 2.10 Streamlines of liquid flowing through a pipe at a steady rate.

At any point in the pipe, the motion of the liquid does not change with time; as a small particle of liquid leaves the point, another reaches it and moves with the same velocity that the preceding particle had there. When the velocity of the liquid is not too great and any changes in shape and area of the pipe are gradual with no sharp edges, a particle of liquid moves through the conductor along a streamline path.

The flow of liquid through a pipe can be mapped by drawing a series of streamlines following the paths of the liquid particles, as illustrated in Figure 2.10. Note that these lines are close together where the liquid is moving more slowly. Since liquids are incompressible (for all practical purposes) and there is no place in the pipe in which the liquid can be stored, the volume of liquid that flows through a given cross-sectional area perpendicular to the streamline, in any interval of time, must be the same everywhere in the pipe. The small volume of liquid ΔV that flows through a small distance Δs past a point in the pipe where the cross-sectional area is A, can be expressed in this relationship:

$$\Delta V = A(\Delta s)$$

The rate of flow Q, the quantity of liquid that flows through this area per unit time t, can be expressed as

$$Q = \frac{\Delta V}{\Delta t}$$

or substituting for ΔV the equivalent $A(\Delta s)$

$$Q = \frac{A(\Delta s)}{\Delta t}$$

where

$$v = \frac{\Delta s}{\Delta t}$$

is the velocity of the liquid at this point.

$$Q = Av$$

Assume two typical areas A_1 and A_2 (refer to Figure 2.10) perpendicular to the streamlines of liquid flow. The volume of liquid passing through area A_1 per unit time is $Q = A_1 v_1$. In this case, v_1 is the velocity of the liquid at this point. Similarly, in a flow condition at a steady rate, the volume of liquid passing through A_2 per unit time is $Q = A_2 v_2$. Because these two quantities must be equal for steady-rate flow, $Q = A_1 v_1 = A_2 v_2$, where Q is flow volume in cubic feet per second (cfs), v is velocity in feet per second (fps), and A is conductor area in square feet (ft²). This is called the *equation of continuity*:

$$\frac{v_1}{v_2} = \frac{A_2}{A_1}$$

This equation states: *The velocity of the liquid at any point in the pipe is inversely proportional to the cross-sectional area of the pipe.* In effect, the liquid will be moving slowly where the area is large and rapidly where the area is small.

▶ **Example** Oil flows out of a horizontal pipe at the steady rate of 2 ft³/min. Determine the velocity of the oil at a point where the diameter of the pipe is (a) 1 in. and (b) ½ in.

Usually the diameter of the bore (also the diameter of the piston) is given in inches. The formula for the area of a circular surface is $A = \pi r^2$. In hydraulics, the bore size is either given or it must be computed. The formula for the area of the bore is expressed in terms of its diameter. The radius (r) is equal to ½ the diameter, or $r = D/2$. The formula now becomes

$$A = \pi \left(\frac{D}{2}\right)^2 = \frac{\pi D^2}{4}$$

Substituting the constant 3.1416 for π, the formula is

$$A = \frac{3.1416 D^2}{4} = 0.7854 D^2$$

Using the formula

$$A = \frac{\pi D^2}{4}$$

determine

$$A_1 = \frac{\pi (1)^2 \text{ in}^2}{4 \times 144 \text{ in}^2/\text{ft}^2} = 0.00545 \text{ ft}^2$$

$$A_2 = \frac{\pi (0.5)^2 \text{ in}^2}{4 \times 144 \text{ in}^2/\text{ft}^2} = 0.00136 \text{ ft}^2$$

$$Q = 2 \frac{\text{ft}^3}{\text{min}} = 2 \frac{\text{ft}^3}{\text{min}} \times \frac{\text{min}}{60 \text{ sec}} = 0.0333 \frac{\text{ft}^3}{\text{sec}}$$

Using the equation $Q = Av$, the velocity through the 1-in. pipe is

$$0.0333 \frac{\text{ft}^3}{\text{sec}} = 0.00545 \text{ ft}^2 v_1$$

$$v_1 = \frac{0.0333 \text{ ft}^3/\text{sec}}{0.00545 \text{ ft}^2} = 6.12 \frac{\text{ft}}{\text{sec}}$$

Since the area A_2 is one-fourth of A_1, the velocity of the oil through the ½-in. pipe is four times that through the 1-in. pipe, or

$$v_2 = 6.12 \frac{\text{ft}}{\text{sec}} \times 4$$
$$v_2 = 24.5 \frac{\text{ft}}{\text{sec}}$$

Conservation of energy

Energy can be neither created nor destroyed, but it can be transformed from one form to another. This statement is known as the *law of conservation of energy.* This law is important to the systems designer in determining and accounting for all forms of energy within a system.

Bernoulli's theorem

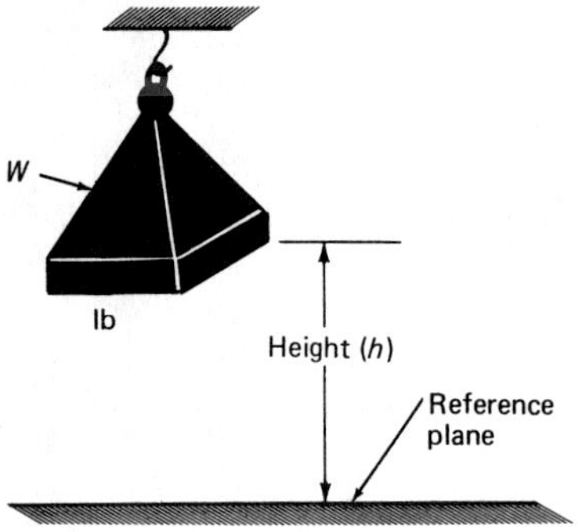

Because Bernoulli's theorem involves the sum total of energy within a system, it is necessary to know the various energy states. There are two basic energy states for Newtonian systems, that is, systems massive enough to behave according to Newton's laws of motion. The first of these states is potential, or stored, energy (Figure 2.11), or, in a system, the ability to do work. Figure 2.11 shows a weight W at an elevation h in relationship to the reference plane. The second energy state is kinetic energy, which is determined by the motion, or velocity, of a body.

$$KE = \tfrac{1}{2}mv^2$$

where *m* is the mass of the body and *v* is its velocity. Fluids, like solids, have mass, weight, and potential and kinetic energy. It follows that the total energy of a fluid system will be the sum of the various energy states within the fluid. This condition applies at any point in a system, and the total energy will be the energies up to the point being considered. Remember that the total energy of a system remains constant throughout the system. The forms of energy may change (Figure 2.12)—potential energy (*static pressure*) may become kinetic (*velocity pressure*) and kinetic energy may become potential—but the total energy (*total pressure*) remains constant (*K*).

In hydraulics potential energy is due to pressure applied to the fluid (called *pressure head*) or to the height (*elevation head*). Work done

Figure 2.11 Stored, or potential, energy.

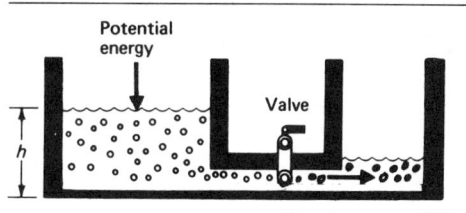

Figure 2.12 Opening the valve allows potential energy (h) to be converted to kinetic energy (KE).

on or by a system is included as a form of kinetic energy. Bernoulli's equation states: *Pressure head plus elevation head plus velocity head equals constant, or pressure energy plus potential energy plus velocity energy equals total energy of the system.*

Pressure $= p$

Elevation pressure (head) $= hw$

Kinetic energy pressure $= \frac{1}{2}\left(\frac{w}{g}\right)v^2$

In the preceding equation, p, hw, and $½(w/g)v^2$ are expressed in pounds per square inch (psi), and w is a unit weight in pounds per cubic inch (lb/in³). Substituting these values, the equation can be expressed mathematically as

$$p + hw + \tfrac{1}{2}\frac{w}{g}v^2 = K$$

If this equation is divided by W, the resulting unit is expressed in inches. Note the difference in units between pressure (psi) and pressure head (in.).

$$\frac{p}{w} + h + \frac{v^2}{2g} = K$$

Bernoulli's equation is the foundation on which systems analyses of energy relationships and power distribution are determined. The expressed equation for K applies to a theoretical steady flow of an incompressible liquid through a frictionless pipe of unvarying cross-section. In actual practice, usable energy is dissipated within a system as a result of friction, restrictions in components and plumbing, and mechanical inefficiencies of actuators.

▶ **Example** Given a reservoir of height $h = 1$ ft, which is open to the atmosphere, determine the velocity of fluid discharging into the atmosphere from a nozzle (1-in. diameter) in the bottom of the reservoir (Figure 2.13). Bernoulli's equation:

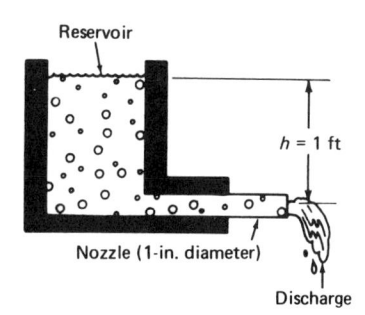

Figure 2.13 Determining fluid velocity.

$$\frac{p_1}{w} + h_1 + \frac{v_1^2}{2g} = \frac{p_2}{w} + h_2 + \frac{v_2^2}{2g}$$

Substituting values

$$\frac{p_1}{w} + h_1 + 0 = \frac{p_2}{w} + 0 + \frac{v_2^2}{2g}$$

$$h_1 = \frac{v_2^2}{2g}$$

or

$$v_2 = \sqrt{2gh_1}$$

$$= \sqrt{2 \times 32.2 \text{ ft/sec}^2 \times 1 \text{ ft}}$$

$$= \sqrt{64.4 \text{ ft}^2/\text{sec}^2}$$

$$= 8.02 \text{ ft/sec}$$

Although Bernoulli's theorem was derived for the steady flow of an incompressible *liquid,* it can also be applied to the steady flow of gases.

Bernoulli's theorem states: *At any two points along a streamline the sum of the pressure, the kinetic energy per unit volume and the potential energy per unit volume has the same value.* This may be expressed by the mathematical statement:

$$p + \rho g h + \tfrac{1}{2}\rho v^2 = K$$

In the motion of a fluid through a horizontal pipe in which changes in potential energy along a streamline are negligible, the preceding equation becomes

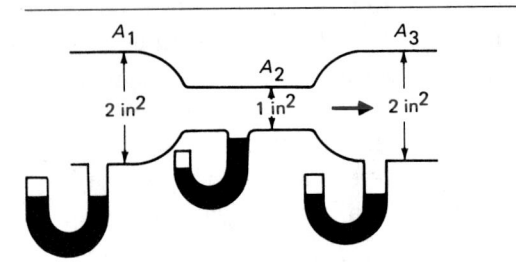

$$p + \tfrac{1}{2}\rho v^2 = K$$

Figure 2.14 The static pressure in the narrow portion is less than in the wide portion.

39 Conservation of energy

This equation shows that in the region where the velocity is small the pressure is large, while in the region where the velocity is large the pressure is small (Figure 2.14).

Torricelli's theorem

Torricelli's theorem states: *Ideally the velocity of the liquid coming out of a tank filled to a height (h) (feet or inches) above the opening is exactly the same as if the liquid had fallen through the same height.* Mathematically, the velocity is equal to the square root of the product of two times the acceleration due to gravity (32.2 ft/sec²) multiplied by the head:

$$v = \sqrt{2gh}$$

If an outlet were directed upward, the liquid discharging into the free atmosphere would rise to a height (h) equal to the level of the liquid within the tank. *Pressure head* is thus defined as the vertical height, in feet or inches, to which a given pressure will elevate a column of fluid. Note the difference in units between pressure (psi) and head (inches or feet).

▶ **Example** Given the information provided in Figure 2.15, determine the velocity at outlets 1, 2, and 3 of an oil tank.

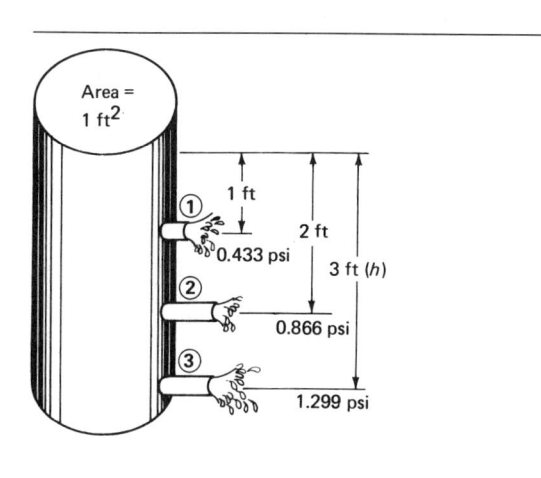

Figure 2.15 Illustration of Torricelli's theorem.

1 $v = \sqrt{2 \times 32.2 \times 1}$

   $= 8.02$ ft/sec

40 Fundamental laws of fluid mechanics

2    $v = \sqrt{2 \times 32.2 \times 2}$

       $= 11.35 \text{ ft/sec}$

3    $v = \sqrt{2 \times 32.2 \times 3}$

       $= 13.90 \text{ ft/sec}$

If the viscosity of the oil is considered, the velocity of the discharging fluid will be less than that determined by the equation.

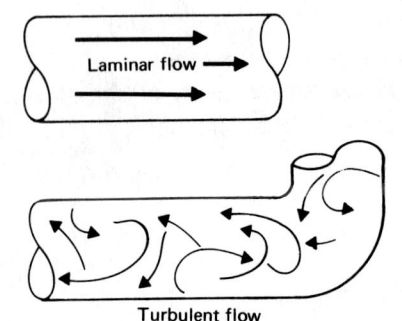

Figure 2.16 Flow patterns of fluid in conductors.

Reynolds number

In his observation of fluid flow characteristics, Osborne Reynolds noted that two basically different forms of flow exist, *laminar* and *turbulent*. Laminar, or streamline, flow occurs when the liquid particles flow smoothly in an orderly manner in even layers (Figure 2.16). Frictional losses are kept to a minimum and are greatest near the surface of the conductor. Turbulent flow occurs when the liquid particles flow in a random pattern. The flow lines are interwoven with each other in an irregular manner. It is possible to determine mathematically whether the flow in any fluid is in the laminar or turbulent range.

A dimensionless number, the Reynolds number (N_R), expresses this particular condition of flow.

$$N_R = \frac{\rho D v}{\mu} \quad \text{or} \quad \frac{Dv}{\nu}$$

where ν = kinematic viscosity
          ρ = mass density (lb-sec^2/ft^4)
          D = pipe inside diameter (ft)
          v = fluid velocity (ft/sec)
          μ = absolute viscosity (lb-sec/ft^2)

Absolute viscosity is the force required to move a flat plate of unit area at unit distance from a fixed plate with unit relative velocity when the space between the plates is filled with the fluid whose viscosity is being measured. Absolute viscosity (μ) is a measure of the value of that quantity, expressed as follows (Figure 2.17):

$$\mu = \frac{(F/A, \text{ lb/ft}^2)}{(v/Y, \text{ ft/sec per ft})}$$

41 Conservation of energy

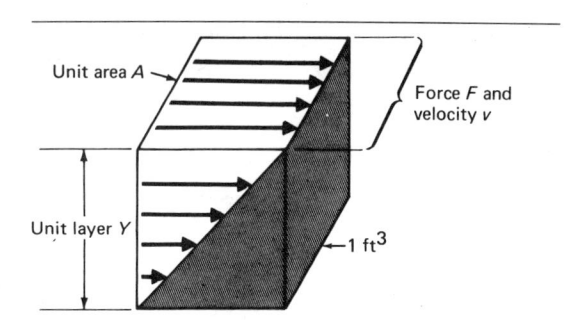

Figure 2.17 Absolute viscosity (μ) is the value of the shearing force per unit area (A) divided by the velocity (v) per unit layer (Y).

$$\mu = FY/vA, \frac{\text{lb-sec}}{\text{ft}^2}$$

Because most viscosity measuring instruments give readings in *kinematic viscosity* (v), which includes mass density (ρ), kinematic viscosity

$$v = \frac{\mu}{\rho} \text{ ft}^2/\text{sec}$$

must be converted to absolute viscosity before applying these units in the equation for the Reynolds number.

Liquid flow is generally laminar when N_R is less than 2000, and changes from laminar to turbulent between the N_R values of 2000 and 4000. It is known that temperature change has a great effect on oil viscosity and particularly on long-chain molecular synthetic fluid viscosity. For example, between the temperatures 70 °F and 160 °F, oil viscosity may change approximately 10:1.

What is the significance of determining the Reynolds number? It is of empirical value to us only for deriving the friction factor (f), which in turn is used in a subsequent equation to determine flow losses in pipes. The friction factor of Darcy's formula is used to determine the pressure drop, or head loss (hf), due to flow friction through a conduit. Mathematically, it may be expressed as

$$hf = \frac{fLv^2}{2Dg}$$

The *friction factor* is a function of the Reynolds number and the roughness of the pipe wall. For conditions in which the flow is laminar (N_R is less than 2000) and the oil is maintained at isothermal (same) temperature, the relationship is

$$f = \frac{64}{N_R}$$

(This ratio is also referred to as the *Hagen-Poiseuille formula*.)

When oil flow remains laminar and the system fluid temperature fluctuates, the friction factor tends to increase as follows:

$$f = \left(\frac{64}{N_R}\right)\left(\frac{L}{D}\right)\left(\frac{v^2}{2g}\right)$$

For turbulent flow determined for smooth steel or brass tubing, the average friction factor tends to follow this relationship:

$$f = \frac{0.3164}{(N_R)^{\frac{1}{4}}}$$

Many of the equations determined by Bernoulli, Torricelli, Reynolds, and other pioneering scientists who studied flow of hydraulic fluids are significant to designers of mathematical models of hydraulic systems. Only the simplest of these equations, however, are used extensively by engineers and technicians.

The Darcy-Weisbach equation expresses a useful relationship for flow losses in pipes.

$$hf = f\left(\frac{L}{D_i}\right)\left(\frac{v^2}{2g}\right)$$

where

hf = head loss (ft)
f = friction factor
L = length of conduit (ft)
D_i = internal diameter of conduit (ft)
v = velocity (ft/sec)
g = acceleration (32.2 ft/sec²)

This equation has limitations in that to assure constant velocity, internal conduit dimensions are assumed to be constant for the length on which calculations are based. Temperature of the fluid affecting the friction factor is also assumed to be constant. For all practical purposes, velocity is important because losses vary as v^2 does. It is therefore common practice to start with some desired or specified velocity and then determine the required size of piping.

▶ **Example** Given a friction factor of 0.016 and a specified velocity of 10 ft/sec, in a 2-in. internal diameter pipe, 100 ft long, determine the head loss.

$$hf = 0.016 \times \frac{100 \text{ ft}}{2/12 \text{ ft}} \times \frac{100 \text{ ft}^2/\text{sec}^2}{64.4 \text{ ft}/\text{sec}^2} = 14.9 \text{ ft}$$

Force-pressure-area relationships

The functioning of a fluid power system depends on the application of Pascal's law: Pressure of a confined and static fluid is transmitted undiminished to all portions of the confining vessel and exerts equal force on all equal areas at right angles to the surfaces involved.

To apply Pascal's law in fluid power, it is important to understand the distinction between the terms "pressure" and "force." *Force* can be defined as a push or pull, exerted against a total area in the case of fluids, and is expressed in pounds. Mathematically, the relationship of pressure, force, and area can be expressed as follows: *Pressure is equal to force divided by the area.*

$$\text{Pressure} = \frac{\text{force}}{\text{area}} \quad \text{or} \quad p = \frac{F}{A}$$

It is apparent from Figure 2.18 that a force acting over a piston area exposed to a fluid creates pressure in the confined fluid. Note

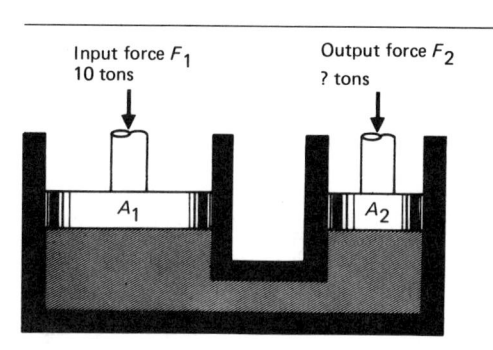

Figure 2.18 Force multiplication via changes in pressure and/or area.

that pressure is caused by any resistance to flow. Conversely, pressure acting on an area creates a force. Therefore, the formula can be expressed as: *Force is pressure times the area.*

$$F = pA$$

Figure 2.19 illustrates a device for recalling these relationships.

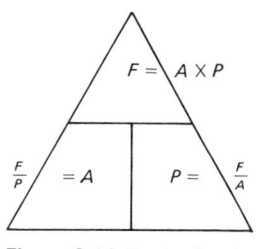

Figure 2.19 Device for determining force, pressure, and area relationships.

▶ **Example** Given a job to raise a 10-ton load using a 12-in. diameter piston A_1 (Figure 2.18) and another cylinder with a piston diameter of 2 in. to supply oil to the large cylinder, find the minimum force that must be applied to the small piston to raise the large one.

Calculation for large cylinder, A_1; plunger, A_2.

$$p = \frac{F_1}{A_1}$$

$$F_1 = 10 \times 2000 = 20{,}000 \text{ lb}$$

$$A_1 = 0.7854 \, D_i^2$$

$$A_1 = 0.7854(12)^2$$

$$= 113 \text{ in}^2$$

$$p = \frac{20{,}000 \text{ lb}}{113 \text{ in}^2}$$

$$= 177 \text{ psi}$$

$$A_2 = 0.7854(2)^2$$

$$= 3.14 \text{ in}^2$$

$$F_2 = pA_2$$

$$p = 177 \text{ psi}$$

$$F_2 = 177 \times 3.14$$

$$= 556 \text{ lb}$$

To overcome the resistance of the load on the large piston, a force in excess of 556 lb must be applied to the small piston.

The preceding example illustrates how force multiplication is used in industrial applications. A force of 556 lb is multiplied to a 10-ton force by utilizing a pressure of 177 psi and stepping from a 2-in. diameter piston to a 12-in. diameter piston.

In Figure 2.20, the effective area of Piston A_2 in a pressure intensi-

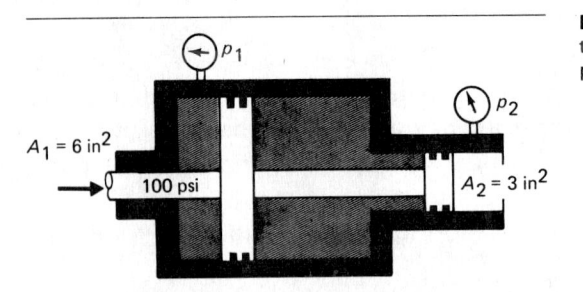

Figure 2.20 Applying Pascal's law to calculate pressure in a two-piston intensifier.

fier is 3 in² while A_1 is 6 in². A pressure of 100 psi is applied to Piston A_1. What will the pressure reading be for p_1 and p_2? Calculation for: force (F) developed by Piston A_1; pressure (p_2) developed by Piston A_2.

$$F = p_1 A_1$$
$$= 100 \times 6$$
$$= 600 \text{ lb}$$

$$p_2 = \frac{F}{A_2}$$
$$= \frac{600 \text{ lb}}{3 \text{ in}^2}$$
$$= 200 \text{ psi}$$

Similarly, it is possible to increase the pressure by reducing the area of a cylinder to perform the same amount of work.

Energy-work-power relationships

Fluid power has been defined as energy transmitted and controlled through a pressurized fluid. A fluid power system offers a relatively simple method of applying a large force with considerable flexibility of control. In fact, there is no comparable system that comes close to the combination of flexibility and dependability that is offered by a fluid power system.

A fluid power system is a planned and functional arrangement of components for getting a job done, that is, doing work. *Work* (G) is the product of a force (lb) times the distance that it moves (ft) (Figure 2.21). Work always involves the actual physical movement

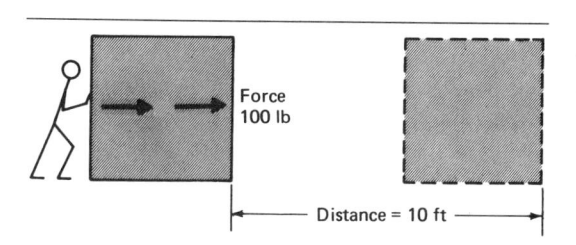

Figure 2.21 Work equals force (lb) times distance (ft) equals foot-pounds.

of an object as a result of an applied force. It is a measure of accomplishment: Unless a sufficient force is applied to move an object, nothing is accomplished by the force and no external work is performed.

Energy can exist in a great many forms—heat, light, electrical, pressurized fluid, potential, kinetic, etc. The many forms that energy can take and their interchangeability are illustrated by a hydroelectric plant (Figure 2.22).

Figure 2.22 Many forms of energy are represented in this figure.

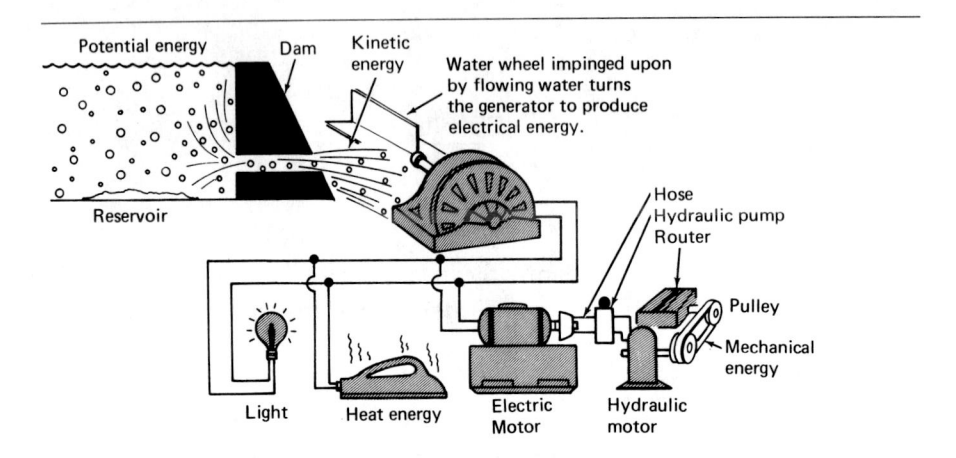

Energy, unlike work, can be at rest and still exist as energy, so long as it is capable of doing work. Work is a transfer of energy or a change in the form of energy (Figure 2.23). When an electric motor

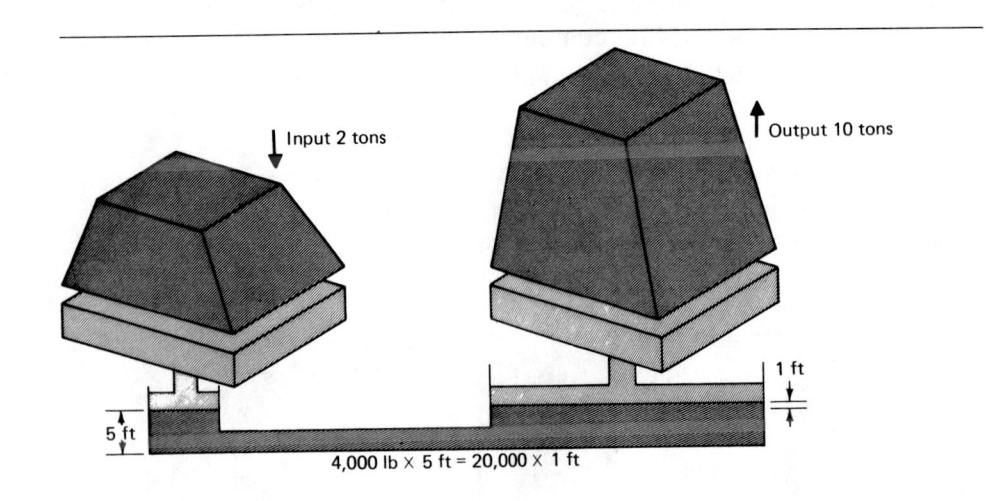

Figure 2.23 Work input equals work output (neglect losses).

turns a pump and fluid is pumped against a load, as in the case of a cylinder, pressure is created within the fluid. Fluid energy in the form of a pressurized fluid acts on the piston, and when the pressure is sufficiently high, the piston moves within the cylinder. Energy is transferred and, as the piston moves, work is accomplished.

In a fluid power system, the simplest and most frequently used form of force is that produced by a cylinder.

Force = pressure (psi) × piston area (in²)
 = lb

▶ **Example** The force exerted by a 5-in² piston (Figure 2.24) at 100 psi will be

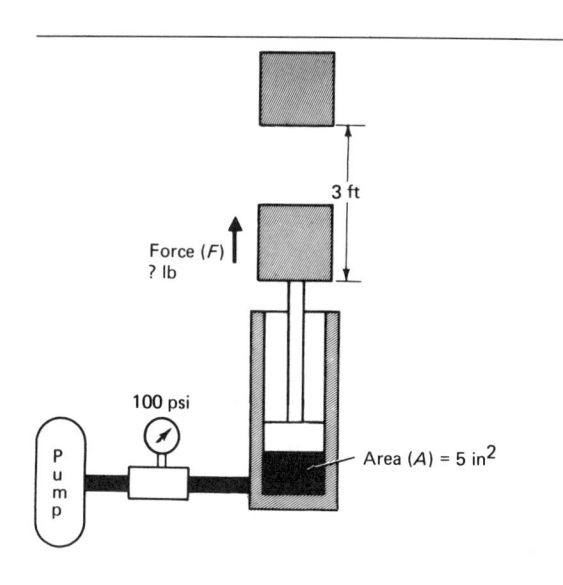

$F = pA$
$= 100 \times 5$
$= 500$ lb

Figure 2.24 The force (F) in lb exerted by the cylinder will be pressure (p) in psi times piston area (A).

A cylinder of this size operating at 100 psi is capable of doing how much work?

Work = force × distance
$G = Fd$
$= 500$ lb $\times$ 3 ft
$= 1500$ ft-lb

The weight or force of 500 lb creates a resistance to the fluid flow until a pressure of 100 psi is reached. At this point (ignoring friction), the pressure acting on the surface of the piston creates a force sufficient to raise the weight.

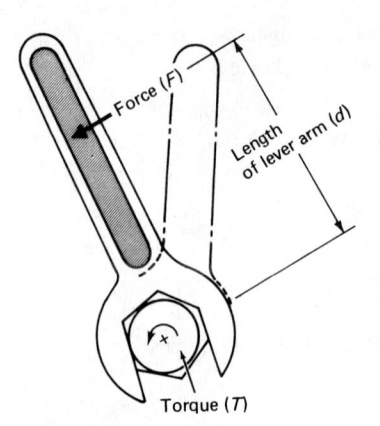

Figure 2.25 Mechanical torque (turning effort).

Figure 2.26 Hydraulic torque produced by a cylinder.

A rotary thrust or turning effort is called *torque*. Mechanical torque (T) may be expressed as *the force (F) multiplied by the length of the turning arm (d)* (Figure 2.25).

$$T = Fd$$

(Force is expressed in pounds, and distance is expressed in inches.) Applying 100 lb to a torque wrench with an arm length of 12 in. will develop a torque of

$$\begin{aligned} T &= Fd \\ &= 100 \text{ lb} \times 12 \text{ in.} \\ &= 1200 \text{ lb-in.} \end{aligned}$$

Hydraulic torque is the measure of the force required to rotate about an axis the body on which it acts, as in the case of a hydraulic cylinder or motor (Figure 2.26). This topic is treated in more depth in Chapter 6.

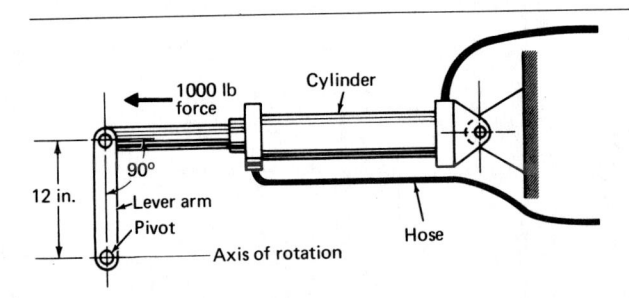

Looked at another way, the pressurized fluid is capable of doing work. With this understanding of the energy-transferring characteristic of a fluid power system, it is now possible to proceed to the definition and calculation of power.

Power is the time rate of doing work and can be calculated by dividing the total work done by the time consumed or required to do the work:

$$\frac{\text{Total work}}{\text{Time}} = \frac{Fd}{t}$$

The units of power used most commonly are

$$\frac{\text{ft-lb}}{\text{min}} \quad \text{or} \quad \frac{\text{ft-lb}}{\text{sec}}$$

49 Energy-work-power relationships

James Watt, in an attempt to rate the power of his steam engine, compared its output to the ability of a horse to lift a 550-lb bale of cotton vertically at the rate of 1 ft in 1 sec. Thus the unit of power was accepted as horsepower (Figure 2.27). One horsepower (hp) is equal to 550 ft-lb of work done per second or 33,000 ft-lb/min.

$$1 \text{ hp} = 550 \frac{\text{ft-lb}}{\text{sec}} \quad \text{or} \quad 1 \text{ hp} = 33,000 \frac{\text{ft-lb}}{\text{min}}$$

Figure 2.27 One horsepower is equal to 550 ft-lb of work done per second, or 33,000 ft-lb/min.

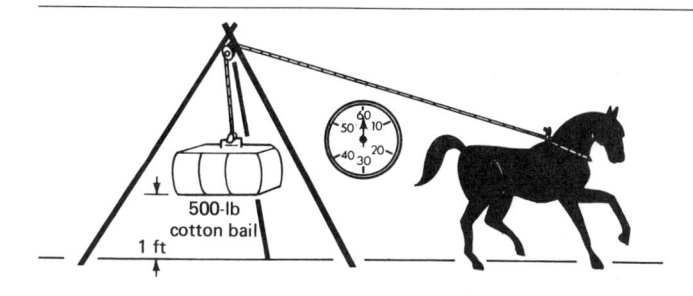

Moving an object through a distance involves work. To pump a quantity (V) gallons of oil against a resistance that causes a pressure of (p) psi, work must be done in the amount of

$$G = \frac{V \text{ (gal)} \times 231 \text{ in}^3/\text{gal} \times p \text{ lb/in}^2}{12 \text{ in/ft}}$$

You will recall that 231 in³ are equal to 1 gal. The quantity (V) is expressed in number of gallons. If the work in the preceding equation must be done in 1 min, the power required is

$$Q = \frac{V}{t} \frac{\text{gal}}{\text{min}}$$

$$\text{hp} = \frac{Q \times 231 \times p}{12 \times 33,000}$$

$$= 0.0005833 \, Qp$$

Stated another way, 0.0005833 hp is required to pump 1 gal/min (gpm) at 1 psi.

▶ **Example** A hydraulic pump has an output of 10 gpm at a pressure of 1000 psi. What horsepower is being consumed?

$$\begin{aligned}
\text{hp} &= 0.0005833 \, Qp \\
&= 0.0005833 \times 10 \times 1000 \\
&= 5.833
\end{aligned}$$

Applications of fluid power theory

Fluid power systems can be designed to perform three specific functions or any combination of these three—pressure control to vary the force; volume control to vary the speed of the actuator; and directional control to start, actuate, or stop the actuator.

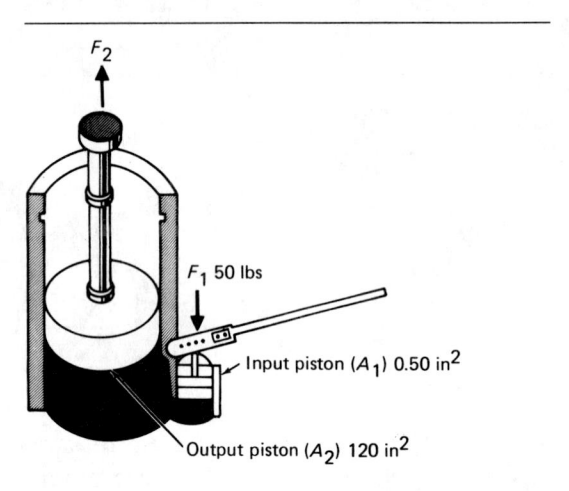

Figure 2.28 Application of Pascal's law.

Multiplication of forces in such applications as hydraulic jacks, rams, and presses is accomplished relatively easily. In Figure 2.28, if a force of 50 lb is applied to the input piston, this force is multiplied by using a larger output piston to exert a force of

$$F_1 = pA_1$$
$$50 = p(0.5)$$
$$p = 100 \text{ psi}$$

$$F_2 = pA_2$$
$$F_2 = 100 \times 120$$
$$= 12,000 \text{ lb}$$

In industrial applications, a hydraulic cylinder can be used in conjunction with a lever to exert great forces, as in presses. Given the force required to press-fit two component parts or to draw-form tubing and knowing the mechanical advantage of the lever as shown in Figure 2.29, it is possible to determine the force that would have to be exerted by the cylinder to produce the output force on the ram. Having determined the necessary force to be exerted by the cylinder, it would be possible to determine the economical size of cylinder and working system pressure to produce this force.

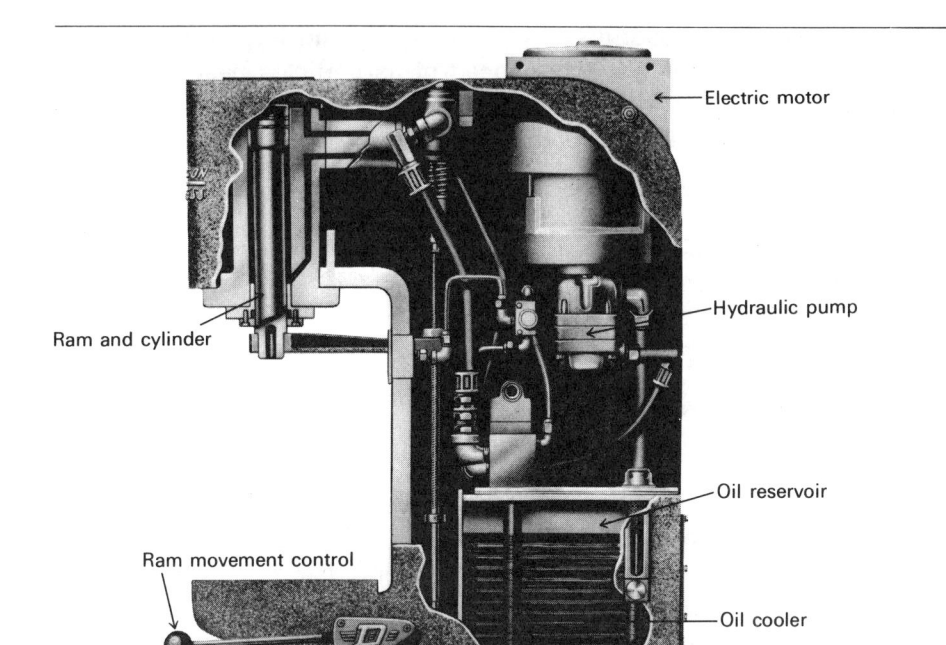

Figure 2.29 Hydraulic press. (Courtesy Denison Division, Abex Corporation)

A hydraulic car lift functions similarly to the hydraulic jack or press. The basic difference between the two is the medium used to transmit the force. In a hydraulic lift, air pressure applied to the surface of oil in a reservoir is transmitted by the oil to the output piston of the hydraulic cylinder (Figure 2.30). The compressed air exerts pressure on the oil, and the oil in turn acts on the piston to raise a vehicle. When the air valve is opened to the exhaust position, the compressed-air supply is blocked and the air from the cylinder exhausts to the atmosphere, thus lowering the vehicle.

Figure 2.30 Hydraulic hoist. Air supplied to the plunger raises it above the oil-filled cylinder.

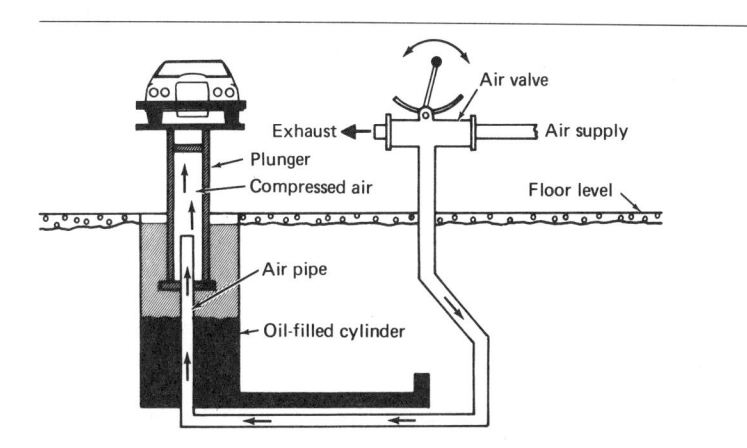

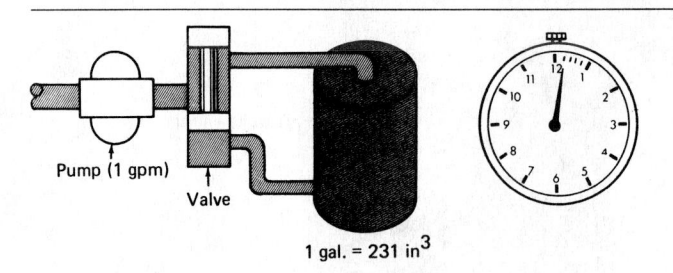

Figure 2.31 A 1-gpm pump would fill a 1-gal container in 1 min.

We have two ways of measuring flow—velocity (v) and flow rate (Q). The velocity of a fluid is the average speed of its particles past a given point. The flow rate is the measure of how much volume (Figure 2.31) of the liquid passes a point in a given time.

We have already mentioned controlling the speed of a piston within a cylinder by varying the volume of fluid flow. The velocity of a piston as it extends or retracts depends (1) on how fast the fluid (oil in the case of hydraulics) is fed into the cylinder and (2) on the area of the piston with which the oil comes into contact. With a fixed pump, delivering 10 gal per minute (gpm) to each of the cylinders in Figure 2.32, it will be noted that the distance traveled by the small cylinder is greater. Stated another way, the velocity of a small-area piston is greater than that of a large-area piston with an equal flow rate. Doubling the flow rate would double the speed of both pistons.

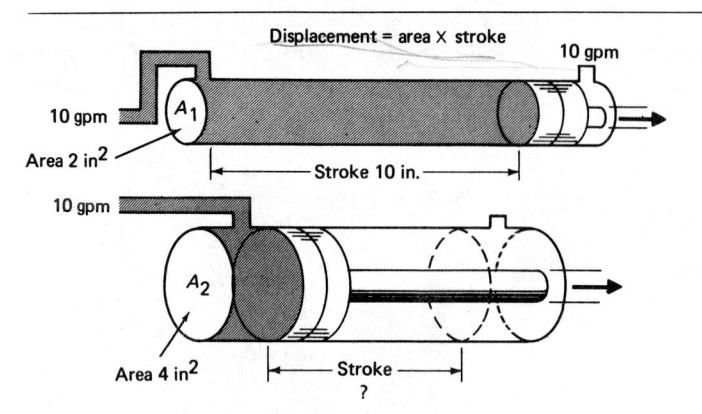

Figure 2.32 The small piston will move faster than the large piston.

Velocity varies directly with flow rate but inversely with the area. Mathematically, this relationship may be expressed as:

$$v = \frac{Q}{A}$$

Velocity is expressed in feet per second (ft/sec) or feet per minute (ft/min). One gallon is equivalent to 231 in³.

To substitute the appropriate units into the formula, it is necessary to make the following conversions:

1. Convert flow rate to equivalent units in cubic inches: 10 gpm × 231 in³/gal = 2310 in³/min.
2. Determine and express the area of the piston: $A = 0.7854 D^2$.
3. Convert the velocity from in/min to ft/min by dividing throughout by 12 (Figure 2.33):

$$v = \frac{10 \times 231}{4 \times 12} = 48.1 \text{ ft/min}$$

A simplified formula is derived by multiplying the original formula by 19.25.

$$v \text{ (ft/min)} = 19.25 \frac{Q \text{ (gpm)}}{A \text{ (in}^2\text{)}}$$

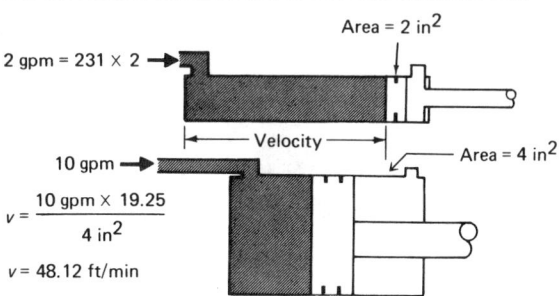

Figure 2.33 Determining the velocity of a piston.

Essentially, there are two ways of increasing the speed at which a piston moves without external controls, by increasing the gpm flow to the actuator or decreasing the size (area) of the piston. Conversely, the speed can be decreased by reducing the flow or increasing the area. The cost of a larger cylinder suggests that, where the power capability is sufficient, it is wiser to decrease the gpm delivery.

More frequently, it is necessary to determine the time it will take a piston to move when the displacement of the pump is known, as well as the distance or stroke the piston has to move. If the cylinder has a piston area A and a stroke s, its displacement volume (V_o) will be

$$V_o = As$$
$$= \text{area (in}^2\text{)} \times \text{stroke (in.)}$$
$$= \text{in}^3$$

Thus the time it will take the piston to move its full stroke can be determined by dividing the volume to be displaced by the flow rate Q of output oil coming from the pump.

$$t = \frac{V_o}{Q}$$

▶ **Example** Given a work cylinder of 23.1 in² and a stroke of 10 in. and a pump with an output of 693 in³/min, determine the time it would take the piston to complete the extension stroke.

$V_o = As$
$V_o = 23.1 \times 10 = 231$ in³

Time for extension stroke (Figure 2.34):

$$t = \frac{V_o}{Q} = \frac{231 \text{ in}^3}{693 \text{ in}^3/\text{min}} = 0.33 \text{ min or } 20 \text{ sec}$$

Figure 2.34 Determining displacement, or time, for the cycle of a piston.

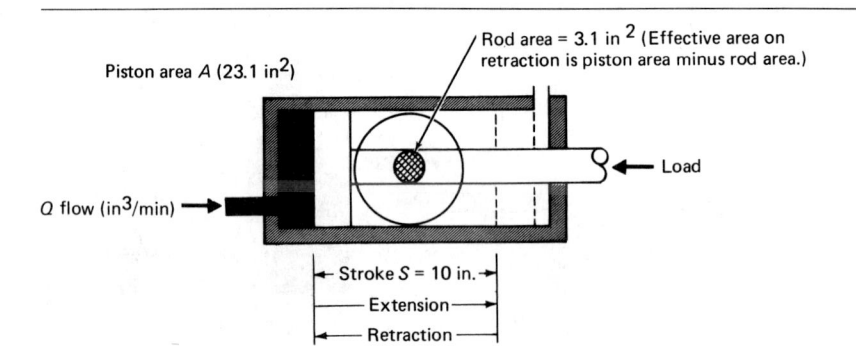

Controlling the direction of cylinders and motors is achieved through the appropriate selection of directional valves. This topic will be treated in Chapter 5. Controlling these three variables—pressure, flow rate, and direction of contained fluids—enables a designer to construct high-power, quick-action, and accurately controlled fluid power systems.

Summary

Fluid mechanics is the study of the physical behavior of fluids and fluid systems and the laws describing their behavior. A fluid is a substance that has a definite mass and volume but no definite shape.

Fluids expand when heated and contract when cooled. The basic gas laws describe the relationships of temperature, volume, and pressure. For mathematical calculation purposes absolute units of temperature and pressure are used in determining these relationships.

Pascal's law explains the relationships of force, pressure, and area of a confined static fluid. The relationship of velocity and cross-sectional areas of a conductor is explained by Bernoulli's theorem. This relationship pertains to a dynamic condition rather than a static one.

For fluid power output device applications, the force exerted is dependent on the available pressure, whereas the speed of the actuator is dependent on volumetric delivery and size of the actuator.

Fluid power systems can be designed to perform one specific function or any combination of these three:

1 control pressure to vary the force,
2 control volume to vary the speed of the actuator, and
3 control the direction of flow to start, actuate, or stop the actuator.

Knowing the basic laws governing the physical behavior of fluids and having the instruments to measure the variables will enable designers to predict, test, and control these variables in a functional fluid power system.

Review questions and problems

1 Define fluid mechanics.
2 On what dimension does the pressure exerted by a column of fluid depend?
3 What happens when fluids are (a) heated? and (b) cooled?
4 The speed of the piston is independent of the load or pressure; on what two variables does it depend?
5 How is the Fahrenheit scale reading converted to the Rankine scale reading?
6 How is the gage pressure reading (psig) converted to an absolute pressure reading (psia)?
7 If a cylinder has an area of 15 in^2 and a stroke of 30 in., what is the displacement on the extension stroke?

8 If the rod area is 5 in² in problem 7, what is the displacement of the cylinder on the retraction stroke?

9 How do you account for the difference between the answers to problems 7 and 8?

10 Oil flows through a 1-in. diameter pipe at a rate of 12 gpm. A valve placed in this line has a passage of $\frac{3}{4}$-in. diameter. What are the flow velocities through (a) the pipe? and (b) the valve?

11 In the cylinder of problem 7, how much work is done by the cylinder (neglect losses) if the pressure on the head end is 1000 psi?

12 Calculate the hydraulic horsepower in a fluid power system requiring 10 gpm at 1000 psi.

Selected readings

- Basal, P. R., Jr. (ed.) *Mobile Hydraulics Manual M–2990–S.* Sperry Rand Corporation, Troy, Michigan, 1967.
- Fitch, Ernest C., Jr. *Fluid Power and Control Systems.* McGraw-Hill, New York, 1966.
- *Fluid Power.* Bureau of Naval Personnel Navy Training Course, NAVPERS 16193–A, U.S. Government Printing Office, Washington, D.C., 1966.
- Hedges, Charles S. *Industrial Fluid Power,* Vol. 1. Nomack Machine Supply Company, Dallas, Texas, 1965.
- Henke, Russell W. *Closing the Loop.* Heubner, Cleveland, 1966.
- ———. *Introduction to Fluid Mechanics.* Addison-Wesley, Reading, Massachusetts, 1966.
- McKinley, James L., and Ralph D. Bent. *Basic Science for Aerospace Vehicles.* McGraw-Hill, New York, 1963.
- Oster, Jon, *Basic Applied Fluid Power: Hydraulics.* McGraw-Hill, New York, 1969.
- Pease, Dudley A. *Basic Fluid Power.* Prentice-Hall, Englewood Cliffs, New Jersey, 1967.
- Semant, Henry. *Fundamentals of Physics.* Holt, Rinehart, and Winston, New York, 1966.
- Stewart, Harry L. *Audel's Practical Guide to Fluid Power.* Theodore Audel, Indianapolis, Indiana, 1966.
- Stewart, Harry L., and John M. Storer. *Fluid Power.* Howard W. Sams, Indianapolis, Indiana, 1968.
- Tinetti, George K. *Fundamentals of Industrial Hydraulics.* General Motors Institute, Grand Rapids, Michigan, 1964.

Chapter 3
Fluids and auxiliaries

This chapter deals with the properties of fluids, as well as their types and additives. It also discusses liquids used in hydraulics and air used in pneumatics, as transmitting media of energy. Filters, as devices for removal of contaminants from the systems, and temperature-controlling devices, for removal and addition of heat, are also included. The chapter then presents reservoirs and accessories as subsystems of a hydraulic power system and discusses fluid conductors (including pipe, tubing, hose, and fittings) as part of the auxiliaries used to conduct fluids from one point to another.

Key terms

- **Additive** A chemical compound added to a fluid to change its properties.
- **Conductor** A system component whose primary function is to contain and direct fluid.
- **Connector** A device for joining a conductor to a component part or to other conductors.
- **Contaminant** Detrimental matter in a fluid.
- **Emulsion** A liquid mixture in which one substance is suspended in minute globules in another (e.g., water in oil or oil in water).
- **Filter** A device whose primary function is the retention by a porous medium of insoluble contaminants from a fluid.
- **Heat Exchanger** A device that transfers heat through a conducting wall from one fluid to another. Includes both coolers and heaters.
- **Manifold** A conductor that provides multiple connection parts.
- **Micron** A unit of length equal to one-millionth of a meter or 0.000039 in.
- **Viscosity** A measure of the internal friction or the resistance of a fluid to flow.
- **Viscosity index** A measure of the relative change in viscosity for a given change in temperature.

Fluids (both liquids and gases) are substances that change shape easily and take on the form of their containers. This characteristic makes it possible for fluids to pass through openings of different shapes and sizes within the components of a fluid power system.

Water was used as a liquid medium in the early hydraulic rams and presses. However, water has a corrosive effect on metals, does not provide adequate lubricity to moving parts, and freezes in cold temperatures. The discovery of petroleum oil and methods of specially compounding it for use in fluid power systems greatly accelerated the widespread use of hydraulics. Control of contamination, closer tolerances of machined components, and the development of better seals have contributed to the improved performance and life expectancy of fluid power components and systems.

Functions of fluids within a system

As previously explained, Pascal's law states that, in a static condition, pressure exerted on a confined liquid is transmitted undiminished in all directions and acts with equal force on all equal areas. This characteristic makes fluid an energy-transmitting medium. However, fluids must serve other important functions. Fluids must also provide good lubrication to precision moving parts to keep friction and wear to a minimum. Fluids must seal moving parts to prevent leakage, pressure loss, and a lower overall efficiency. By removing contaminants, fluids clean bearing surfaces and thus prolong the performance of equipment. Additives specially compounded with fluids provide the desired original qualities for given applications and help to retain those qualities.

Properties of fluids

To perform their functions well in hydraulic systems service, petroleum-base oils must be well refined and specially processed for such systems. Because of the varying operating requirements of hydraulic systems, additives are included to achieve the necessary performance properties and to retain these desirable properties for the longest time possible. Basic property requirements for hydraulic systems follow.

Viscosity

Viscosity is probably the most important physical property of a hydraulic fluid and is the characteristic most commonly used when specifying a fluid for a particular system application. It is defined as a measure of the internal friction or the resistance of a fluid to flow (Figure 3.1).

Figure 3.1 Viscosity is a measure of the internal resistance of a fluid to flow.

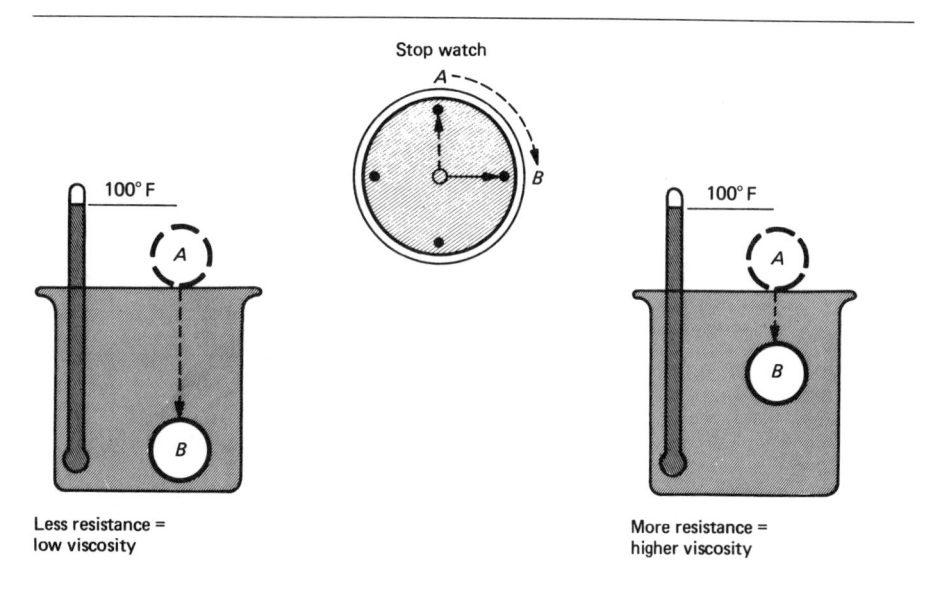

Less resistance = low viscosity

More resistance = higher viscosity

Measuring viscosity The viscosity of a petroleum-base oil and lubricant is measured with a viscosimeter according to an ASTM standard (Figure 3.2). This instrument measures the number of seconds it takes for a fixed quantity of liquid (60 cm^3) to flow through a standard orifice under a standard head and at a given temperature. Both 100 °F and 210 °F are common temperatures for determining viscosity. The time flow is recorded in seconds, and the viscosity reading is indicated by Saybolt Seconds Universal (SSU). Thus viscosity is a measure of flowability at a definite temperature. An oil of high viscosity, 850 SSU [Society of Automotive Engineers (SAE) 40], has more resistance to flow than an oil of 202 SSU (SAE 10-W) at 100 °F. The overall viscosity requirements of oils for hydraulic systems depend on component and systems design and size, operating pressures, and temperatures.

Effect of viscosity on fluid power systems Viscosity varies with the actual temperature of the fluid and the operating pressure. If the viscosity of the oil is too high, the following undesirable results may occur:

Figure 3.2 Saybolt viscosimeter for measuring viscosity of fluids.

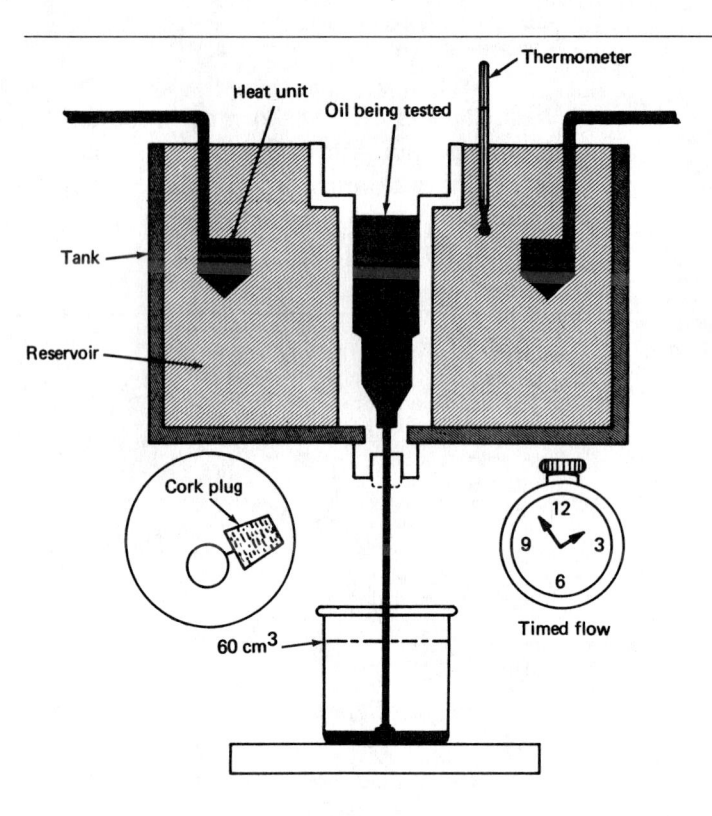

- The internal friction of the fluid will increase, which in turn will increase the flow resistance through the clearances of the pumps and valves.
- The temperature will increase.
- The operation will become sluggish.
- Pressure drop throughout the system will increase.
- Power consumption will increase.

Should the viscosity of the oil be too low, the following results may occur:

- Internal and external leakage will increase.
- Pump slippage will increase, which will reduce pump efficiency and increase oil temperature.
- Rate of wear of moving parts will increase.
- The system will lose pressure.
- There will be a loss of precision control.

The hydraulic pump, which is a high-precision performing device, is the most critical component within a hydraulic system with respect to viscosity. It is important that the pump manufacturer's specification be observed in the selection of appropriate fluids.

Petroleum oils, including the naphthenic and the paraffinic, tend to become thin (low viscosity and high flowability) at higher temperatures. Such oils also thicken (high viscosity and low flowability) as the temperature decreases.

Hydraulic fluid suppliers, pump manufacturers, system designers, plant equipment purchasers, and plant supervisors are all concerned with viscosity.

Viscosity index The rate of change in viscosity with a change in temperature is denoted by the viscosity index (VI), which is a numerical value determined at 100 °F or 210 °F. A high VI value indicates a low rate of change in viscosity with respect to temperature and is therefore desirable where considerable temperature changes do occur (Figure 3.3). For example, a fork lift may operate alternately indoors at room temperature and outdoors at a much colder temperature. A petroleum-base fluid having a viscosity index of about 90 is generally selected for hydraulic applications.

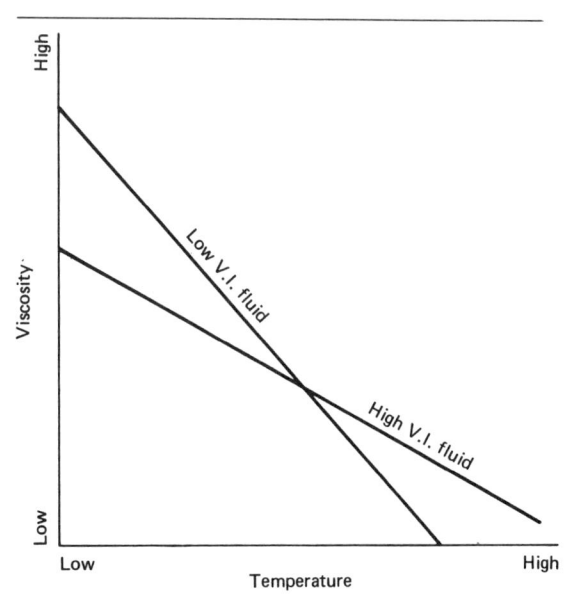

Figure 3.3 Viscosity index: The higher the VI, the smaller the change in viscosity due to temperature changes.

Paraffinic oils as a group generally have much higher viscosity indices than naphthenic oils. By solvent-refining an oil from a paraffin-base crude, viscosity indices in the range of 80–110 can be produced. Additives such as polymers and long-chain hydrocarbons are used as VI improvers in aircraft hydraulic fluids, which are exposed to a wide range of temperatures.

It is important to note that the VI's of the aqueous-base fire-resistant fluids are quite different from those of petroleum oils. Any change in the water content will change the fluid's viscosity. Whenever it is necessary to change from a petroleum oil to a fire-resistant fluid, the viscosity index should be known and considered. This is particularly critical if the hydraulic system operates in extreme temperature ranges. In such cases, it is essential that a high-VI oil be used. If the oil is too thick when the hydraulic system is started at a low temperature, it won't flow, or it may be so thin at the highest temperature as to produce all of the unfavorable results previously discussed.

Viscosity indices for phosphate esters are generally very low: Some are very low, approximately 150 for water-containing fluids, and some are very high for silicones and halogenated hydrocarbons. Most of the oil companies and fluid manufacturers have excellent bulletins that provide technical data and performance specifications.

Oxidation stability

Oils, which are hydrocarbons, are susceptible to *oxidation,* which is the reaction of oxygen with oil which can cause numerous undesirable effects. The chemical reaction will ultimately produce insoluble gum, sludge, and varnish deposits. Oils vary widely in their oxidation stability, depending on their type and whether oxidation inhibitors are present. In addition to refining and blending the oil, factors such as operating temperatures, pressures, contaminants, metal surfaces of components, flow rates, and cycle patterns all affect the rate of oxidation.

Temperature is the prime accelerator of oil oxidation. The rate of chemical reaction, including oxidation of hydrocarbons, will approximately double for every 18 °F increase in temperature. Resulting contamination tends to speed up the oxidation process. Below 135 °F the oxidation rate is relatively low; however, for every 15° above 140 °F the life of the oil is estimated to decrease approximately 50%. Fluids developed for aerospace and military applications may be operated with normal upper limits of 240 °F. Currently, research is under way in the development of silicones and ester fluids for operation at temperatures up to 400° and 450 °F.

Contaminants such as dirt, moisture, thread compounds of hose, paint, and entrained air all accelerate the rate of oil oxidation. Such contaminants should be reduced to a minimum level. High-quality hydraulic oils contain inhibitors that decrease the influence of factors contributing to oil oxidation.

Even though laboratory tests for viscosity and neutralization are available, the length of satisfactory service obtained in an actual installation depends on the equipment and the environment in which the fluid is functioning.

Hydraulic systems, being made of ferrous metals, are susceptible to *rusting,* which is oxidation of the surface of a ferrous metal. Rust occurs in a hydraulic system because of moisture due to condensation either from air entering through a vent of a reservoir or from air leaks on the inlet side of the pump. As ferrous metal rusts it gains weight and enlarges. Precision-fitted parts cannot function efficiently if they are too large or if flakes of rust get between precision moving parts of pumps, valves, or cylinders.

In mobile and outdoor applications of hydraulic equipment, condensation of appreciable amounts of water in the reservoir may occur. Periodic draining or installation of water separators is necessary in such systems. Because it is practically impossible to completely eliminate moisture from the system in all prevailing working conditions, chemists have developed chemical inhibitors that keep moisture away from the metal's surface. Rust-inhibiting hydraulic oils form a protective layer several molecules in thickness, which is sufficient to retard rusting and contamination.

Foaming

In addition to normal amounts of entrained air in hydraulic fluid introduced when the oil comes in contact with the atmosphere within the reservoir, it is possible for excessive air to be in the fluid. Some of the common causes of excessive air are: too low a level of oil within the reservoir, permitting the pump inlet pipe to be exposed to air; excessive turbulence of return oil due to improper design; and leaks in defective components of the system. Air entering oil not only results in erratic motion of component parts but also increases oil oxidation and contamination. When air trapped in cylinders is compressed, it creates heat, which in turn causes deterioration of the oil. Depressants of the foaming ability of oil tend to speed up the separation of air from the oil. It is necessary to bleed air out of all lines and components when first starting up a system or after a system has been serviced.

Pour point

The *pour point* may be defined as the lowest temperature at which a fluid will pour or flow under specified laboratory procedures (Figure 3.4). Pour point becomes a consideration if low starting temperatures are encountered. For hydraulic installations on equipment required to operate in cold temperatures, it should be 20–30 °F below the lowest operating temperature. If it is too close to the operating temperature, the oil may thicken sufficiently to prevent it from flowing readily into the inlet side of the pump. Pour point depressants can be added to hydraulic oils to render them suitable for use at lower temperatures.

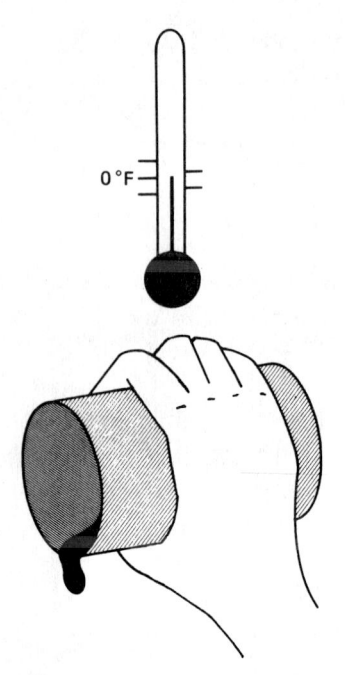

Figure 3.4 Pour point is established at the temperature oil stops to flow.

Antiwear

One of the two important functions of a hydraulic fluid is to lubricate component parts in motion. A high-quality hydraulic fluid must have good lubricity to prevent undue friction and wear of the system's moving parts under operating conditions of given temperature, pressure, and speed. Premium quality hydraulic fluids that incorporate antiwear additives provide significant reductions in wear during laboratory tests.

Water separation

Any moisture that enters a hydraulic system, whether through condensation or contamination, will emulsify with hydraulic oils under the dynamic conditions that exist within a functioning system. These emulsions further promote the collection of dirt, dust, and grit, which cause wear and oxidation that adversely affect the proper functioning of valves and actuators. Highly refined mineral oils tend to separate readily from water. Additives mixed in the fluid must be chosen to have no adverse effect on the water-separating properties. Keeping condensation to a minimum, using high-quality refined oils, and periodically draining the system will reduce wear due to friction and oxidation.

Understanding the properties of hydraulic oils and their relative importance under varying conditions should enable designers to select more intelligently the oil for a system to operate within given conditions. Because the chemistry of oils is complex, and systems are designed to operate under specified conditions, it is best to rely on the specifications of the manufacturer of the equipment and your oil supplier. Maintenance of a system should ensure maximum performance under the operational conditions for which the equipment was designed.

In general, superior grades of hydraulic power transmission oil should have the following characteristics:

- exceptionally high stability,
- high viscosity index,
- good rust and corrosion prevention,
- low carbon-forming tendency,
- good water-separating tendency,
- nonfoaming capability, and
- low pour point.

Types of hydraulic oils and fluids

There are several basic types of hydraulic oils and fluids. The more common are: (1) straight-distilled oils from various mineral crudes; (2) straight-distilled oils that are treated with additives; (3) solvent-extracted oils; and (4) oils from the above groups to which chemical inhibitors have been added. The characteristics of petroleum-base hydraulic oils are controlled by the type of crude oil used, the method of refining, and the additives included.

Properly refined naphthenic oils have good resistance to emulsification and have a high resistance to foaming. Naphthenic oils are suitable for systems working at temperatures as high as 150 °F, but their relatively low viscosity index makes them unsuitable for use in systems where the oil temperature varies too widely.

Paraffinic oils have a higher viscosity index than naphthenic oils and are thus more suitable for use in systems where temperature variations are widely separated. Oils of this type will perform satisfactorily in the presence of high moisture, contaminants, and other accelerators of oxidation.

Fire-resistant

When a hydraulic system is located near high-temperature equipment or sources of ignition that could endanger personnel or equipment, the use of a fire-resistant hydraulic fluid is important. Hazards of fire exist particularly in industries such as die casting, foundries, heat treating, mining, steel mills, and welding. Fire-resistant hydraulic fluids fall into three types:

1 water-in-oil emulsions,
2 water-glycols, and
3 synthetic fluids.

Water Water was used as the fluid medium in early hydraulic systems. It is still suitable for a number of large industrial hydraulic installations or processes that require high pressures and low operating speeds such as forges, foundries, die casting, and welding. As a hydraulic liquid, water presents many problems such as rust and corrosion, poor lubrication, temperature variation intolerance, and abrasive contaminants. On the credit side, water has ideal fire resistance and is available in large quantities at relatively low cost, as opposed to petroleum-base fluids. When water is used as a hydraulic fluid, it is often combined with certain oils or glycols to provide the necessary properties for resistance to fire, lubricity, and rust prevention.

Water-in-oil emulsions These fluids are mixtures of water, petroleum oil, and an emulsifying agent (soluble oil-type). Each drop of water is encased in a skin of oil that breaks at elevated temperatures to release fire-smothering steam. With their higher viscosity, these fluids don't leak too much and their distinctive yellow color makes leaks easier to detect. Operating temperatures are lower with emulsions, freezing does not materially affect them, and stability is adequate. Their lubricating property and corrosion resistance are fair to good. Water-in-oil emulsions are compatible with most seals except those made of natural rubber.

Emulsion fluids have excellent cooling characteristics and very low viscosity. Such oils are highly refined, with good emulsion, and are inhibited against oxidation, rust, and foaming. Water-in-oil emulsions cost generally twice as much as petroleum-base fluids.

Several manufacturers and suppliers have concentrates of ready-mix base oil, emulsifier, and additives. Distilled water, free from harmful iron, lime, salts, and other contaminants, is mixed with the concentrate.

Water-glycol These fire-resistant fluids are mixtures of water, glycol, lubricant, inhibitors, and a thickener. Their fire resistance is excellent, provided the water content is kept at 35 to 50% of the solution. Water-glycol mixtures have many of the desirable lubricating properties of oil-in-water emulsions, particularly at temperatures below 120 °F.

Water-glycol fluids cost more than emulsions, have a higher specific gravity that may cause pump starvation, and have a solvent action on most paints, enamels, and varnishes. Buna "N" elastomers, neoprene, treated leather seals and packings are compatible with water-glycol fluids. Other materials, including butyl rubber, nylon, Teflon, and Viton, are also nonreactive with water-glycol fluids.

Perhaps the most important precaution when using water-glycol fluids is to check the fluid in a system at regular intervals (3–6 months) to determine the rate of water evaporation so that proper viscosity is maintained.

Synthetic The phosphate esters, phosphate-ester-base, and chlorinated-hydrocarbon-base fluids have outstanding lubricating quality at high pressures and they are particularly suitable where pumps have heavily loaded antifriction bearings. These fluids provide excellent fire resistance and are stable in continuous operation up to 300 °F. Unless fortified with rust inhibitors, phosphate esters do not protect against rusting caused by water contamination. Synthetic fire-resistant fluids help to increase efficiency by reducing maintenance. They are also excellent detergents and prevent build-up of sludge on the hydraulic components.

A major factor in selecting straight synthetic fluids is that they are not compatible with the seals normally used for petroleum fluids. Seals of butyl rubber, Viton, ethylene-propylene (EP) rubber, silicone

68 Fluids and auxiliaries

rubber, Teflon, and nylon are suitable. Seals must be changed when a system is being converted to synthetic fluids. Straight synthetic fluids have a strong solvent action on most paints, enamels, and varnishes. Cured phenolic and epoxy paints, as well as nylon-base paints, are compatible with these fluids.

Silicone-base fluids are being researched for operation in more critical and elevated temperatures. These fluids are compatible with nitrile rubber, Fluorel, and braid impregnated with Teflon suspensoid.

Manufacturer's recommendations should be closely followed in changing over from a petroleum-base fluid to a fire-resistant fluid or from one fire-resistant fluid to another. Thorough draining, cleaning, flushing, and refilling procedures are usually required. In most cases, it is necessary to disassemble components and change seals and gaskets.[1]

Fluids of the future

Fluid manufacturers and suppliers are working closely with component and system manufacturers to gain acceptance of their products. As new and more critical requirements are called for in numerically controlled machines, aircraft systems, and servo controls, the search for compatible fluids goes forward. Greater extremes of operating temperatures, higher working pressures, new seal materials, better safety standards, and closer machine tolerances are all changing the requirements of fluid types and properties.

Improved additives, better conditioning, high-quality refining of base oil, and improved testing will contribute to safe and efficient fluids for a wide variety of applications.

The record of fluid power progress clearly indicates that the development of mechanical equipment is simultaneously affected by the development of specific fluids for each intended new application. Breakthroughs in high-temperature tolerant fluids have provided evidence in the past that when equipment required fluid power, satisfactory systems and fluids were made available. For example, in space vehicles hot gases are obtained by bleeding off rocket engines' exhaust gases or by using a liquid propellant gas generator.

---

[1] For a thorough discussion of compatibility factors of fluids, see Dudley A. Pease, *Basic Fluid Power,* Prentice-Hall, Englewood Cliffs, New Jersey, 1967, pp. 47–52.

69 Types of hydraulic oils and fluids

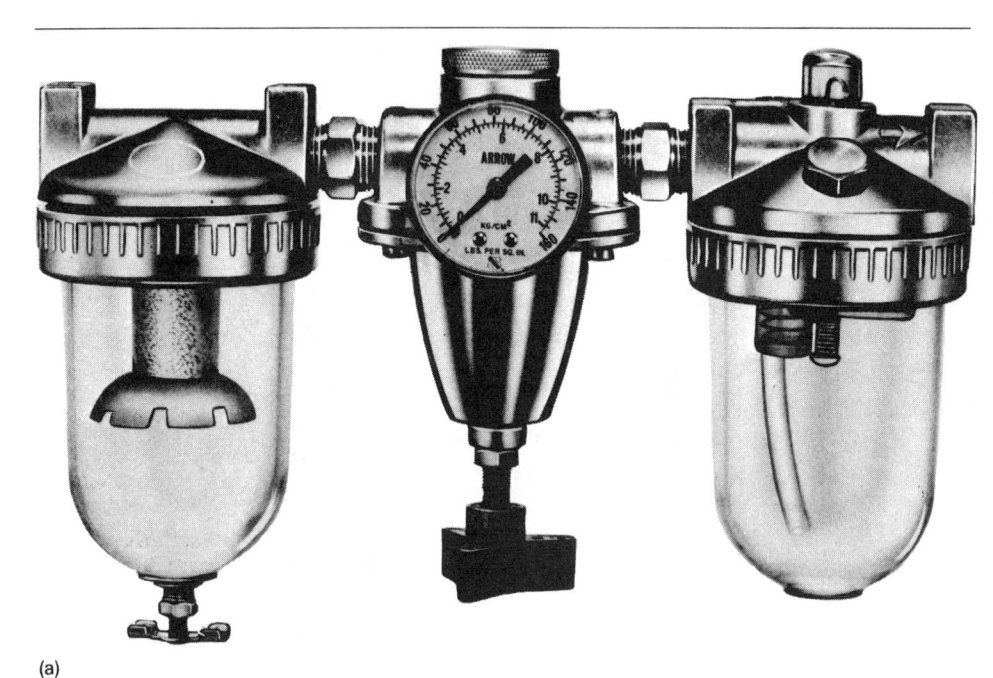

Figure 3.5 (a) Filter-regulator lubricator (FRL) unit for conditioning and regulating air pressure. (Courtesy Arrow Sintered Products Co.) (b) Construction and operation of an air-line filter. (Courtesy Arrow Sintered Products Co.) (c) A balanced piston regulator. (Courtesy Arrow Sintered Products Co.) (d) Components of a drop lubricator (Courtesy Arrow Sintered Products Co.) (e) Compressed air aftercoolers circulate compressed air through finned tubes and cool it by blowing atmospheric air around fins and tubes. (Courtesy Young Radiator Company, Racine, Wisconsin) (f) Heat transfer compressed-air drier. (Courtesy Kellogg-American, Inc.) (g) Air purifier: Compressed air is dried during compression cycle to meet critical moisture content in air lines. (Courtesy Aeroquip Corporation) (h) Air flow is reversed through dessicant canister during the regeneration cycle. (Courtesy Aeroquip Corporation)

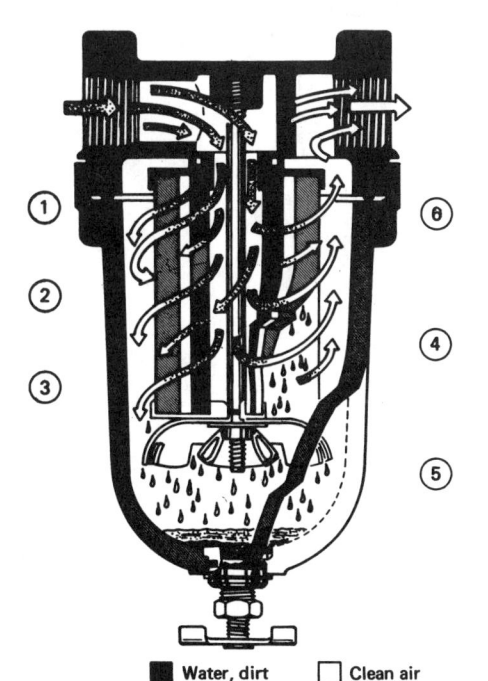

This is how it works:

1 Contaminated air flows to the inside of the element assembly.

2 Larger particles are filtered by the inner core of ribbon layers. Core's high strength supports entire assembly. Its continuous series of tiny orifices distributes air evenly to next filter media.

3 Air flow through a $\frac{1}{8}$-in. thick wall composed of glass fibers during which the minute particles are trapped, and vapors are coalesced into droplets.

4 A wall of foam completely encircles the filter walls and collects the contaminants to prevent reentry into the cleaned air path.

5 Separated liquids flow by gravity into quiet zone in bottom of filter bowl where they are discharged through the drain.

6 Clean clear air flows out through discharge port.

70 Fluids and auxiliaries

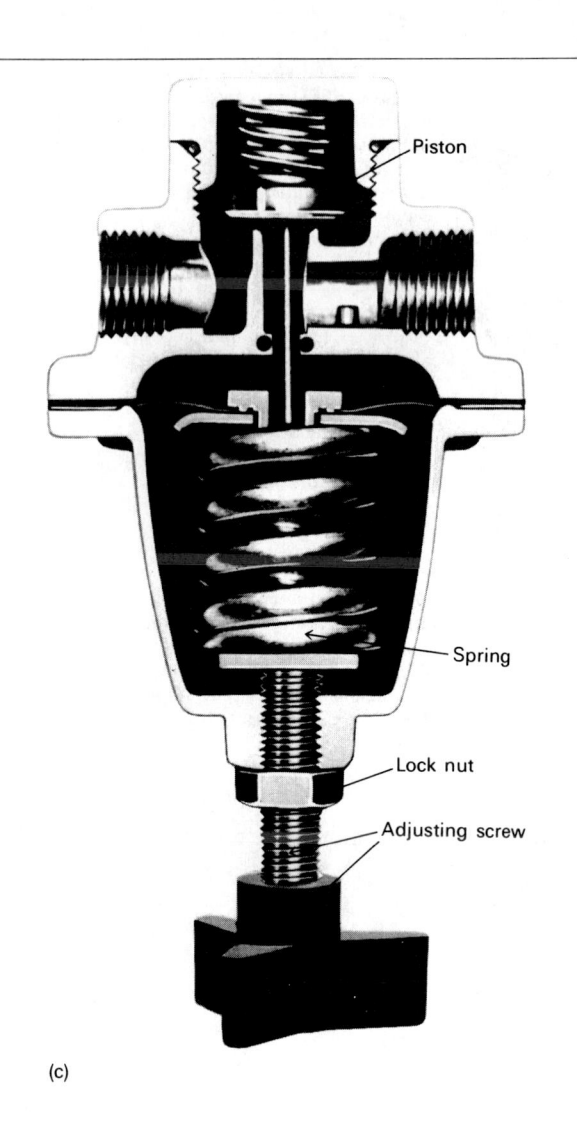

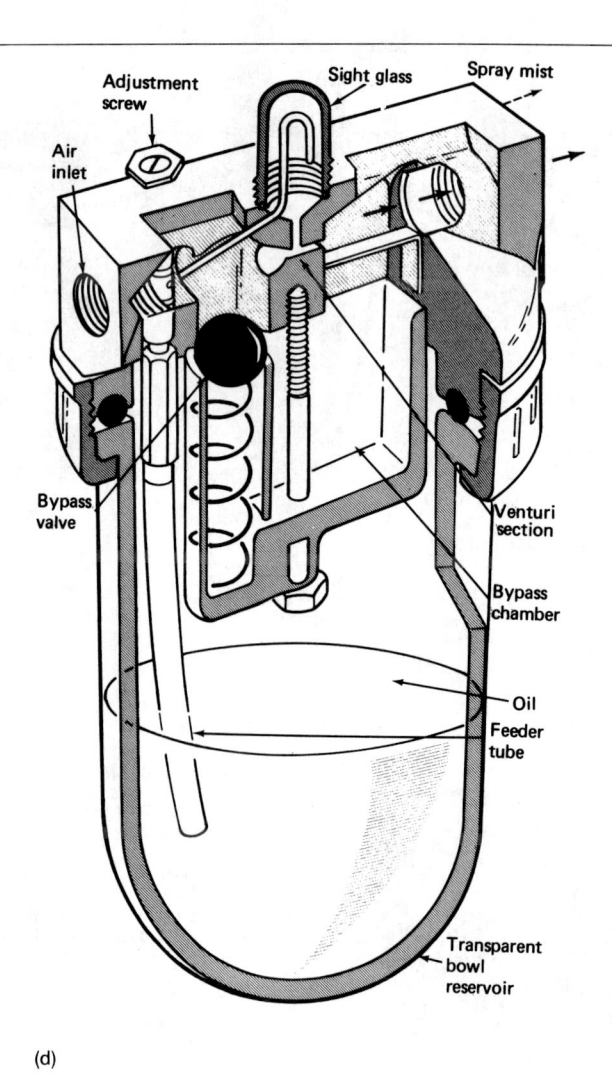

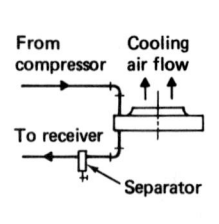

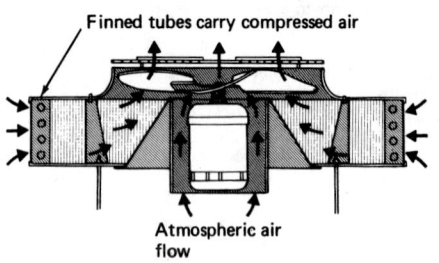

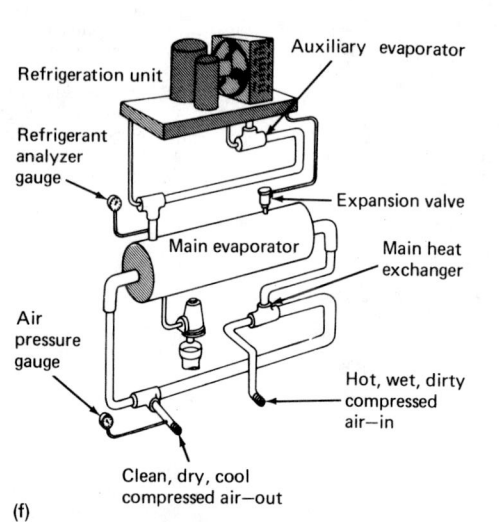

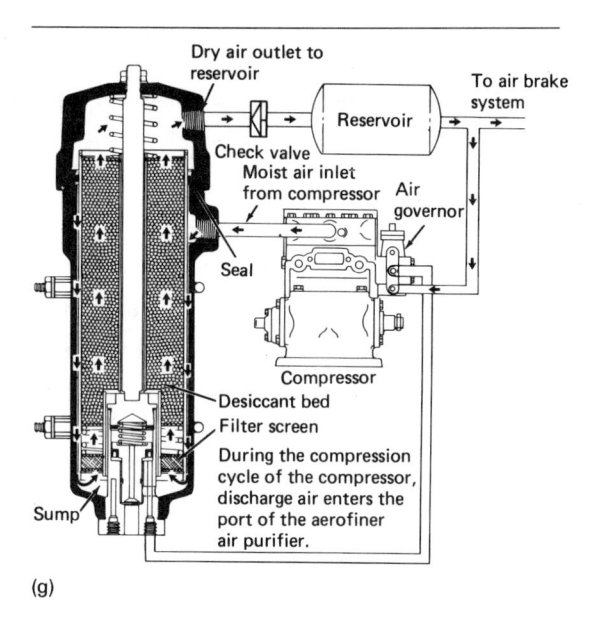

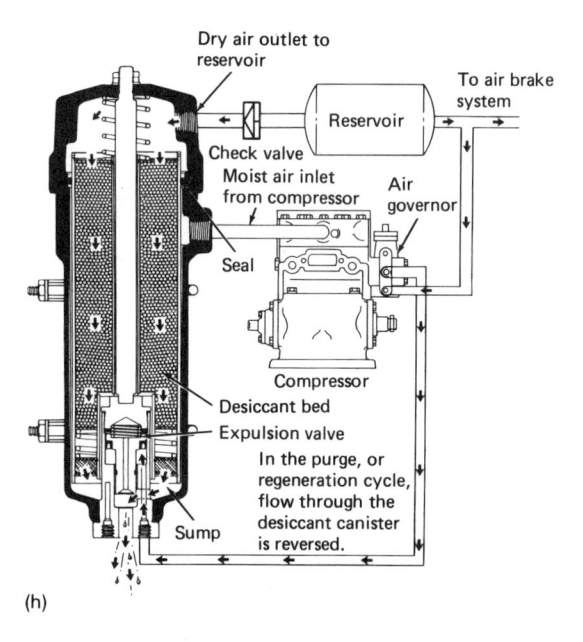

(g) *[Diagram: Aerofiner air purifier during compression cycle, showing dry air outlet to reservoir, check valve, moist air inlet from compressor, air governor, compressor, seal, desiccant bed, filter screen, sump. Caption: During the compression cycle of the compressor, discharge air enters the port of the aerofiner air purifier.]*

(h) *[Diagram: Aerofiner air purifier during purge/regeneration cycle, showing dry air outlet to reservoir, check valve, moist air inlet from compressor, air governor, compressor, seal, desiccant bed, expulsion valve, sump. Caption: In the purge, or regeneration cycle, flow through the desiccant canister is reversed.]*

Fluid conditioning

Gases are conditioned by four means: filtration, lubrication, heat transfer, and drying (Figure 3.5). Air filters are used to remove both liquid and solid contaminants from the compressed air and should be placed as close to the working components as possible. Lubricators are used to add a lubricant to the compressed air for applica-

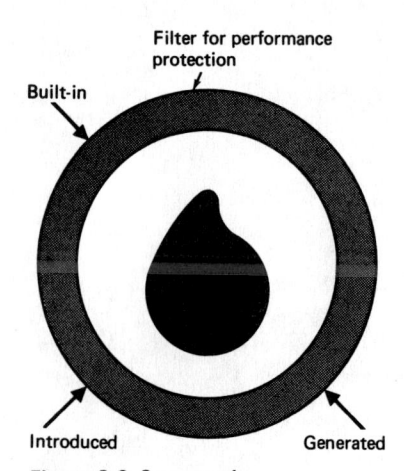

Figure 3.6 Sources of contamination.

Figure 3.7 Typical forms of contamination.

tions where this is necessary. Heat exchangers are used between stages of a multistage compressor or at the output of the compressor. Dryers are used wherever excessive moisture must be removed from compressed air. Conditioning of hydraulic liquids is accomplished by two methods, namely, filtration, or removal, of contaminants and heat transfer.

Contamination is defined as the presence of any substance in a fluid system that is detrimental to that system (Figure 3.6). Hydraulic system contamination may be classified into three types:

Built-in Contaminants are left in the components or systems during manufacturing or fabrication.

Introduced A system operating in a polluted environment may take in airborne contaminants, or replenished oil may introduce contaminants.

Generated Wear particles are continually generated and permitted to flow through the system.

Common contaminants include liquid, metallic, nonmetallic, and fibrous particles (Figure 3.7). For optimum system performance the three types of contamination may be controlled as follows:

1 To safeguard against built-in contaminants, the system is flushed with clean fluid prior to operation.
2 Introduced contaminants are frequently the result of neglect and poor housekeeping. An air filter breather on the reservoir and a strainer on the filler cap will also prevent any foreign particles from entering the reservoir.
3 Generated contaminants such as water condensation, corrosion particles, acids, sludges, and bacteria can be removed from circulation through full-flow filtration, use of magnetic plugs, and periodic draining and flushing of the system.

The effects of contaminants include not only a loss of function but also a gradual loss of accuracy and reliability as a result of wear. Excessive noise, leaks, sluggishness, and wear are all indicators of poor performance and suggest that maintenance is required. In many well-organized and carefully operated maintenance programs, the components are inspected for wear at scheduled times. A schedule for regular inspections should be established in conjunction with careful analysis of the environment, the system application, and

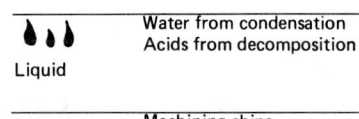

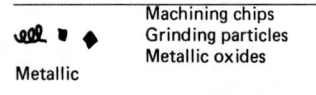

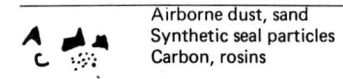

73 Fluid conditioning

recommendations of the manufacturer. Preventive maintenance, good housekeeping, and adherence to manufacturer's specifications will ensure longer and more efficient system performance.

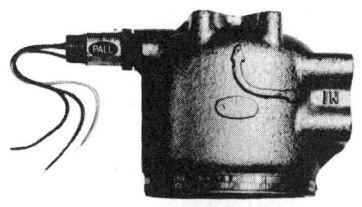

Filtration

Properly designed and maintained filter systems can eliminate about 75% of the potential causes of fluid power system failure. In addition, the life of system components is increased considerably.

Materials used in filters Hydraulic filters and strainers are usually located near to or in the system reservoir. A *filter* is a device that retains, by some porous medium, insoluble contaminants from a fluid [Figure 3.8(a) and (b)]. A *strainer* is a device that removes coarse contamination from a hydraulic fluid. Strainers are often used in inlet lines and reservoir fill pipes. They are usually classified by the number of meshes per square inch (for example, 60) or the degree of fineness of the weave. Filters are rated in microns [Figure 3.8(c)]. A micron is a unit of length equal to one-millionth of a meter; 25 microns equal approximately 0.0001 in.

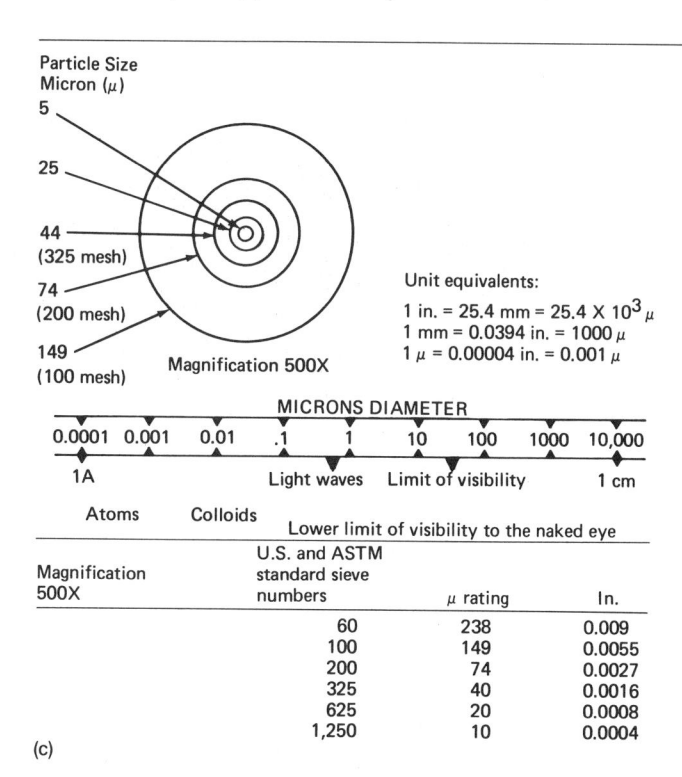

(a)

(b)

Figure 3.8 (a) A magnetic-shield filter that picks up ferrous particles. (Courtesy Marvel Engineering Corporation) (b) A mechanical cartridge filter. (Courtesy Pall Corporation) (c) Relative size of particles and comparison of dimensional units. A micron is a very small unit of measure; few people realize that visible red light is half a micron. Metrologists have selected a red wavelength of light as the international standard of measurement.

Particle Size Micron (μ)
5
25
44 (325 mesh)
74 (200 mesh)
149 (100 mesh)

Magnification 500X

Unit equivalents:
1 in. = 25.4 mm = 25.4 × 10^3 μ
1 mm = 0.0394 in. = 1000 μ
1 μ = 0.00004 in. = 0.001 μ

MICRONS DIAMETER

0.0001	0.001	0.01	.1	1	10	100	1000	10,000
1A				Light waves	Limit of visibility			1 cm
Atoms		Colloids						

Lower limit of visibility to the naked eye

Magnification 500X	U.S. and ASTM standard sieve numbers	μ rating	In.
	60	238	0.009
	100	149	0.0055
	200	74	0.0027
	325	40	0.0016
	625	20	0.0008
	1,250	10	0.0004

(c)

A large percentage of contaminants in a reservoir is removed when larger particles settle to the bottom of it. However, due to the precision clearances in hydraulic components, the presence of contaminants cannot be tolerated. Consequently, filters should be used because they allow oil to flow through their pores but trap solid particles.

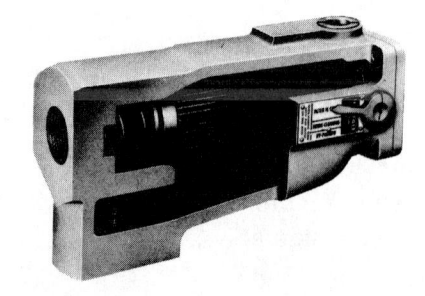

There are three main types of filter materials, mechanical, absorbent (inactive), and adsorbent (active). *Mechanical* filters use finely woven metal-wire screens or disks. Filters of this construction are generally used to remove fairly coarse insoluble particles. *Absorbent (inactive)* filters are constructed of a porous, permeable material such as cotton, wood pulp, diatomaceous earth, cloth, paper, or asbestos [Figure 3.9(a) and (b)]. As the fluid permeates the material, the absorbent action causes the contaminant particles to be trapped by the walls of the cartridge medium [Figure 3.9(c)]. Absorbent media filters make it possible to filter extremely small particles. When the elements of this type of filter become sufficiently clogged, the filter should be replaced.

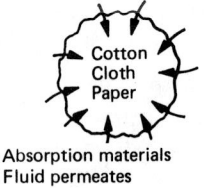

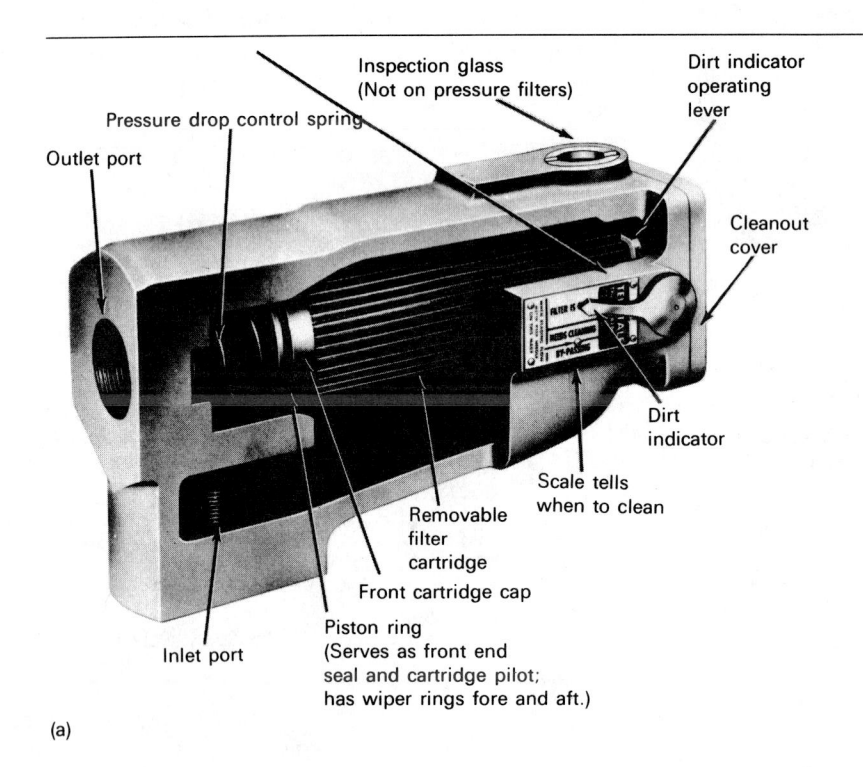

Figure 3.9 (a) Intake-line filter with telltale device. This full-flow filter is located above the reservoir, making it accessible for cleaning and replacement. (Courtesy Rosaen Filter Division, Parker-Hannifin) (b) A clogged filter element would compress the coil spring and move the element to the bypass position. (Courtesy Rosaen Filter Division, Parker-Hannifin) (c) Adsorbent and absorbent filter-element materials.

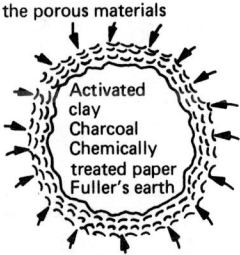

Adsorbent (active) filter materials adhere to the surface of certain solids or liquids. Adsorption is a *surface phenomenon,* which means that the capacity of a solid to adsorb depends on the extent of its surface exposed to the particles of another substance as well as on its chemical nature. Materials such as Fuller's earth, charcoal, activated clay, and chemically treated paper are used in the elements of adsorbent filters. These filter elements remove contaminants mechanically as well as by the ionic attraction inherent in the adsorbent material. One possible disadvantage is that adsorbent filters have a tendency to remove antifoam and other additives. This type of filter element is compatible with straight mineral oil.

There are two basic system designs for hydraulic fluid filtration: (1) *full-flow filtration* with the entire volume of oil in the circulating system passing through the filter element [Figure 3.10(a)]; and (2) *proportional filtration* with only a portion of the fluid passing through the filter element during a given cycle and the remainder bypassing it [Figure 3.10(b)].

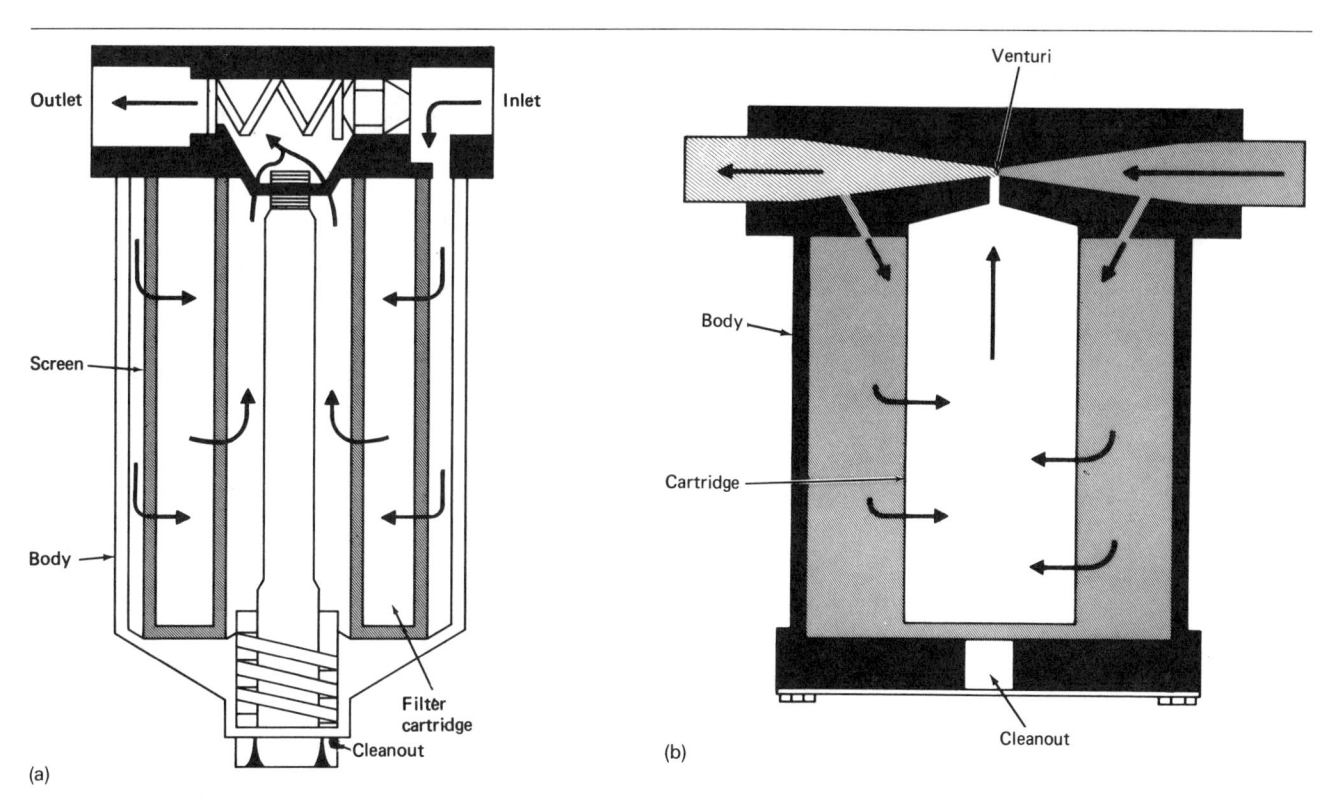

Figure 3.10 (a) Full-flow filter.
(b) Proportional-flow filter.

Location of filters and strainers The primary purpose of a filter or strainer is to prevent the introduction of contaminants and thus promote trouble-free performance of the hydraulic system components (Figure 3.11). Therefore the first consideration is to have a strainer

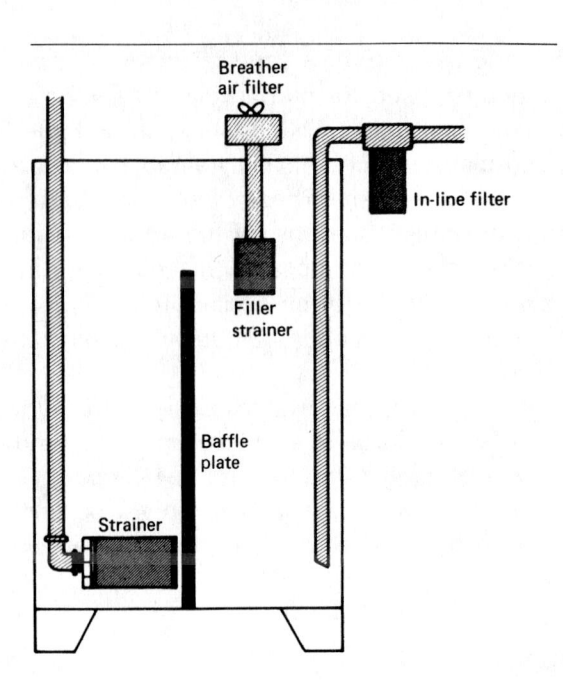

Figure 3.11 Reservoir filler strainers and filters.

in the fill pipe. Generally, a 50–60 mesh or finer is satisfactory for this purpose.

Proper operation of the pump depends on the presence of atmospheric pressure within the reservoir. The main function of the reservoir's vent cap is to let air in and out of the reservoir. Systems operating under severe air contaminant conditions should have an oil-bath air cleaner or pleated-paper air filters. Protection against contaminants in the pump is critical because of the precision and speed of the pump's moving parts. A strainer on the inlet line will provide protection from larger particles and offer minimum resistance to flow. Approximately 60-mesh strainers with a capacity rated at four times pump flow will provide adequate protection to the system.

Reservoir return-line filtration is used to incorporate full-flow filtration in the return line using filter elements of 25 to 35 microns. In highly sophisticated systems, using servo-valves, it may be necessary to get as low as 10 microns. Return-line filter elements should be properly sized with a capacity of two to three times the pump volume delivery.

Selection of filters In selecting a filter for a hydraulic system application, the following factors should be considered:

- type and degree of filtration,
- flow rate and capacity,

- pressure drop,
- system pressure,
- compatibility with fluid,
- temperature, and
- service interval recommended.

Filtration is one of the most neglected areas of hydraulic system design and maintenance, but one that frequently contributes to downtime. With many different fluids on the market and systems operating at higher temperatures and pressures, the user must determine from the manufacturer's specifications whether the filter is compatible with the fluid and what are the service-interval recommendations.

Lubrication and heat transfer

Temperature regulation of hydraulic fluids is important in maintaining narrow limits to viscosity changes and leakage [Figure 3.12(a)]. This is particularly essential with hydraulically controlled machines working to precision tolerances. When oil gets too hot, it loses its lubricating property and tends to oxidize more rapidly. Hydraulic systems designers favor the use of operating temperatures as high as possible without encountering fluid breakdown and excessive wear of moving parts. Whereas a decade ago 120 °F was the highest recommended temperature, today MIL–H5606 hydraulic fluid has an upper normal limit of 240 °F and 275 °F maximum (Class II system). Class III systems require components and fluids that will operate at temperatures up to 400 °F and 450 °F [Figure 3.12(b)].

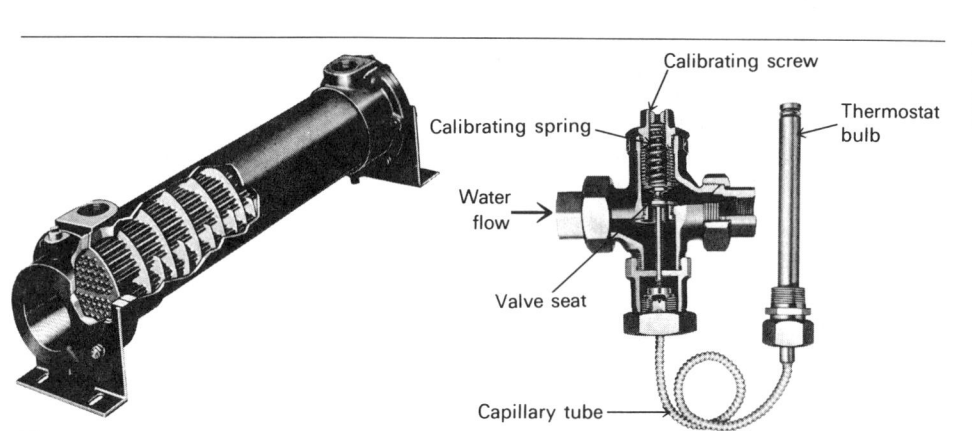

Figure 3.12 (a) A fluid cooler. (Courtesy Young Radiator Company) (b) A temperature-control valve. (Courtesy Sterling Inc.)

Heat exchangers

Coolers or heaters are considered heat exchangers and are used to maintain the optimum temperature in an operating hydraulic system. For many applications, normal heat losses from lines, components, and the reservoir are adequate to maintain hydraulic system temperatures within the normal limits as specified by a fluid manufacturer. When high temperatures are encountered, as when machinery is in continuous use near furnaces, heat exchangers may be needed to aid in the dissipation of fluid heat from the system.

Oil coolers (water-type) for hydraulic circuits consist of cooling plates or multiple tubes in a metal housing through which water is circulated. Baffles break up the oil stream to provide more cooling surface. Oil coolers may be located in the return line to the reservoir or in the inlet line to the pump. Also it is practical to draw the hot incoming oil from the reservoir, pass it through the cooler, and return it to the reservoir. Many of the water-type coolers have a maximum recommended working pressure of 125 psi or less and therefore should be installed on the low-pressure side of the hydraulic circuit.

Air coolers serve the same function as water coolers—to transfer heat from the hot oil to the surrounding cooling medium (Figure 3.13). Air coolers are used more frequently where water is not readily avail-

Figure 3.13 Air-cooled oil coolers. [(a) Courtesy Young Radiator Company; (b) Courtesy Continental Machines, Inc.]

(a)

(b)

able or where a water-type cooler is considered to be too expensive. Air coolers consist of a core of tubes and plate fins. Several designs of cores and shapes of tubes are available. Essentially, the fluid passes through the tubes with the large fin surface providing a high rate of heat transfer. The oil-to-air type of unit is more economical to operate, but it is not effective unless the hydraulic fluid enters the cooler at a temperature 10 °F or more above the ambient temperature. Automatic thermal controls with heat-exchanger circuits make it possible for complex machinery to operate positively according to predetermined performance characteristics.

Heaters are used where mobile and other hydraulically equipped machinery must operate in subzero temperatures. Heating units employing hot water, steam, or electricity are restricted to start-ups. Once the fluid is sufficiently warm to circulate through the system, the heat-generating characteristics of the operating system are generally sufficient to maintain efficient operation. Both extreme cold and extreme heat have adverse effects on system efficiency and component life.

Fluid reservoirs

A *reservoir* is a container for storing liquid in a fluid power system. Although the primary function of a reservoir is to supply an adequate amount of fluid to the entire hydraulic system, it is more than a storage vessel (Figure 3.14).

A properly designed reservoir that conforms to the American National Standards Institute (ANSI) specifications incorporates several important features. A vertical baffle plate, extending about two-thirds the height of the oil level, separates the return oil from the oil on the pump inlet side. This baffle permits slower, more complete oil circulation to release heat, air, dirt, water, and other contaminants. The breather assembly used to vent the reservoir contains a filter to ensure that only clean atmospheric air enters the reservoir. In large-capacity systems it is important to have a large-volume breather to permit rapid discharge of air displaced by the rapid and large volume of fluid returning to the reservoir.

Other features include cleaning accessibility, a filler opening with a strainer, legs to improve air circulation for dissipation of heat, a rigid mounting plate for the electric motor and pump, dishing of the

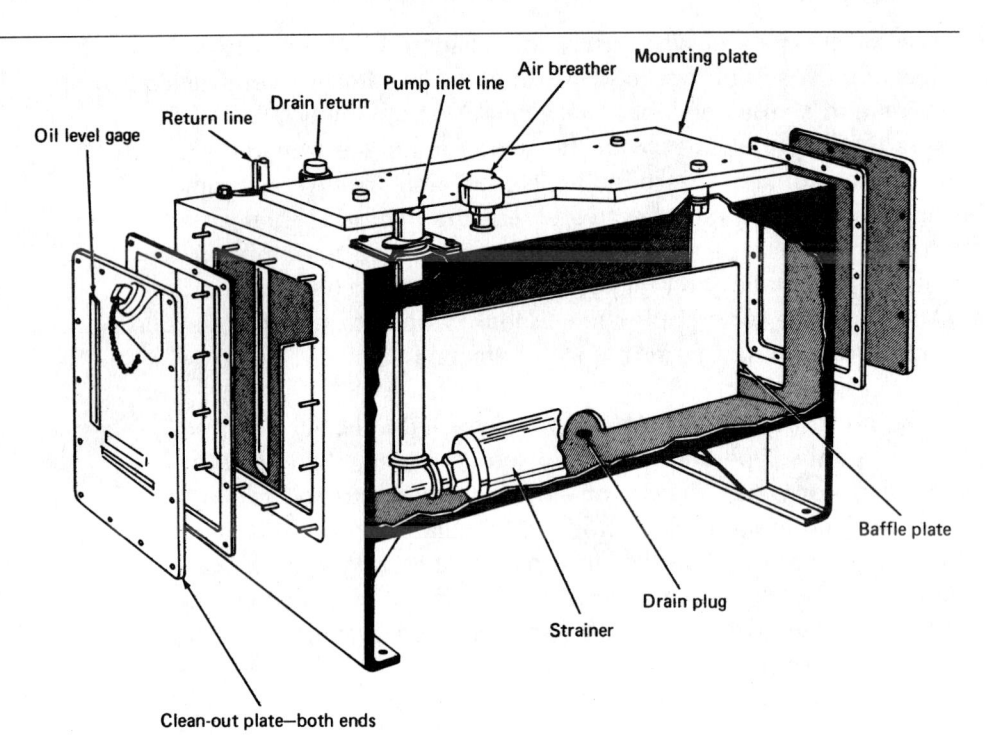

Figure 3.14 Major components of a reservoir. (Courtesy Vickers, Sperry Rand)

bottom to provide for proper drainage, provisions for bolting the reservoir to the floor or other surface, and a protected fluid-level indicator. It is evident that an oil drum does not meet the requirements outlined above. A well-designed reservoir should be completely enclosed and self-contained. It may be an integral part of a machine or a separate tank. The shape should be such that the pump inlet pipe is not overtaxed by atmospheric lift.

Sizing a reservoir

The reservoir should be large enough to store more than the largest volume of fluid that the system will require at any one moment. As a general rule, the volume of the oil in the reservoir should normally be equal to at least two to three times the capacity in gallons of rated pump delivery for 1 min of operation. Where the system cycle frequency is high, as in many industrial applications, three times the rating of the pump will give the oil a longer period to cool and settle contaminants before being recirculated. In mobile applications, where the cycle frequency of operation is considerably less, a reservoir two times the pump rating would be adequate.

The most critical consideration in deciding reservoir capacity is the heat-dissipation factor, which usually sets the size minimum. Heat dissipation depends on the demands of the circuit. To dissipate the equivalent of 1 hp of heat energy requires approximately 90 ft^2 of steel surface for active circulation of 30 sec. Two important considerations in capacity design are that the fluid never (1) approaches the level of the pump inlet, which could cause pump cavitation, or (2) rises so high that there is inadequate space above the oil to allow air escaping from the oil to move to the atmosphere. Thermal expansion has also to be taken into account in sizing large reservoirs.

Accessories

In addition to the many features to be considered in designing and sizing a reservoir, a number of accessories contribute to an effective hydraulic system operation. Most reservoirs include a submerged filter on the pump inlet line to prevent dirt, grit, sludge, lint, rust, and other contaminants from entering the pump and other hydraulic components (Table 3.1). This type of filter is designed to provide maximum active filtering area in a minimum of space. Generally, the filter has twice the oil flow of the pump capacity. Return-line filters,

Table 3.1 Sources of component contaminant

	Oxide scale	Plastic elastomer	Oil additive	Metal particle	Airborne dirt	Silica sand	Lapping compound	Process residues	Fibers
Oil	X	—	X	—	X	X	—	—	X
Tank	X	X	—	X	XX	X	—	—	X
Pressure-relief valve	—	X	—	XX	X	X	X	—	X
Accumulator (bladder and piston types)	X	X	—	X	X	—	—	X	X
Filter	X	X	—	X	XX	—	—	—	X
Piping, fittings, and rubber tubing	X	X	—	X	XX	—	—	X	X
Control valves	—	X	—	X	X	—	X	—	X
Actuators	—	X	—	X	X	—	—	—	X
Pump	—	X	—	XXX	X	X	—	—	X

Key: X = noticeable contamination contribution potential
XX = medium contamination contribution potential
XXX = strong contamination contribution potential

generally of the bypass type, are installed in the incoming line. Only about 10 to 20% of the return oil is filtered. Such filters prevent contaminants from components from entering the reservoir and ultimately the pump.

The more practical and efficient water-type oil coolers can be attached as an accessory to the reservoir. Temperature-control valves are used to achieve a constant oil temperature for accurate control of feed rates through automatically controlled heat exchangers. Pressure gages are installed as a means of controlling and protecting the heat-exchange units. Hydraulic lines should be large enough to conduct the maximum flow from the pump without turbulence or excessive friction losses. They should be as short as possible with minimum bends to avoid pressure losses.

Package units are beginning to predominate in industrial applications. A *package unit* includes a power unit and operating hydraulic components fabricated into one integral unit, completely piped and providing a terminal connection block. These units are usually individually designed to the customer's specifications for a particular job. Package units offer convenience, portability, minimum installation time and expense, and optimum performance achieved through compatibility of components and sound engineering (Figure 3.15).

Figure 3.15 This package unit is a circuit master power unit. Accessories provided include filtration devices, controls, and conductors. (Courtesy Double A Products Co.)

Large-demand systems in some cases may have one large central reservoir with several independent pressure systems. Multiple pressure systems can utilize independent pumps. These systems increase flexibility, reduce maintenance, and save space in large plants.

Fluid conductors

Fluid conductors interconnect the various components of a system. Conductors can be pipes, tubing, hose, or manifolds. The operation of the system depends on the efficiency and serviceability of the vital lines connecting the units. These lines have to be capable of withstanding the working pressures, including peak shock pressures, created within a system. Conductors must be sized properly for transmitting the required volumetric flow rate demanded by the system. A practical method of assembly and disassembly must be provided that is not only leakproof but also avoids excessive pressure drop in the line.

Types of conductors

Fluid conductors are divided into three classes: (1) *rigid,* or piping; (2) *semirigid,* or tubing; and (3) *flexible,* or hose and flexible conduit. Selection of a specific type of conductor is determined by the requirements of the installation. Flexible lines must be used in installations where interconnected units need to be free to move relative to each other. Where movement of the components containing fluid connections is not required, rigid or semirigid conductors can be installed. In some applications, machine motions may require fluid lines to pivot or swivel. In other applications, such as mobile and farm hydraulically equipped machinery, it may be necessary to use quick-disconnect hose couplings.

Selection of conductors

The proper selection and installation of the lines are critical in the efficient operation of a system. Factors that should receive careful study in selecting a particular type of conductor are as follows:

- Lines must be strong enough to contain the fluid at the calculated working pressure and be able to withstand momentary surge peak pressures as high as four times the working pressure.
- Lines should be strong enough to support in-line components.
- Conductor size should be large enough to prevent pressure drops of more than 10% of the initial pressure and large enough to transmit required flow rate.

84 Fluids and auxiliaries

- The lines should have a smooth interior surface to reduce turbulence and frictional losses.
- Line material must be compatible with the fluid used within the system.
- Terminal fittings must be available at all junctions with other lines and components that may require removal for repair and replacement.
- Weight is a factor in such applications as aerospace and aircraft.

Several materials including steel, iron, aluminum, copper, stainless steel, and plastic are readily available for the manufacture of conductors.

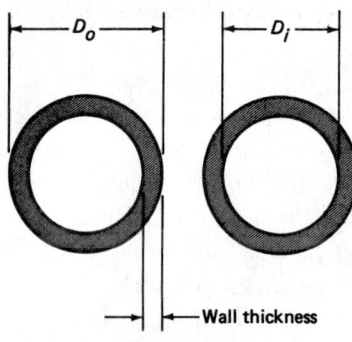

Figure 3.16 Tubular dimensions.

Sizing pipes

The inside diameter of a conductor is important in determining flow-rate capacity for that particular size of conductor (Figure 3.16). The

Figure 3.17 Nomograph of flow capacity of pipes at recommended velocities. (Courtesy Stratoflex)

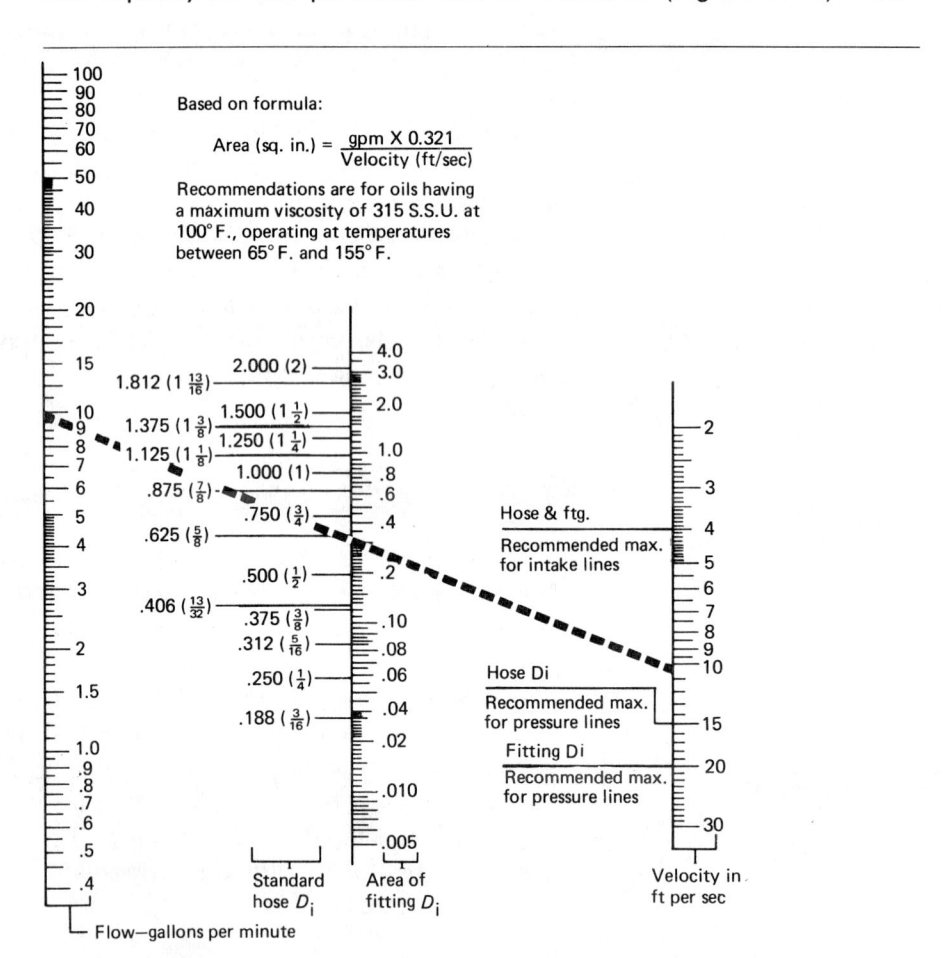

Based on formula:

$$\text{Area (sq. in.)} = \frac{\text{gpm} \times 0.321}{\text{Velocity (ft/sec)}}$$

Recommendations are for oils having a maximum viscosity of 315 S.S.U. at 100°F., operating at temperatures between 65°F. and 155°F.

Flow—gallons per minute

Standard hose D_i

Area of fitting D_i

Hose & ftg. Recommended max. for intake lines

Hose D_i Recommended max. for pressure lines

Fitting D_i Recommended max. for pressure lines

Velocity in ft per sec

velocity of a known flow through a conductor varies inversely as the square of the inside diameter. A flow rate of 15 ft/sec is recommended by most fluid systems designers. This rate is subject to pressure, continuous versus intermittent use, and fluid characteristics.

A nomograph is a helpful chart for selecting the size of pipe when the other two variables—flow velocity and flow capacity—are known (Figure 3.17).

▶ **Example** What is the required inside diameter of a pipe that will conduct 14 gpm at a flow rate of 10 ft/sec?

With a straightedge, connect the mark "10 ft/sec" on the velocity scale on the right to the mark "10 gpm" on the flow scale on the left. Note that this line crosses the center scale (diameter) at approximately $\frac{3}{4}$ in.

There are three important dimensions for tubular products—outside diameter (D_o), inside diameter (D_i), and wall thickness. In the past, the thickness of pipe wall was classified as "standard," "extra strong," and "double-extra strong." In recent years, however, a trend has developed toward the use of a scheduled number designation. These scheduled numbers were established by the American Standards Association; they range from 10 to 160 and cover ten different pipe-wall thicknesses and pipe fittings. The schedule number 40 is equivalent to standard (STD); schedule 80 to extra strong (XS); and schedule 160 to double-extra strong (XXS).

Pipe sizes are determined and specified by the nominal inside diameter (D_i) of the pipe. A nominal dimension is close to, but not identical with, the actual measured inside diameter. As indicated in Table 3.2, a pipe with a nominal size of 3 in. has a 3.068-in. D_i for schedule 40 and only a 2.624-in. D_i for schedule 160.

Hydraulic installations that utilize pipe tend to create problems due to shock or vibration. In some instances it is advisable to use an accumulator as a shock-absorbing device.

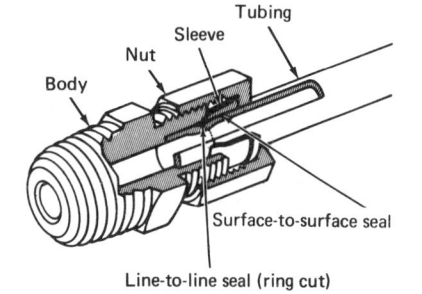

Figure 3.18 Steel tubing with appropriate connector. (Courtesy Weatherhead)

Semirigid conductors

There are two types of tubing manufactured for hydraulic systems, (1) seamless (Figure 3.18) and (2) electrically welded. Tubing is

Nominal size	Pipe D_o	Sched. 10	Sched. 20	Sched. 30	Sched. 40	Sched. 60	Sched. 80	Sched. 100	Sched. 120	Sched. 140	Sched. 160
$\frac{1}{8}$	0.405				0.269		0.215				
$\frac{1}{4}$	0.540				0.364		0.302				
$\frac{3}{8}$	0.675				0.493		0.423				
$\frac{1}{2}$	0.840				0.622		0.546				0.466
$\frac{3}{4}$	1.050				0.824		0.742				0.614
1	1.315				1.049		0.957				0.815
$1\frac{1}{4}$	1.660				1.380		1.278				1.160
$1\frac{1}{2}$	1.900				1.610		1.500				1.338
2	2.375				2.067		1.939				1.689
$2\frac{1}{2}$	2.875				2.469		2.323				2.125
3	3.500				3.068		2.900				2.624
$3\frac{1}{2}$	4.000				3.548		3.364				
4	4.500				4.026		3.826		3.624		3.438
5	5.563				5.047		4.813		4.563		4.313
6	6.625				6.065		5.761		5.501		5.189
8	8.625	8.125	8.071	7.981	7.813	7.625	7.439	7.189	7.001	6.813
10	10.750	10.250	10.136	10.020	9.750	9.564	9.314	9.064	8.750	8.500
12	12.750	12.250	12.090	11.934	11.626	11.376	11.064	10.750	10.500	10.126

Table 3.2 Wall thickness schedule designations for pipe (in terms of the inside diameter)

available in three basic materials: steel, aluminum, and copper. More specifically, these materials can be identified as follows:

- **Steel seamless** (SAE 1010), fully annealed and suitable for bending
- **Stainless steel seamless** (18–8), chrome-nickel (corrosion-resistant) fully annealed, suitable for bending and flaring, for use in high-pressure sections (3000 psi and above)
- **Aluminum-alloy seamless** (5052–0), used for low-pressure systems (1500 psi) and (6061) used for high-pressure systems (3000 psi)
- **Copper seamless,** fully annealed, not recommended with petroleum-base fluids, suitable for bending and flaring

Tubing, unlike rigid pipe, is specified by its actual outside diameter. Thus $\frac{3}{8}$-in. tubing has an outside diameter of $\frac{3}{8}$ in. As indicated in Table 3.3, tubing is available in a variety of wall thicknesses. It is made in incremental sizes of $\frac{1}{16}$ in. for diameters $\frac{1}{8}$ through $\frac{3}{8}$ in., $\frac{1}{8}$ in. for diameters $\frac{1}{2}$ to 1 in., and $\frac{1}{4}$ in. for diameters 1 in. and beyond.

Manufacturers of hose, tube, and fittings use a dash numbering system to identify the size of conductors. The dash number represents the D_o of the tube in sixteenths of an inch. The numerator of the fraction is listed as the dash size. For example, converting a $\frac{3}{8}$-in. tube size into the fraction $\frac{6}{16}$ produces a numerator of 6; hence this size is designated as a −6.

The type of tubing used is determined by the system pressure, fluid velocity, flow, and type of fluid. Manufacturers of hose, tubing, and

Tube D_o	Wall thickness	Tube D_i	Tube D_o	Wall thickness	Tube D_i	Tube D_o	Wall thickness	Tube D_i
$\frac{1}{8}$	0.028	0.069	$\frac{5}{8}$	0.035	0.555	$1\frac{1}{4}$	0.049	1.152
	0.032	0.061		0.042	0.541		0.058	1.134
	0.035	0.055		0.049	0.527		0.065	1.120
				0.058	0.509		0.072	1.106
				0.065	0.495		0.083	1.084
$\frac{3}{16}$	0.032	0.1235		0.072	0.481		0.095	1.060
	0.035	0.1175		0.083	0.459		0.109	1.032
				0.095	0.435		0.120	1.010
$\frac{1}{4}$	0.035	0.180						
	0.042	0.166	$\frac{3}{4}$	0.049	0.652	$1\frac{1}{2}$	0.065	1.370
	0.049	0.152		0.058	0.634		0.072	1.356
	0.058	0.134		0.065	0.620		0.083	1.334
	0.065	0.120		0.072	0.606		0.095	1.310
				0.083	0.584		0.109	1.282
$\frac{5}{16}$	0.035	0.2425		0.095	0.560		0.120	1.260
	0.042	0.2285		0.109	0.532		0.134	1.232
	0.049	0.2145						
	0.058	0.1965	$\frac{7}{8}$	0.049	0.777	$1\frac{3}{4}$	0.065	1.620
	0.065	0.1825		0.058	0.759		0.072	1.606
				0.065	0.745		0.083	1.584
				0.072	0.731		0.095	1.560
$\frac{3}{8}$	0.035	0.305		0.083	0.709		0.109	1.532
	0.042	0.291		0.095	0.685		0.120	1.510
	0.049	0.277		0.109	0.657		0.134	1.482
	0.058	0.259						
	0.065	0.245	1	0.049	0.902	2	0.065	1.870
				0.058	0.884		0.072	1.856
$\frac{1}{2}$	0.035	0.430		0.065	0.870		0.083	1.834
	0.042	0.416		0.072	0.856		0.095	1.810
	0.049	0.402		0.083	0.834		0.109	1.782
	0.058	0.384		0.095	0.810		0.120	1.760
	0.065	0.370		0.109	0.782		0.134	1.732
	0.072	0.356		0.120	0.760			
	0.083	0.334						
	0.095	0.310						

Table 3.3 Tubing size designations

pipe usually supply charts, graphs, or tables that aid in the selection and sizing of fluid power lines (Figure 3.19).

Conductors should be kept as short and as free of bends as possible. Tubing, however, should not be assembled in a straight line because a bend tends to eliminate strain by absorbing vibration and compensates for thermal expansion and contraction. Eliminating strain prevents cyrstallization and ultimate failure. Lines should be anchored to prevent vibration and minimize stress on the lines and fittings.

In tubing fabrication, installation, and maintenance, a technician needs to pay particular attention to the radii of bends, the alignment of tubing, and fittings. It is general practice to make a bend radius nine or more times the inside diameter of a tube to minimize friction and turbulence (Figure 3.20).

Figure 3.19 Hose-selection nomograph. (Courtesy Synflex-Samuel Moore)

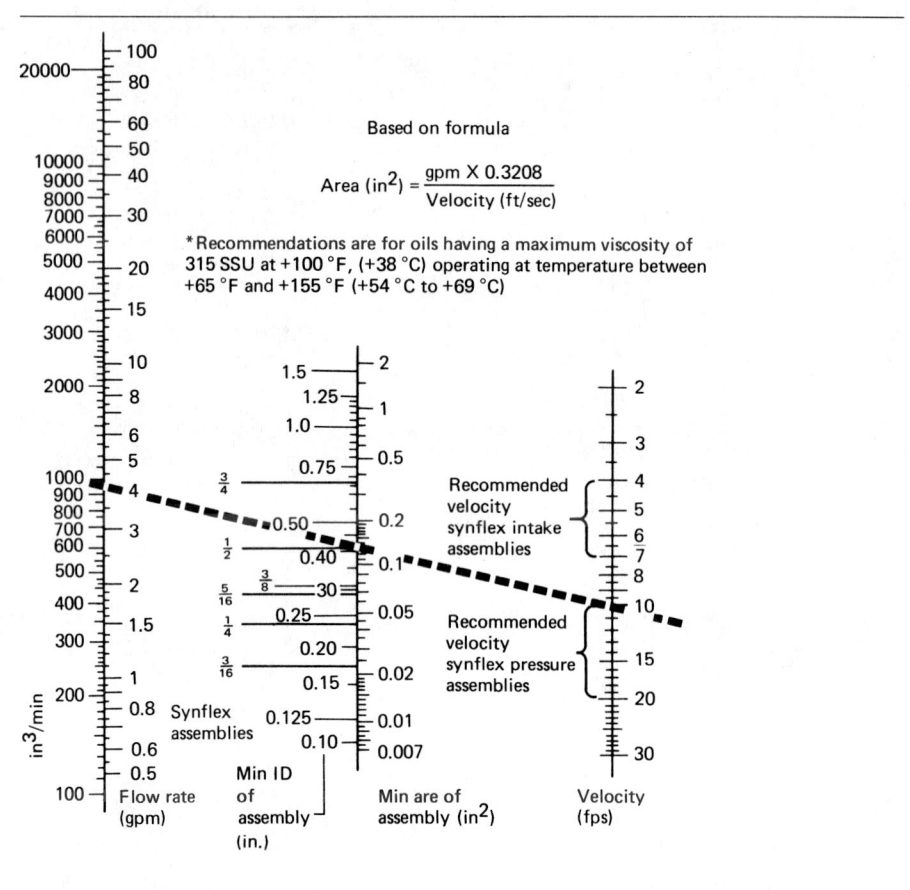

Based on formula

$$\text{Area (in}^2) = \frac{\text{gpm} \times 0.3208}{\text{Velocity (ft/sec)}}$$

*Recommendations are for oils having a maximum viscosity of 315 SSU at +100 °F, (+38 °C) operating at temperature between +65 °F and +155 °F (+54 °C to +69 °C)

Figure 3.20 (a) Recommended tube-fitting practices. (Courtesy Weatherhead) (b) Hand-operated tube bender. (Courtesy Rosaen Filter Division, Parker-Hannifin)

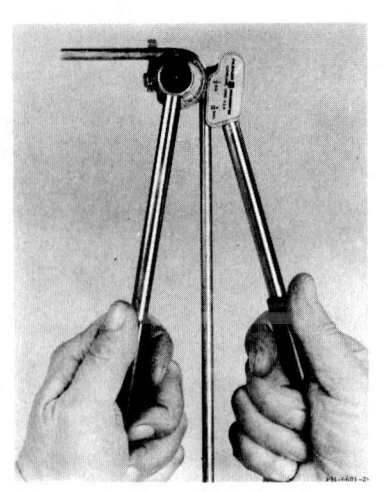

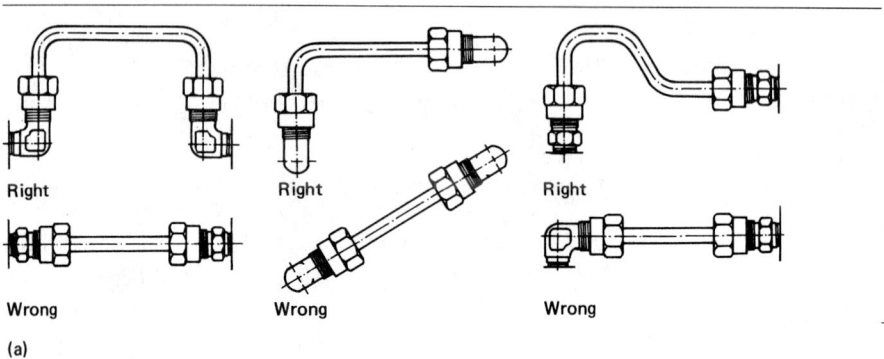

Flexible hose

It is often necessary to use hose instead of rigid tubing or pipe for the transmission of oil or air under pressure because of the need to bend lines in vehicle and machine construction. Flexible hose is manufactured from a variety of elastomeric materials, fibers, and wire. Hoses are available in a variety of pressure ratings and sizes. The hose size is specified by its nominal inside diameter. The pressure rating is governed by the number of layers and type of materials and how the hose is constructed. Three pressure ratings are usually specified to describe hose pressure capability: (1) the recommended *working pressure* which is the maximum safe pressure at which a given hose can be operated continually for satisfactory service; (2) the *test pressure* which is the pressure a hose is warranted to withstand; and (3) *burst pressure* which is the pressure at which the hose will rupture.

Hose is commonly used because it can be easily installed, will absorb shock, requires less installation skill than pipe or rigid tubing, and is readily available in a wide range of pressure ratings; also, many fittings, angle tubes, and quick-disconnect assemblies are readily available.

Flexible hosing must not be twisted as this may cause a fitting to loosen and puts stress on the hose. It should be protected from rubbing on another surface. This can be done by suspending and supporting the hose correctly. It needs to be installed with a minimum of flexing and be supported by appropriate clamps. Hose should be slightly longer than the maximum distance between outstretched limits. While hose is stored or installed it should not be kinked. This is particularly important with Teflon hose (Figure 3.21).

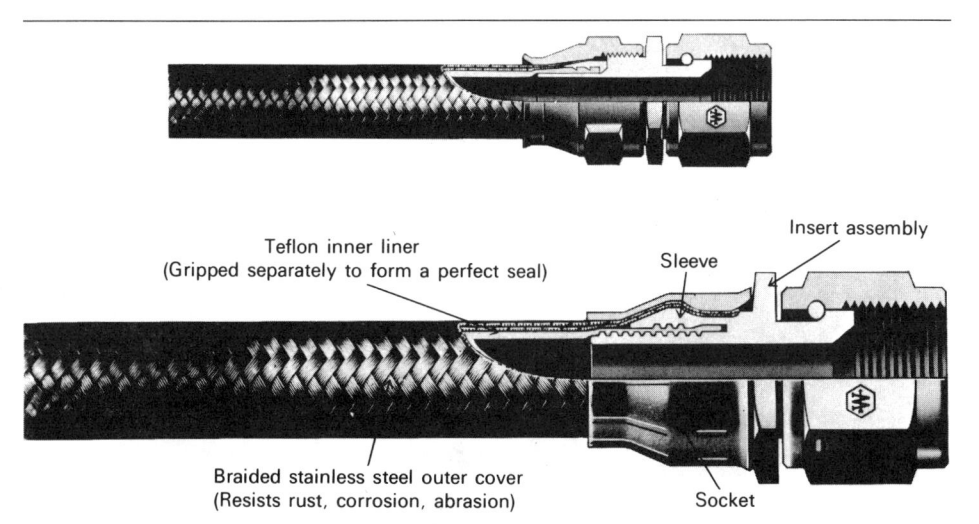

Figure 3.21 Medium-pressure Teflon hose. (Courtesy Weatherhead)

Fittings

Tubes and pipes are connected to other tubes and pipes or to the components of an installation by some type of connector. There are many different types of connectors manufactured for fluid power systems. The type of connector that is ultimately selected for a given system will be determined by the type of conductor (pipe, tubing, or flexible hose), the fluid used, and the maximum operating pressure of the system.

Many manufacturers of conductors and fittings publish catalogs and bulletins that provide up-to-date technical data in the form of tables, charts, nomographs, and descriptive specifications that will be most helpful for more advanced study purposes.

Threaded

Threaded connectors [Figure 3.22(a)] are generally used in low-pressure fluid power systems such as inlet, return, and drain lines. Standard pipe threads [Figure 3.22(b)] are tapered slightly to ensure tight connections. Straight threads generally require a seal of some kind. To avoid corrosion on threaded portions and to facilitate assembly, a protective compound of some type is used to seal and protect

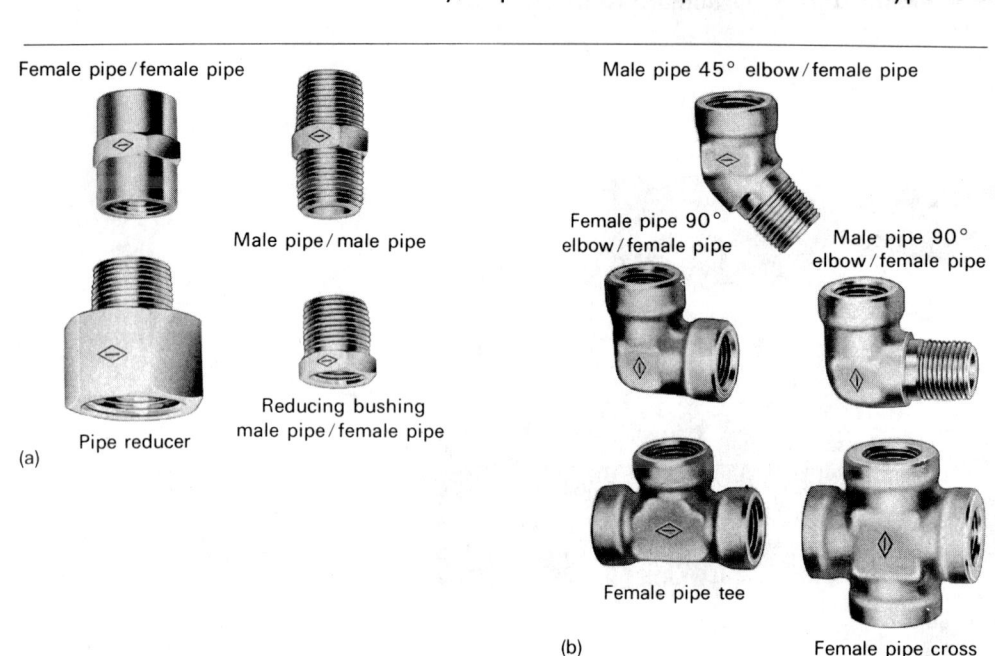

Figure 3.22 (a) Threaded pipe connectors. (Courtesy Imperial Eastman) (b) Standard pipe threads. (Courtesy Imperial Eastman)

Tube

Tube fittings are available in standard design, sizes, and types. A vital concern in designing a circuit is to select and apply fittings that offer a minimum pressure drop, permit adequate rate and volume flows, and are easy to install. Fittings are made of steel, aluminum alloy, or bronze. They are available in a variety of shapes to form unions, elbows, tees, nipples, caps, plugs, couplings, and crosses. Unlike iron pipe, which is threaded on the outside, tubing is not generally threaded, but instead is connected with fittings (Figure 3.23).

Figure 3.23 Tube fittings. (Courtesy Rosaen Filter Division, Parker-Hannifin)

Flared

Fittings can be divided into two classifications, flared and flareless. In flared connections, the tubing end is flared, and a matching surface on the fitting is tightened against it. Fittings of this type are made with a 45° or 37° flare angle. The 37° flare fitting is used for high-pressure hydraulic circuits. It is a three-piece reusable fitting that consists of a body, sleeve, and nut (Figure 3.24).

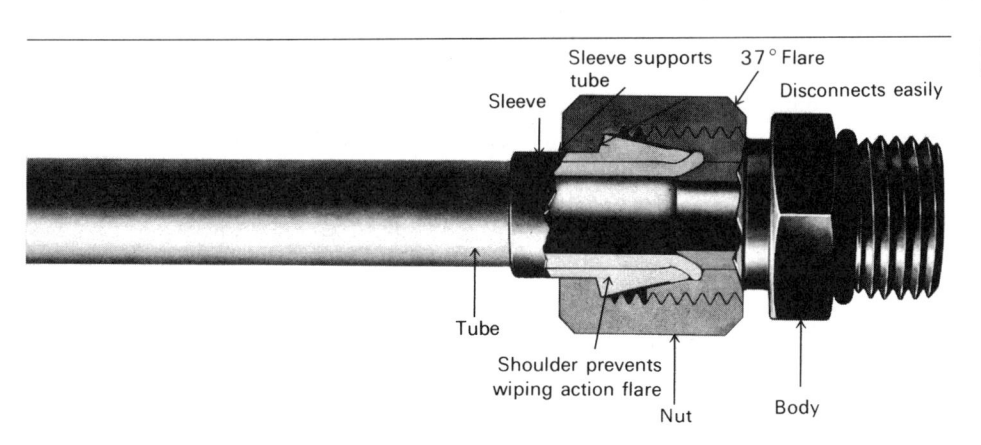

Figure 3.24 A flared fitting. (Courtesy L&L Manufacturing)

Flareless

The flareless, or compression, fitting is attached to the tube by compressing a metal sleeve or ferrule (or elastomeric seal) around the tubing. The wedging action, resulting from compression applied by the nut, forms a leakproof seal between the ferrule and the body (Figure 3.25).

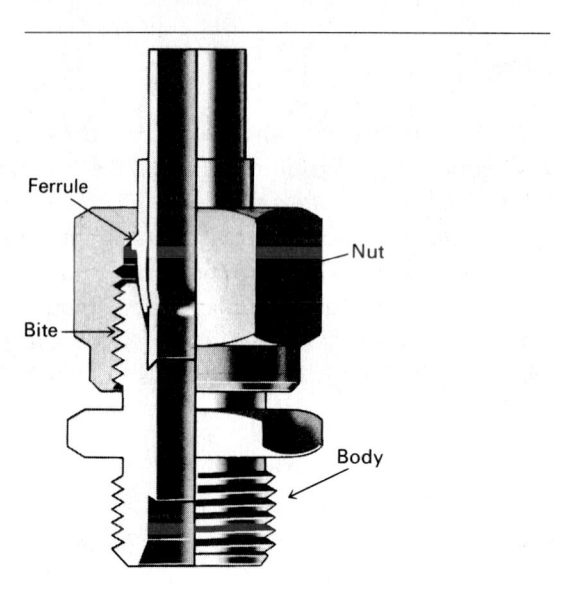

Figure 3.25 A flareless fitting. (Courtesy L&L Manufacturing)

Welded

Welded tube connections are used for very high-pressure systems. The disadvantage of welded joints is that they cannot be disassembled. Fittings are generally welded to the tubing according to standard specifications that define the materials and techniques. The assembly is thoroughly cleaned for scale and other contaminants before it is assembled into a circuit. Stainless steel tubing is welded in some aircraft applications that require high pressure, high temperatures, and high velocities. Welded joints have an advantage over mechanical joints in terms of system maintenance.

Quick-disconnect couplings

Quick-disconnect couplings are becoming popular on fluid power equipped machinery, particularly where there is frequent need to uncouple the lines for separating units, maintenance, testing, and safety (Figure 3.26). Many quick-disconnect couplings have double checks that can be readily disconnected without any loss of fluids.

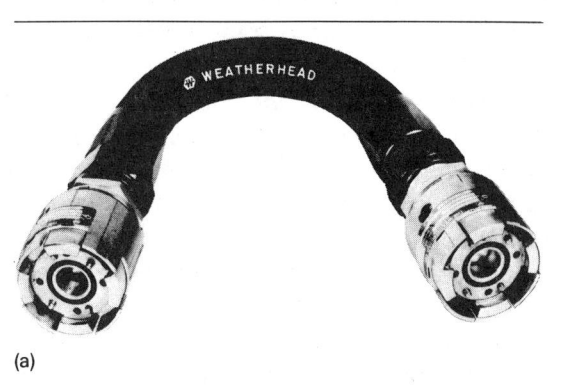

Figure 3.26 (a) Quick-disconnect coupling. (Courtesy Weatherhead) (b) Cutaway view of quick-disconnect coupling. (Courtesy Weatherhead)

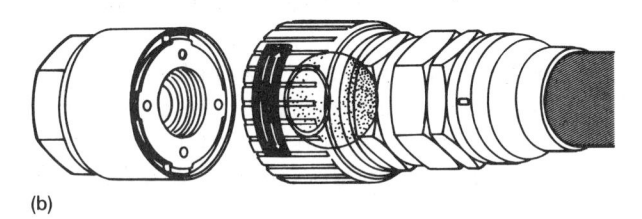

Future trends in conductor development

The future of conductors and fittings is difficult to predict. However, as operating pressures and heat and variety of fluids increase, manufacturers will develop and test systems that will perform safely and efficiently. Extra-high-pressure hose of seamless extruded tetrafluoroethylene (TFE), with an outer braided stainless steel cover, will withstand temperatures from $-67\ °F$ to $+400\ °F$ and pressures of 10,000 psi. Teflon keeps its flexibility, chemical structure, and strength at temperatures in the range of $-100°$ to $+450\ °F$. It has an extremely low friction coefficient, is corrosion-resistant, and is extruded in seamless tube form. Fittings using metal-to-metal seals that are unaffected by vibration are being designed and machined.

Summary

The primary function of fluid power systems, whether they use liquids or air, is to transmit energy through a pressurized fluid. Other functions include providing lubricity to moving parts, sealing moving parts to prevent leakage, and cleansing bearings and precision-machined parts by removing contaminants. To perform their functions well, fluids are specially processed and compounded with additives to obtain and retain desirable qualities.

Viscosity is probably the most important physical property of a hydraulic fluid since it is the measure of a fluid's ability to flow in excessive heat or cold. Water introduced by condensation is detrimental to pneumatic and hydraulic components. Lubricity is the most important additive to air for pneumatics. Water contamination is the greatest enemy of compressed air systems.

Adequately maintained filters will remove the majority of harmful contaminants. Contaminants such as water accelerate oxidation whereas gritty substances tend to wear precision-machined parts.

Designers should follow manufacturer's recommendations for matching fluid requirements with the seals and performance of system components. Fire-resistant fluids must be compatible with the seals, environmental temperature, and painted surfaces of a reservoir.

To retain the properties compounded in most fluids, it is essential to consider the micron rating of the filter. For example, when additives are present in a fluid, the filter must not remove them if the chemical properties of the fluid are to be retained. One of the most neglected areas of a hydraulics system, and one that contributes to downtime and costly repairs, is filtration.

Another element very destructive to fluids is excessive heat that may be generated and not dissipated effectively. Heat exchangers are necessary to maintain the optimum temperature in an operating hydraulic system and the chemical stability of the hydraulic fluid. Proper sizing and installation of conductors will prevent excessive velocities that can create pressure losses, heat generation, and expensive maintenance costs.

Review questions and problems

1 What is the primary function of a fluid in a fluid power system?
2 What other functions does the fluid serve?
3 Define the terms "viscosity" and "viscosity index."

4 What results may occur in a hydraulic system if the viscosity of the oil in the system is too low?
5 What is the prime accelerator of oil oxidation?
6 List four inhibitors that can be added to a hydraulic oil to improve its performance characteristics.
7 Name three types of fire-resistant hydraulic fluids.
8 In selecting filters for a hydraulic system, what factors should be considered?
9 List three classes of fluid conductors.
10 How are the sizes of the following conductors specified: (a) pipe? (b) tubing? and (c) hose?
11 Calculate the velocity of fluid flow passing through a pipe with a cross-sectional area of 5 in^2 at the rate of 20 gpm.
12 What would be the velocity of fluid flow through a tubing with 0.402-in. D_i, if the pump delivery were also 20 gpm?

Selected readings

- Atland, George. *Practical Hydraulics.* Vickers Division of Sperry Rand Corporation, Troy, Michigan, 1968.
- Basal, P. R., Jr. (ed.) *Mobile Hydraulics Manual.* Sperry Rand Corporation, Troy, Michigan, 1967.
- Cellulube Safety Series, *Fire-Resistant Hydraulic Fluids and Synthetic Lubricants in Controlled Viscosities.* Celanese Chemical Company, New York, 1961.
- *Compressed Air Power in Manufacturing.* Compressed Air and Gas Institute, Cleveland, 1961.
- Farris, John A. *The Meaning of Filter Ratings.* Pall Corporation, New York, 1965.
- *Filtration for Hydraulic Fluid Power Systems.* Technical Manual, T3.10.65.2. National Fluid Power Association, Thiensville, Wisconsin, 1965.
- *Fluid Power.* Bureau of Naval Personnel Navy Training Course, NAVPERS 16193-A, U.S. Government Printing Office, Washington, D.C., 1966.
- *Hydraulic Fundamentals and Industrial Hydraulic Oils.* Sun Oil Company, Philadelphia, 1963.
- *Hydraulic Power Transmission.* Engineering Bulletin no. HP-221. Standard Oil, American Oil Company, Chicago, 1966.
- McKinley, James L., and Ralph D. Bent. *Basic Science for Aerospace Vehicles.* McGraw-Hill, New York, 1963.
- *Operation and Care of Machinery.* Texaco, Inc., New York, 1965.

- Oster, Jon. *Basic Applied Fluid Power: Hydraulics.* McGraw-Hill, New York, 1969.
- Pease, Dudley A. *Basic Fluid Power.* Prentice-Hall, Englewood Cliffs, New Jersey, 1967.
- Stewart, Harry L., and John M. Storer. *Fluid Power.* Howard W. Sams, Indianapolis, Indiana, 1968.
- Wheeler, H. L. *Filtration in Modern Fluid Systems.* Bendix Filter Division, Madison Heights, Michigan, 1964.

Chapter 4
Energy input devices

This chapter discusses the operating principle of hydraulic system pumps and pneumatic compressors. It describes the performance and operation of energy input devices and introduces pump and compressor mechanisms, including gears, vanes, and pistons, with the intent of relating these mechanisms to their applications and characteristics. The chapter concludes with methods of calculating the efficiency of such energy input devices.

Key terms

- **Axial-piston pump** A pump having multiple pistons disposed with their axes parallel.
- **Compressor** A device that converts mechanical force and motion into pneumatic fluid power.
- **Cycle** A single complete operation that consists of progressive phases starting and ending at the neutral position.
- **Delivery** The volume of fluid discharged by a pump in a given time; usually expressed in gallons per minute (gpm).
- **Displacement** The quantity of fluid that passes through a pump or motor in a single revolution.
- **Fixed-displacement pump** A pump in which the displacement per cycle cannot be varied.
- **Mechanical efficiency** The ratio between the theoretical input hydraulic horsepower to the pump and the actual input horsepower.
- **Multiple-stage compressor** A compressor having two or more compressive steps, in which the discharge from each step supplies the next in series.
- **Overall efficiency** The product of volumetric efficiency and mechanical efficiency.
- **Port** The open end of a passage. Can be within or on the surface of a component housing.
- **Power unit** A combination of pump, prime mover, reservoir controls, and conditioning components.
- **Pump** A device that converts mechanical force and motion into hydraulic fluid power.

- **Radial-piston pump** A pump having multiple pistons disposed radially and actuated by an eccentric element.
- **Variable-displacement pump** A pump in which the displacement per cycle can be varied.
- **Volumetric efficiency** The ratio between the actual output from the pump at a given pressure and the theoretical output determined by the geometric displacement.

Historically, the term "hydraulics" referred to the flow of water unretarded by opposing forces of friction. It is interesting to note what Mansfield Merriman wrote at the turn of the century:

> The force pump is a device for raising water by means of pressure exerted on it by a piston. The syringe, which has been known from very early times, is an example of this principle, but the first true force pump was invented in Egypt about 250 B.C., by Ctesibus, a Greek hydraulician, and the description of it given by Vitruvius indicates that it was used to some extent by the Romans.[1]

The early force pumps were placed with their cylinders below the level of the water to be lifted and had valves that closed under the back pressure of the water.

The same author also states, "By placing the cylinder above the water level and utilizing the principle of suction, the suction and force pump originated."[2] This technological advance was accomplished in France some time after 1732. Today we recognize that pumps operate on the principle of pressure differential rather than suction.

Pump performance and rating

Designers are no longer concerned only with hydraulic pumps designed to create fluid motion or transfer. Today fluid power circuits use hydraulic pumps as devices that convert mechanical force and motion into hydraulic fluid power. Thus hydraulic pumps are capable of both moving fluids and also inducing fluids to do useful work.

Figure 4.1(a) shows that a pump requires a *prime mover* such as an electric motor, an internal combustion engine, a power takeoff, or some other mechanical device that can impart force and motion to operate the pump. The work applied to the pump shaft [Figure

[1] Mansfield Merriman, *Treatise on Hydraulics,* John Wiley & Sons, New York, 1906, p. 496.
[2] *Ibid.*

Figure 4.1 (a) Principal components of a hydraulic circuit. (b) Work applied to the pump shaft is converted to either kinetic or potential energy in the fluid.

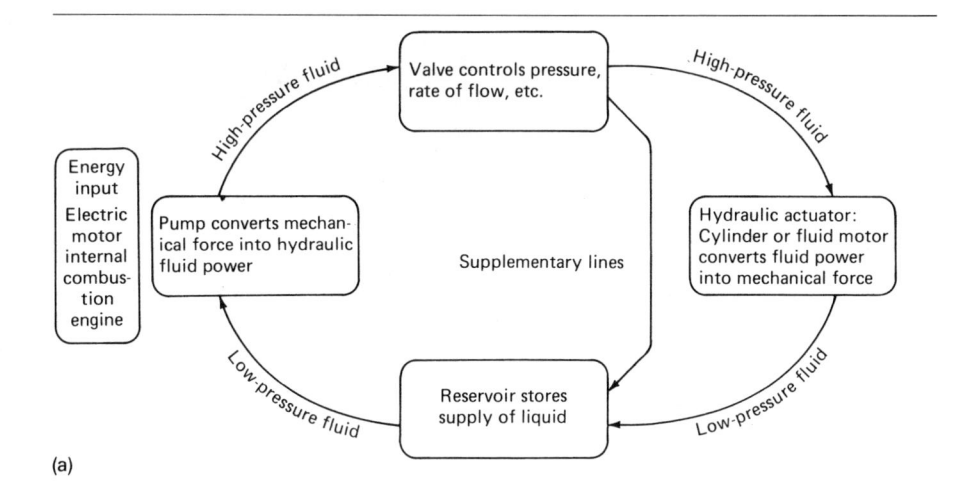

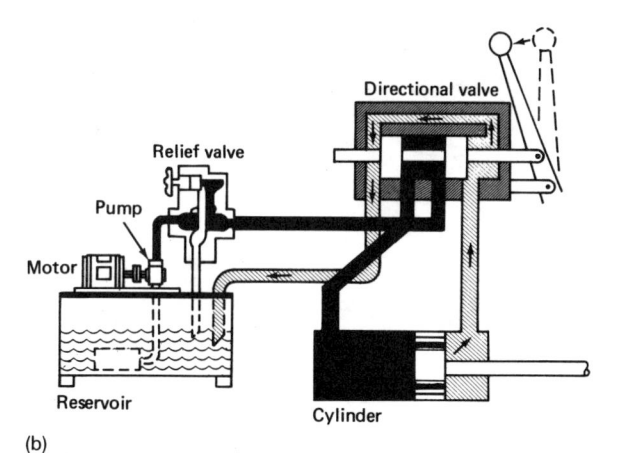

4.1(b)] is converted, with some mechanical losses, into either kinetic or potential energy in the fluid. This energy, measured as pressure and volume, is capable of inducing fluid to do useful work with the aid of a cylinder or hydraulic motor.

Hydraulic pumps create flow (Figure 4.2); they do not pump pressure. Pressure is the result of resistance to flow encountered by the confined fluid. The pressure is controlled by (1) the load imposed on the system or (2) the restriction created by the pressure-regulating device.

Pumps are classified by several different methods. In general, the fluid power industry broadly classifies them as either *positive-displacement* or *nonpositive-displacement* pumps (Figure 4.3). *Displacement* is the volume of oil moved, or displaced, during each cycle of a pump. Nearly all pumps used in hydraulic system applications—

Figure 4.2 Hydraulic pumps create flow.

Figure 4.3 Displacement pumps.

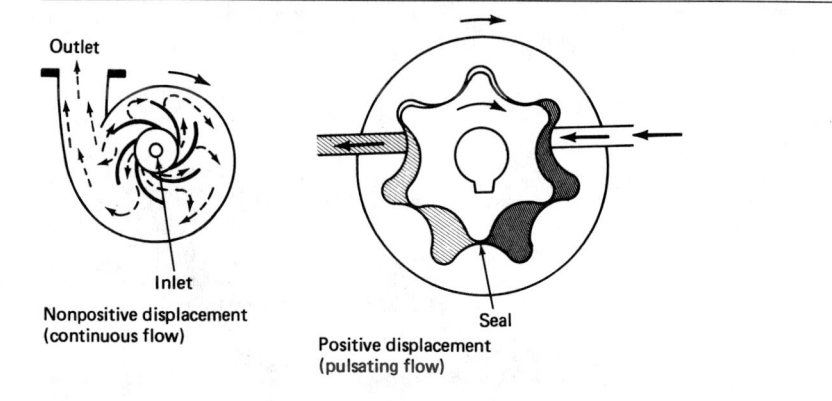

whether on industrial machinery, mobile vehicles, or aircraft power-actuated systems—are of the positive-displacement type.

Nonpositive-displacement pumps

A nonpositive-displacement pump is primarily a velocity type of unit. Because of its design and the lack of an internal seal between the stationary and rotating parts of the pump, internal leakage is more likely to occur. The leakage causes the volume of liquid delivered for each cycle to vary, depending on the resistance to flow. The higher the resistance, the greater the internal leakage. Without a positive seal between the inlet and outlet sides of the pump, the displacement between the inlet and the outlet is not positive. The volume of oil displaced depends on its speed and resistance at the discharge side. These pumps are generally used for low-pressure, high-volume applications.

Positive-displacement pumps

A positive-displacement pump is able to deliver a definite volume of liquid for each cycle of pump operation, at any resistance encountered, because there is an internal seal against leakage between the inlet and outlet sides of the unit. All positive-displacement pumps operate on a common principle. The mechanical action of the pump element creates a pressure differential in the pump cavity at the inlet side compared with the higher atmospheric pressure exerted on the surface of the fluid in the reservoir. When this mechanical action creates a sufficient vacuum (a pressure reduction of 2 to 3 psi) inside the pump cavity at the inlet side, the atmospheric pressure acting on the surface of the fluid in the reservoir forces the fluid up the pump inlet line and into the inlet side of the pump cavity. The pump element then delivers this fluid to the outlet, forcing the fluid into the hydraulic system.

Variable- and fixed-delivery pumps

Positive-displacement pumps can be further subdivided into fixed- or variable-delivery types. The delivery of a pump depends on the working relationships and geometry of its internal operating elements. In a fixed-delivery pump, these relationships cannot be physically altered, and the volumetric output remains constant for a given speed of the pump. The output can be varied only by varying the speed of the pump. A fixed-delivery pump incorporated in a system requires a pressure-relief control valve to protect it against overloads.

In a variable-delivery pump, the delivery can be varied by altering the physical relationships of the pump elements at the same speed (Figure 4.4). Note that as pressure rises or parts wear, the internal leakage in positive-displacement pumps will increase.

Volumetric output and pressure capabilities

If the geometric design and displacement of a pump are known, it is possible to determine its theoretical displacement—100% of pump capability, assuming there is no internal leakage. Even with a mechanical seal a certain amount of internal leakage occurs. This is

Figure 4.4 Fixed- and variable-displacement pumps.

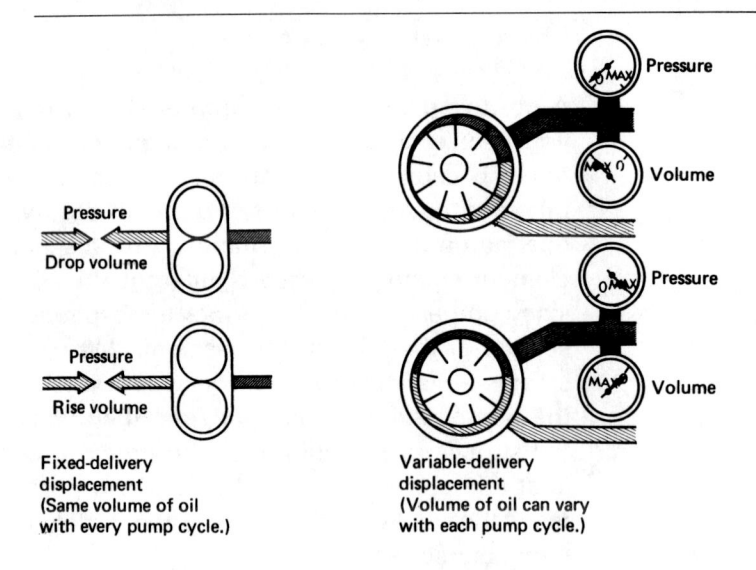

particularly true if the pressure, speed, and heat increase. The difference between the theoretical displacement and the actual displacement is the result of leakage due to pressure, viscosity, and speed. A typical example is shown in Figure 4.5, which illustrates that, as the system (output) pressure increases, the volumetric output (flow rate) decreases. Internal leakage is called *pump slippage* and is a factor that must be considered in the selection and application of all pumps.

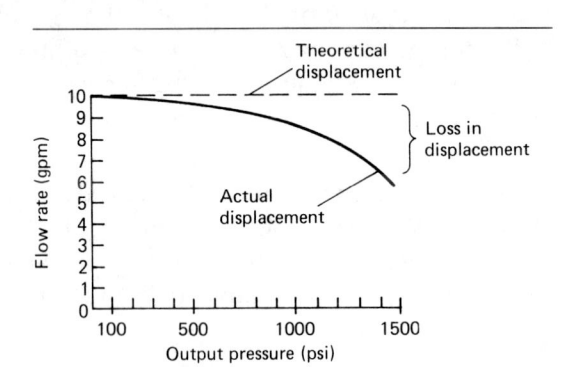

Figure 4.5 Relationship of theoretical and actual displacement. As the system pressure increases, the volumetric output decreases.

Pump slippage

Because slippage is a common factor in all pump designs, most pumps are rated in terms of volumetric output at a given pressure. For example, a pump rated at 10 gpm using a fluid of 100 SSU

(Saybolt Seconds Universal) at 140 °F would theoretically displace 10 gpm. In practice, under test conditions, this same pump delivers only 9.5 gpm with a system pressure of 600 psi and only 5.5 gpm with a pressure setting of 1500 psi. The decrease in pump output in relationship to a given increase in output pressure is governed by the pump's volumetric efficiency. Some pumps have more internal slippage than others. The *volumetric efficiency* is defined as the ratio of actual displacement to theoretical displacement. While the volumetric efficiency of the pump performance reported in Figure 4.5 may be 95% at 600 psi, it drops to only 55% at 1500 psi with the same number of revolutions per minute and the same fluid.

Pump ratings for continuous and intermittent use

Manufacturers of hydraulic pumps may also rate a pump for continuous and intermittent use. For example, a pump may be rated at 1000 psi continuous and 1500 psi intermittent. At 1500 psi there would be considerable internal slippage, which creates heat and other undesirable effects on the fluid. With intermittent use, the performance loss would not be as costly as it would be with continuous use.

Because pumps operate best between specific minimum and maximum speeds, manufacturers also specify the drive-speed rating of a pump. Excessively low speeds can be as damaging to pump elements as excessively high speeds.

Hydraulic pumps are manufactured with volumetric capacities ranging from only a few cubic inches per minute to hundreds of gallons per minute. Pressure ratings range from 150 psi or less to 10,000 psi or higher, and pump operating speeds may be as high as 3600 rpm. It should be noted that pump operating speeds increase as the pressure ratings go up, while volumetric capacities decrease. It is not uncommon to find pumps used on today's aircraft that operate at 1200 rpm, with a capacity of 10 gpm and a pressure rating of 3000 psi. The weight of such a pump might vary from 0.25 to 0.5 lb/hp. It would be capable of producing the equivalent of approximately 20 hp and even at 0.5 lb/hp would weigh only 10 lb. (See Table 4.1.)

Manufacturers of hydraulic pumps are continually seeking performance improvements in terms of greater efficiency and dependability

Table 4.1 Pump comparison

Pump type	Pressure rating (psi)	Overall efficiency (%)	Weight (lb/hp)	Cost ($/hp)	Rated speed (rpm)
External gear	2000 to 3000	80 to 90	0.5	4 to 8	1200 to 2500
Internal gear	500 to 2000	60 to 85	0.5	4 to 8	1200 to 2500
Vane	1000 to 2000	80 to 95	0.5	5 to 30	1200 to 1800
Axial piston	2000 to 10,000	90 to 98	0.25	7 to 50	1200 to 3600
Radial piston	3000 to 10,000	85 to 95	0.35	5 to 35	1200 to 1800

and reduction of size, weight, and noise level. This effort is evidenced by their research, testing, and reported data.

Classification of pumps by design

Pumps may also be classified according to the specific design of the element used to create flow of liquid. The types commonly used to develop pressures for hydraulic power transmission systems are *centrifugal, reciprocating,* or *rotary.* Rotary pumps, however, are the most extensively used today.

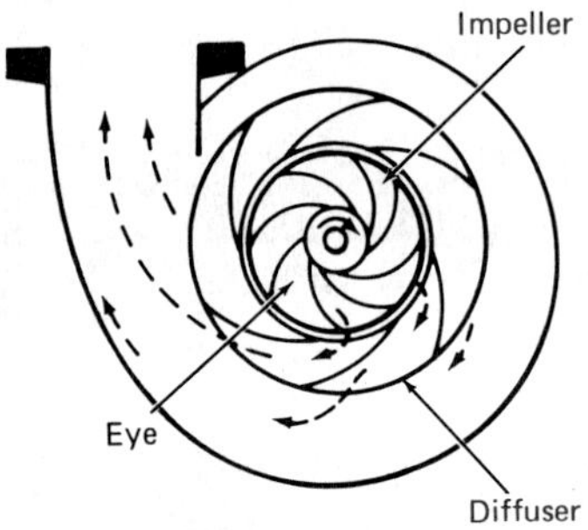

Centrifugal pumps

Figure 4.6 Centrifugal pump.

The centrifugal pump (Figure 4.6) has a normal pressure rating of only 150 psi and for this reason is generally not used for hydraulic power transmission systems. Such pumps are more commonly used for transfer of any type of liquid. Centrifugal pumps can be staged, that is, coupled in series to produce pressures up to 3000 psi. Centrifugal pumps are used in some applications to precharge other hydraulic pumps.

The operation of a centrifugal pump depends on the centrifugal force induced to the liquid pumped by the rotation of the impeller at high speeds. The induced force imparts a high velocity to the liquid and creates fluid transfer. As the shaft turns, a low-pressure area is created in the center opening (eye) of the pump (Figure 4.6). Atmospheric pressure acting on the surface of the liquid in the reservoir forces it to the eye of the pump. The blade of the impeller then starts rotating the liquid. It can be observed that the liquid between the blades moves out to a continuously increasing radius. In so doing, it acquires an increasing tangential velocity.

As the outward radius increases, the liquid must travel a greater distance with each revolution of the shaft. Eventually, the liquid reaches the tip of the blades and enters the volute space. Two actions are imparted to the fluid: (1) owing to centrifugal force, it tends to continue to move directly outward from the center of the pump cavity because of inertia; (2) it acquires *tangential* velocity, which is the speed at which a free body tends to go off on a tangent (at a right angle) to the circular path it was previously following (Figure 4.7). The two actions—centrifugal and tangential—combine to cause the liquid to flow out of the discharge port.

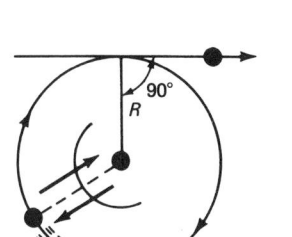

Figure 4.7 Tangential velocity.

The volute shape of the pump cavity continually expands in volume in a definite ratio to provide for the capacity of liquid escaping at all points and creating a larger flow stream as it moves toward the discharge port.

Single-stage impeller pumps have only one impeller. Staged or multistage impeller pumps have two or more impellers in one pump housing. Generally, each impeller acts separately, discharging to the inlet of the next impeller. It is possible to create higher pressure-capacity ratings through the use of multistage centrifugal pumps than are possible with a single-stage pump.

Centrifugal pumps are nonpositive-displacement pumps. They are used occasionally in hydraulic systems that require a large volume of flow and operate at relatively low pressures. In some instances they are used in conjunction with positive-displacement pumps.

Reciprocating pumps

A reciprocating pump depends on a reciprocating-piston motion to transmit liquid from its inlet side to its outlet side. The basic components of the pump are a cylinder, a reciprocating piston, and valves that direct fluid to and from the cylinder. Figure 4.8 illustrates a double-acting reciprocating pump.

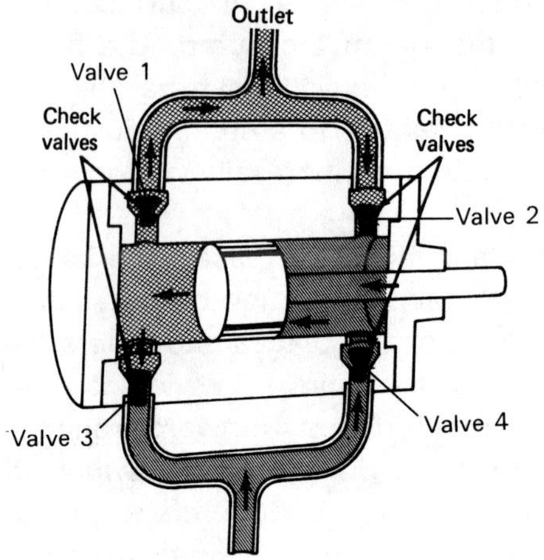

Figure 4.8 Reciprocating pump with a double-acting piston.

Reciprocating pumps are practical for extremely high-pressure applications because of low leakage losses. These pumps are manufactured with pressure ratings that range from 150 psi or less to 15,000 psi. One large-capacity reciprocating pump can supply hydraulic power to several machines. An inherent characteristic of a reciprocating pump is that the flow pulsates too greatly to provide the smooth feed demanded by some hydraulically equipped machines. Pumps with two or more pistons on the same shaft are designed to reduce the pulsations in flow by arranging the compression stroke of one cylinder to overlap with the next. Most reciprocating pumps are of the fixed-displacement design. On some models, however, it is possible to vary fluid flow by varying the length of the piston stroke. Hand pumps with reciprocating pistons are used in such devices as automobile jacks, small presses, and tensile testers.

The design of the pump shown in Figure 4.8 isolates the inlet from the discharge port by means of four check valves. When the piston moves to the right, a partial vacuum is created in the cap-end chamber of the cylinder. This reduced pressure enables the atmospheric pressure exerted on the fluid surface in the reservoir to unseat Valve 3 and push the oil into the cylinder. Valve 1 remains seated during this stroke. While the piston moves to the right, Valve 4 remains seated and the pressure created by the restriction of ball check Valve 2 is exceeded to unseat this ball and permit the oil to flow to the discharge. As the piston moves to the left, the action of each valve is reversed and liquid is forced from the rod-end chamber to the discharge port.

Rotary pumps

Most pumps used in hydraulic system applications today are the rotary type, in which a rotating assembly carries the fluid from the pump inlet to the outlet. Rotary pumps operate with a continuous rotary motion instead of a reciprocating movement. The three most common pumping mechanisms used in rotary pumps are the *gear, vane,* and *piston.*

Gear

Rotary gear pumps (Figure 4.9), which consist of two or more meshing gears enclosed in a close-fitted housing, are positive-displacement pumps. All gear pumps can be classified into three general categories:

1 gear-on-gear (*external gear*),
2 gear-within-gear (*internal gear*), or
3 axial-flow (*screw*).

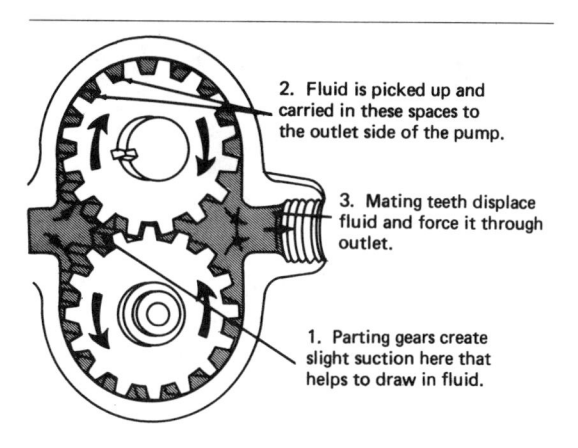

2. Fluid is picked up and carried in these spaces to the outlet side of the pump.

3. Mating teeth displace fluid and force it through outlet.

1. Parting gears create slight suction here that helps to draw in fluid.

Figure 4.9 Rotary gear pump. (Courtesy Brown & Sharpe Manufacturing)

Three significant facts to be noted about the operation of a gear pump are: (1) when oil is transferred away and gears are unmeshed on the inlet side, a pressure drop occurs that helps the atmospheric pressure acting on the surface of the oil in the reservoir to force it up the pipe to fill the void; (2) the oil is transferred around the periphery of the pump housing; and (3) as the gear teeth mesh again they form a seal that prevents oil from backing up to the inlet.

External-gear pumps External-gear pumps having two gears, as shown in Figure 4.10, have one gear mounted on the drive shaft (making it a drive gear), while the second is the driven gear. These gears rotate in opposite directions (Figure 4.10) and mesh at a point in the housing between the inlet and outlet ports. As the gear teeth become unmeshed on the inlet side, the volume on the inlet increases, thereby decreasing the pressure and making it possible for the atmospheric pressure exerted on the surface of the oil in the reservoir to push oil up the inlet pipe and into the pump inlet port.

Figure 4.10 External-gear pump. (Courtesy Webster Electric Company)

Rotation of the gears from the inlet port toward the outlet traps oil in each tooth space and transfers the oil peripherally to the outlet side. A mechanical metal seal between the end of the gear tooth and the pump housing prevents oil slippage. In addition, a seal is provided between the sides of the gears and the wear plates (Figure 4.11). As the gears rotate, the teeth meshing on the outlet side also form a seal to prevent oil from backing to the inlet. Oil is continually delivered to the outlet chamber with the rotation of the gears. Since oil is not compressible, it is forcibly ejected out of the discharge port.

It is possible to obtain quieter and somewhat smoother operation than that achieved with a spur gear pump by using helical or herringbone external gears [Figure 4.12(a) and (b)]. Larger quantities of oil can be delivered with less pulsation by rotating pumps with these gear designs. Another adaptation of the external-gear pump is the

Figure 4.11 (a) Wear plates on a gear pump. (b) Mechanical seals and wear plates of a rotary gear pump. [(b) Courtesy Warner-Motive Hydraulics]

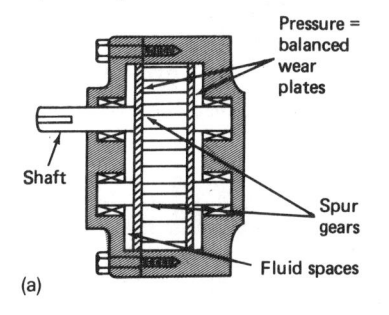

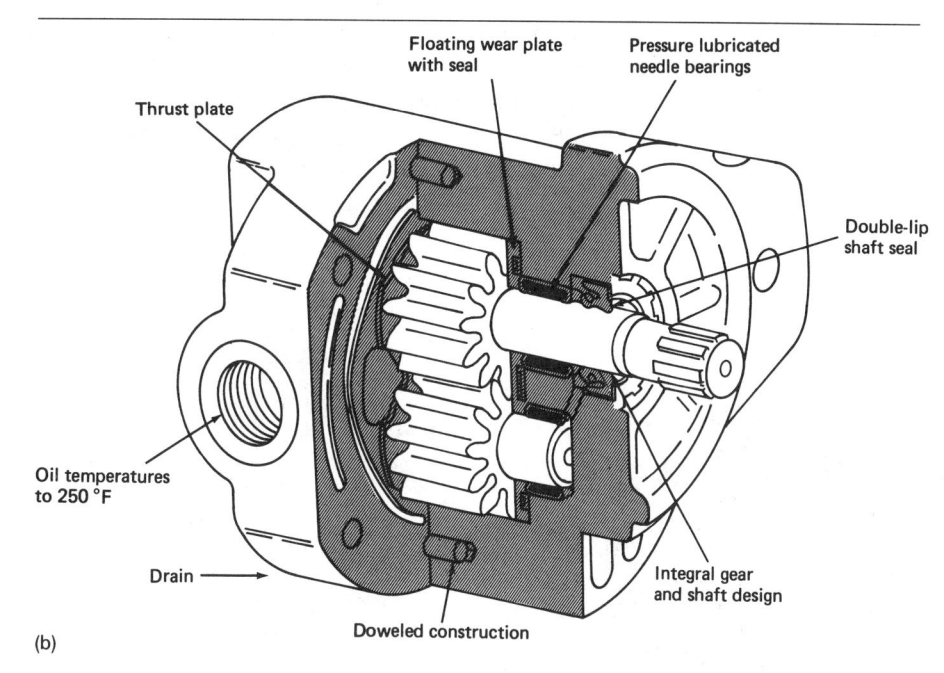

lobe-shaped rotating element [Figure 4.12(c)]. Unlike the spur gear, the lobe-shaped and helical mechanisms are both driven through suitable external timing gears.

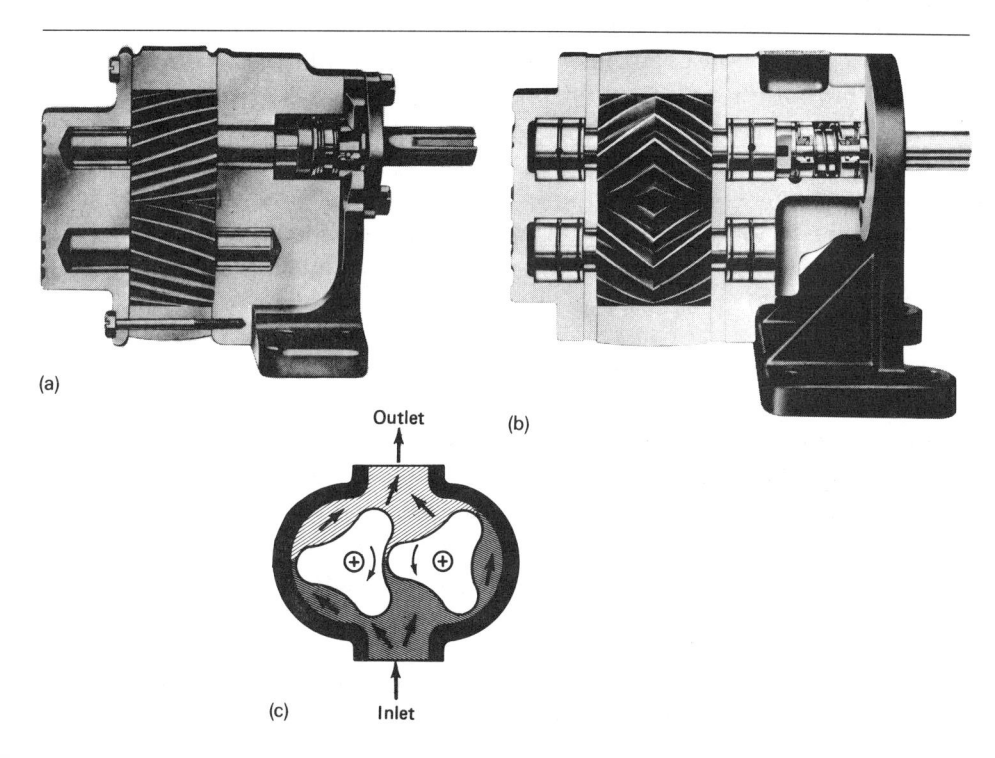

Figure 4.12 (a) Helical-gear pump. (Courtesy Brown & Sharpe Manufacturing) (b) Herringbone-gear pump. (Courtesy Brown & Sharpe Manufacturing) (c) Rotating-lobe pump.

Internal-gear pumps The internal-gear pump also uses two gears, but it is a modification of the external-gear pump [Figure 4.13(a)]. The spur gear is mounted inside a larger ring gear. The smaller spur gear is in mesh with one side of the larger gear and kept apart by a crescent-shaped separator on the other side, as illustrated in Figure 4.13(b). The separator isolates the inlet port from the outlet port. The prime source of power may be applied to either the spur gear or the outer ring gear.

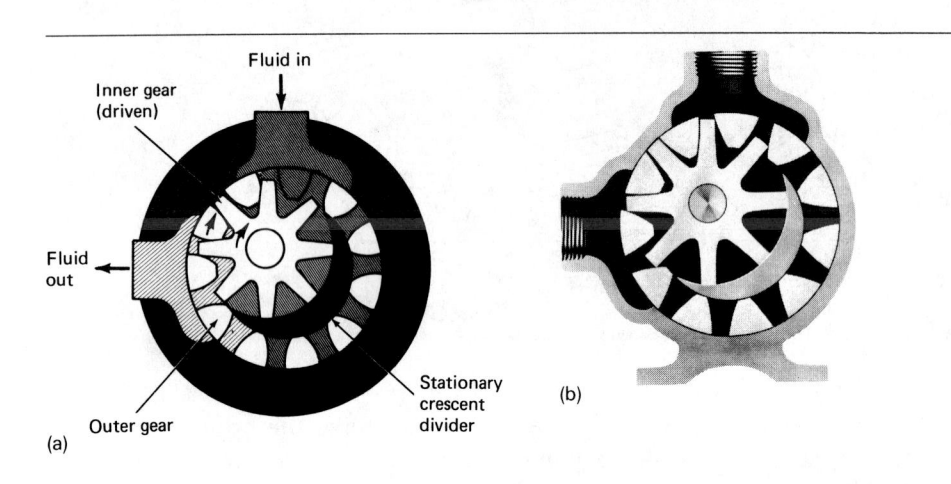

Figure 4.13 (a) Internal-gear pump. (Courtesy Brown & Sharpe Manufacturing) (b) Cutaway view of internal-gear pump. (Courtesy Viking Pump)

The principle of operation is essentially the same as for the external-gear pump. One major difference is that both gears rotate in the same direction. As the gear teeth are unmeshed, a partial vacuum is created on the inlet side. Atmospheric pressure forces oil from the reservoir into the space created, and with the rotation of the gears oil is carried around the periphery of the gears and the separator, as illustrated in Figure 4.13(a), until it reaches the outlet port. As the gear teeth mesh again, a seal is formed which prevents a back-up of the oil. A continuous flow of oil pushes out through the outlet port.

Another variation of the internal-gear pump, the internal-lobe gear pump, is shown in Figure 4.14(a). This design makes use of a special tooth form. The inner gear (driven) has one tooth less than the outer, or ring, element. Geometrically, the two elements are sized so that part of the periphery of the inner gear maintains contact with the surface of the outer gear element at all times. There is always a seal between the inlet and outlet port, as shown in Figure 4.14(b).

Screw pumps There are two types of screw pumps available, single screw and multiple screw. A single-screw pump consists of a helical gear (screw) that rotates eccentrically in an internal element (Figure

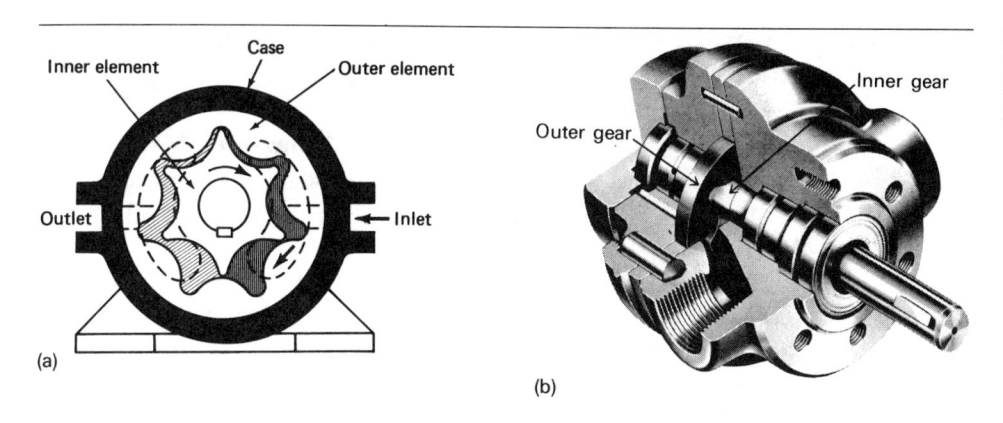

Figure 4.14 Internal-lobe gear pump. The outer element has one lobe more than the inner element. [(b) Courtesy Viking Pump]

4.15). Multiple-screw pumps consist of two or more helical gears that mesh as they rotate in a close-fitting housing. As the screws rotate, a volume of fluid from the inlet is trapped between the contact points of the screws and the space between the screws and the housing. Rotation of the screws causes the volume of trapped oil to move linearly along the screws' axis until it is pushed through the outlet of the pump, as shown in Figure 4.15. Flow through a screw pump is axial and in the direction of the driving screw.

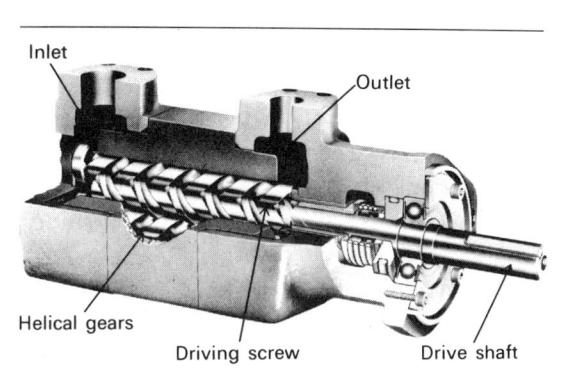

Figure 4.15 Constant-displacement, rotary-screw pump. (Courtesy De Laval Turbine)

Screw pumps produce extremely smooth, nonpulsating delivery with a very low noise level. These types of pumps are used in elevators, submarines, and other applications requiring a quiet environment.

Output capabilities of gear pumps Gear pumps are available with output capabilities ranging from a fraction of a gallon per minute to over 100 gpm. Most production gear pumps have a capacity of 1 to 25 hp. The average pressure rating is 500 psi; however, some models have pressure-loaded wear plates and close tolerances that permit operational pressures ranging from 1000 to 3000 psi.

Gear pump efficiency is quickly reduced by wear. Volumetric efficiency is rapidly reduced by the following leakage losses: (1) around the periphery of the gears, (2) across the faces of the gears, and (3) at the points of contact of gear teeth.

The major uses of gear pumps are in mobile equipment, machine tools, and aircraft. Their relative simplicity of design, low cost, rugged construction, large capacity in a small space, and tolerance for contaminants have made them the most common type of pump. However, low efficiency, severe wear characteristics, and short life expectancy contribute to relatively high maintenance costs.

Vane

All vane pumps have a rotor driven within a ring by a drive shaft coupled to a prime mover. A cylindrical rotor with sliding rectangular vanes, generally in radial slots, rotates within the ring. The center of the rotor is offset from the center of the ring, as illustrated in Figure 4.16, leaving a crescent-shaped chamber between the rotor and the ring. In a simple vane pump (Figure 4.16), as the rotor turns, the vanes are forced outward against the inner surface of the cam ring by centrifugal force. In some pumps, fluid under pressure is directed to the lower end of the vanes to force them outward. This outward radial movement of the vanes and turning of the rotor causes the chamber between the vanes to increase as the vanes pass further away from the inlet port. This increase in volume results in a lowering of pressure until the atmospheric pressure is sufficient to force oil from the reservoir into the inlet chamber. Oil from the inlet is

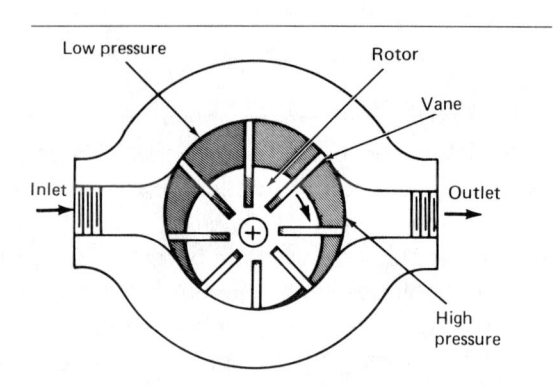

Figure 4.16 Vane pump (unbalanced).

swept away by the vanes toward the outlet port through a decreasing series of chambers until it is forced through the outlet port. System pressure is exerted against the leading side of the vanes, assuring their sealing contact with the ring during the pumping operation. Slippage is reduced to a minimum by the back pressure on the vanes.

The pump illustrated in Figure 4.16 is hydraulically unbalanced because high pressure is generated on only one side of the rotor shaft. A vane pump of balanced design has an elliptical cam ring so that two pumping chambers are formed. The pump illustrated in Figure 4.17 is in hydraulic balance, since the two intake and two outlet ports are diametrically opposed to each other. In balanced-design pumps, the side loads cancel out, thereby increasing bearing life and permitting high operating pressures.

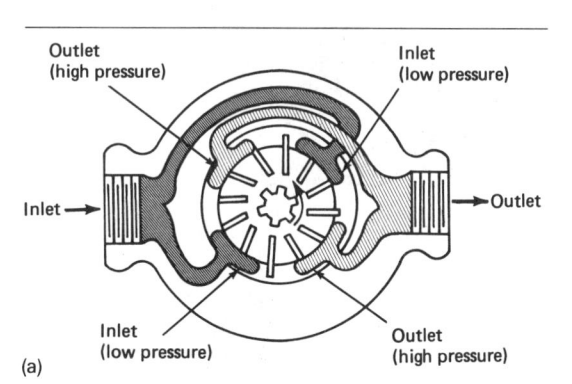

Figure 4.17 Hydraulically balanced vane pump. [(b) Courtesy Webster Electric Company]

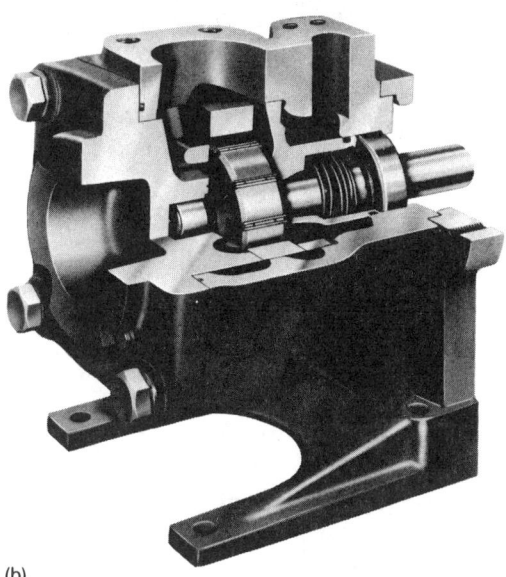

114 Energy input devices

Vane pumps are either fixed- or variable-delivery types. Fixed-delivery vane pumps can be unbalanced; however, they are generally balanced by being ported symmetrically to equalize the reaction forces. The eccentricity of the rotor with respect to the ring determines the displacement of a vane pump.

The single variable-delivery vane pump is controlled to deliver maximum or reduced flow to meet circuit requirements (Figure 4.18). Pumps of this design can provide an infinitely variable volume from zero to the maximum volumetric capacity. The basic purpose of the pressure compensator on the variable-delivery pump is to limit the maximum system pressure [Figure 4.18(a)]. Delivery is automatically

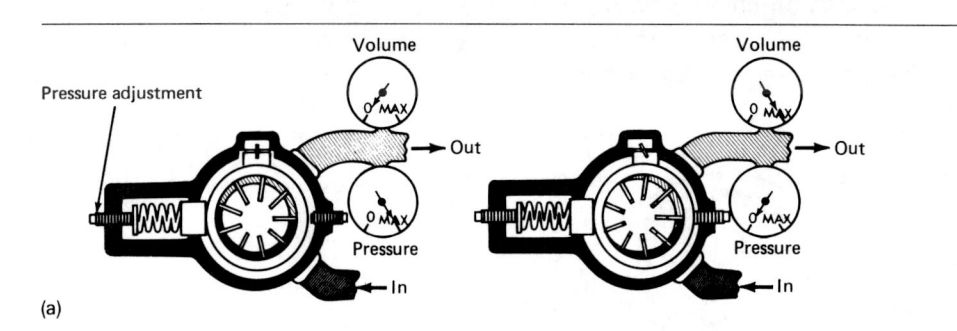

Figure 4.18 Variable-delivery vane pump. [(b) Courtesy Racine Hydraulics and Machinery]

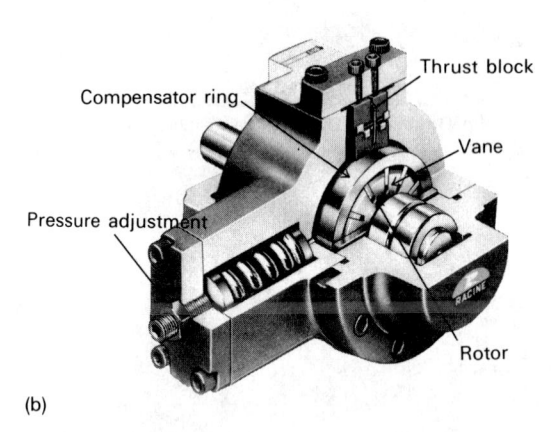

changed to provide the exact flow rate demanded by the system. System pressure is maintained at a nearly constant value regulated by the compensator setting. When the hydraulic circuit no longer requires flow, the pump ring returns to a nearly neutral (concentric) position, and only minimal flow takes place through leakage at the set pressure. When the circuit demands full flow, the pressure ring is offset to its full-flow position as a result of the decrease in circuit pressure. The reaction pressure of the circuit and the compensator spring bias force are matched so that any flow rate from zero to maximum capacity is automatically delivered to meet circuit demands.

One advantage of a pressure-compensated pump is that power consumption is reduced as the flow rate decreases. Stated in another way, a pump consumes power to deliver oil at the outlet. Pressure-compensated pumps also reduce the need for complex valving. Some manufacturers pressurize the control spring chamber in pressure-compensated pumps. In such designs, a pressure-control valve governs the maximum pump pressure. Hydraulic pressure in the spring chamber, when added to the spring force, allows the pressure ring to remain in a full-flow position for a higher pressure setting before shifting the pressure-control valve. Both the low- and high-pressure settings are adjustable and may be switched by means of a solenoid valve.

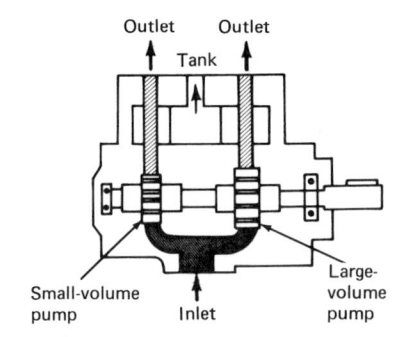

Figure 4.19 Double vane pump.

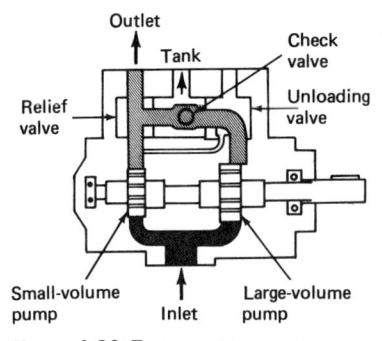

Figure 4.20 Two-pressure vane pump.

Vane pumps are available as single or multiple pumps. Figure 4.19 shows two pumps mounted on a common shaft; this is commonly referred to as a *double pump.* A double pump consists of two cartridges that may or may not be the same size on the same shaft and that can supply two circuit branches in situations where separate hydraulic circuits with different requirements are needed. Each pump is controlled by its own relief valve.

Combination pumps are used extensively to provide fluid at more than one pressure. Pumps connected in tandem will deliver greater volumes of oil than either pump alone, and multistage pumps will produce pressures higher than the component pumps alone. Figure 4.20 shows two pumps on the same shaft. Their combined deliveries provide large-volume, low-pressure oil for a rapid approach to work. A small volume of high-pressure oil for the slow movement in the feed cycle is delivered by the small pump, while the large pump unloads to the tank.

The two-pressure pump shown in Figure 4.20 could be used in a circuit application such as a drill press, first providing rapid advance to work and then slow feed during the drilling operation. The operating pressure of the smaller combination pump is controlled by the relief-valve setting.

The two-stage pump illustrated in Figure 4.21 consists of two vane

116 Energy input devices

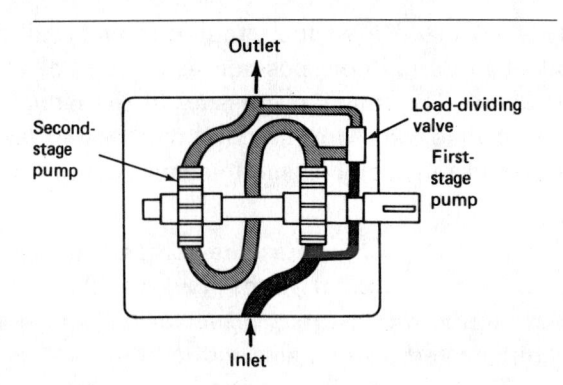

Figure 4.21 Two-stage vane pump.

pumps operating in series, mounted on the same shaft. Much higher pressures can be generated by a two-stage pump. The volumetric displacement is determined by the first-stage pump.

Vane pumps, because of their design, can handle large volumes of oil. Pumps with capabilities of 100 gpm exist, but the more common capacities are from 5 to 40 gpm. Vane pumps are classified as medium-pressure pumps, with the common operating range from 500 to 1500 psi; however, high-performance vane pumps for mobile

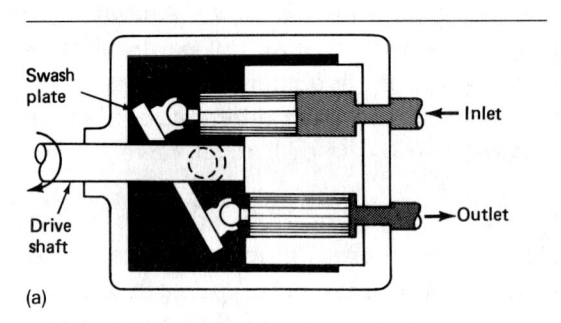

Figure 4.22 (a) Axial-piston pump. (b) Radial-piston pump.

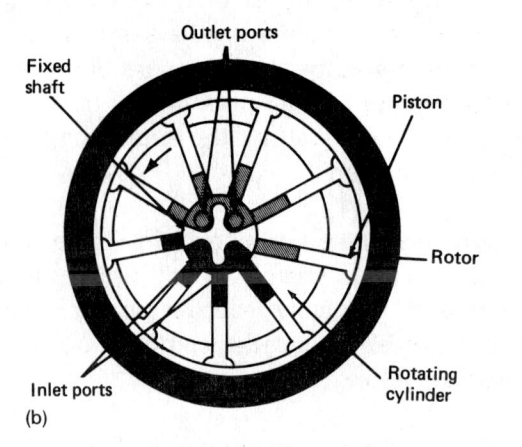

applications are manufactured with ratings as high as 2500 psi. Multistage vane pumps may have a rating of 3000 psi.

Industrial vane pumps are designed to run at speeds corresponding to synchronous electric motor speeds of 1175 to 1750 rpm. Mobile-application pumps are engine driven so that they can operate at varying speeds up to 2400 rpm.

In general, vane pumps are efficient and durable if operated within a clean hydraulic system and to the manufacturer's specifications. Volumetric efficiency for vane pumps ranges from 85 to 90% at rated pressure and speed. They have good mechanical efficiency as a result of hydraulic balancing, but overall efficiency is somewhat lower, on the order of 75 to 87%. Losses in vane pumps are due to clearance leakage across the faces of the rotor and between the bronze wear plates and the pressure ring. They are generally quiet, but will whine at higher speeds, and are compact in relation to their output. Satisfactory performance of vane pumps depends on clean oil, good lubricity, and appropriate shaft speed.

Piston

Rotary-piston pumps are classified according to their piston arrangements, either axial or radial. Axial-piston pumps convert rotary shaft motion to a reciprocating motion of a piston [Figure 4.22(a)]. Radial-piston pumps convert rotary shaft motion to a radial reciprocating motion of the pistons [Figure 4.22(b)].

Axial-piston pumps The pistons in an axial pump operate along the axis of the pump drive shaft (Figure 4.23). The cylinder block has a series of cylinder bores (always either seven or nine) with pistons that move in and out. Rotation of the drive shaft causes the pistons and the cylinder block to rotate at the same speed. As the block rotates, each piston moves in and out of its cylinder, the length of stroke depending on the angle of the cylinder block with reference to the drive plate. As each piston moves outward, fluid is drawn into the cylinder bore through the valve plate. On the return stroke, fluid is forced out through the valve plate under pressure due to restriction of flow.

Several basic designs are available in axial-piston pumps. The bent-axis, fixed-delivery type has a fixed angle of the cylinder relative to

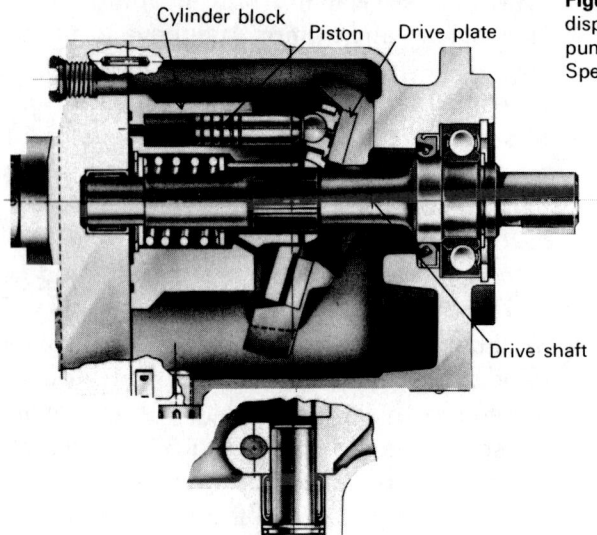

Figure 4.23 Inline, fixed-displacement, axial-piston rotary pump. (Courtesy Vickers Division, Sperry Rand Corporation)

the housing (Figure 4.24). There is no provision for altering this angle. The bent-axis, variable-displacement type has the cylinder block mounted in a yoke that can be positioned at various angles.

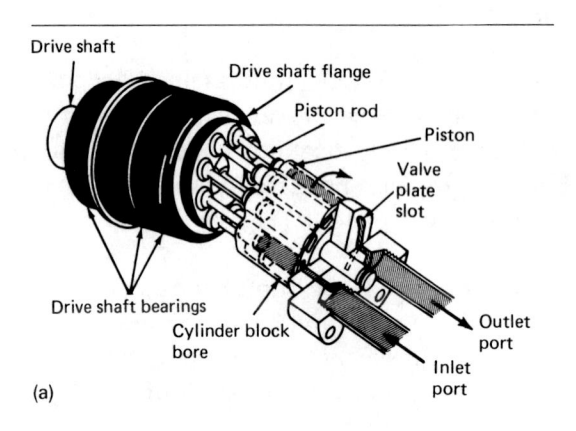

Figure 4.24 Bent-axis, fixed-displacement pump. (Courtesy Vickers Division, Sperry Rand Corporation)

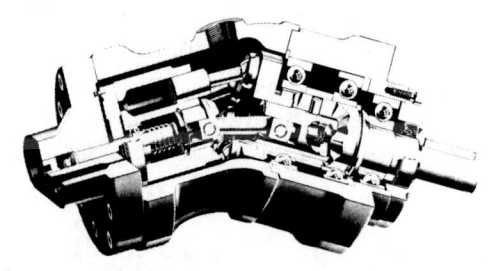

The relative positions of the cylinder block and the drive shaft determine the displacement of this type of pump. When the cylinder block centerline is parallel to the drive shaft, no flow is produced. Displacement is regulated by moving the yoke to offset the cylinder block. Mechanical linkages are used to make the precision adjustment.

Inline axial-piston pumps, although they are similar to bent-axis pumps, differ in that the cylinder block is parallel to the drive shaft. The stroke length of the piston is determined by the angular position of the swash plate, as illustrated in Figure 4.23. Inline axial-piston pumps are available in fixed- or variable-displacement models. A variable-displacement unit has a swash plate cradled so that its angle can be altered by any of several methods (Figure 4.25). A fixed-displacement unit has the swash plate mounted at a fixed angle within the housing. An axial-piston pump always has some leakage, which must be drained internally or externally.

Figure 4.25 Variable-delivery, axial-piston pump. (Courtesy Abex Corporation, Denison Division)

Radial-piston pumps Radial-piston pumps have a cylindrical element that rotates about a stationary central pintle (Figure 4.26). This

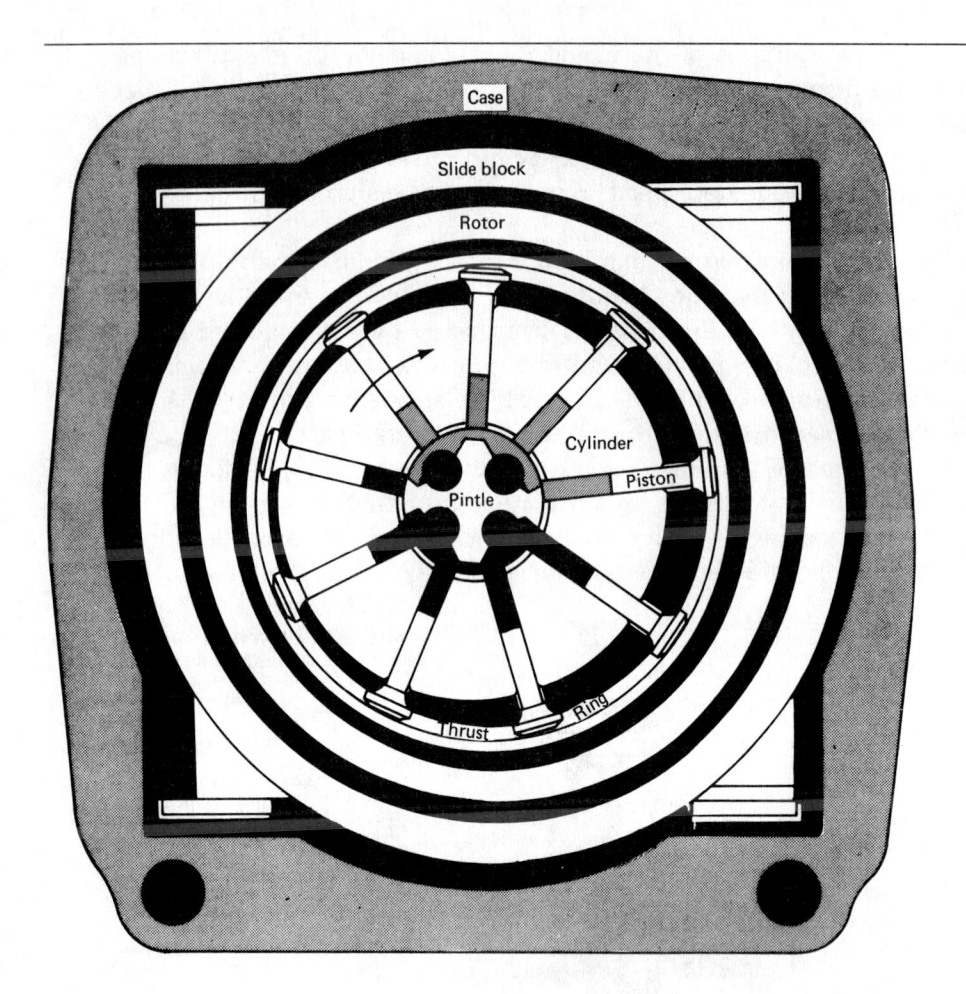

Figure 4.26 Major parts of a radial-piston pump. (Courtesy Oilgear)

cylinder contains seven or nine radial bores fitted with pistons that move in or out as the cylinder block turns. The pintle includes inlet and outlet ports that connect with the inner openings of the cylinder bores. A rotor and its support, called a slide block, move eccentrically with respect to the cylinder block. The pistons in the cylinder block are equipped with shoes at their outer ends, which bear against the freely rotating reaction rotor.

As the drive shaft rotates the cylinder block, the individual pistons travel outward while the cylinder bores pass the inlet ports of the pintle, drawing in oil (Figure 4.27). As the oil-filled bores pass toward the outlet ports of the pintle, the pistons are moved inward by the rotor, forcing pressurized oil through the outlet ports to the system. The piston heads are held against the rotor by centrifugal force. Pump delivery is determined by the degree of eccentricity between the rotor and the cylinder.

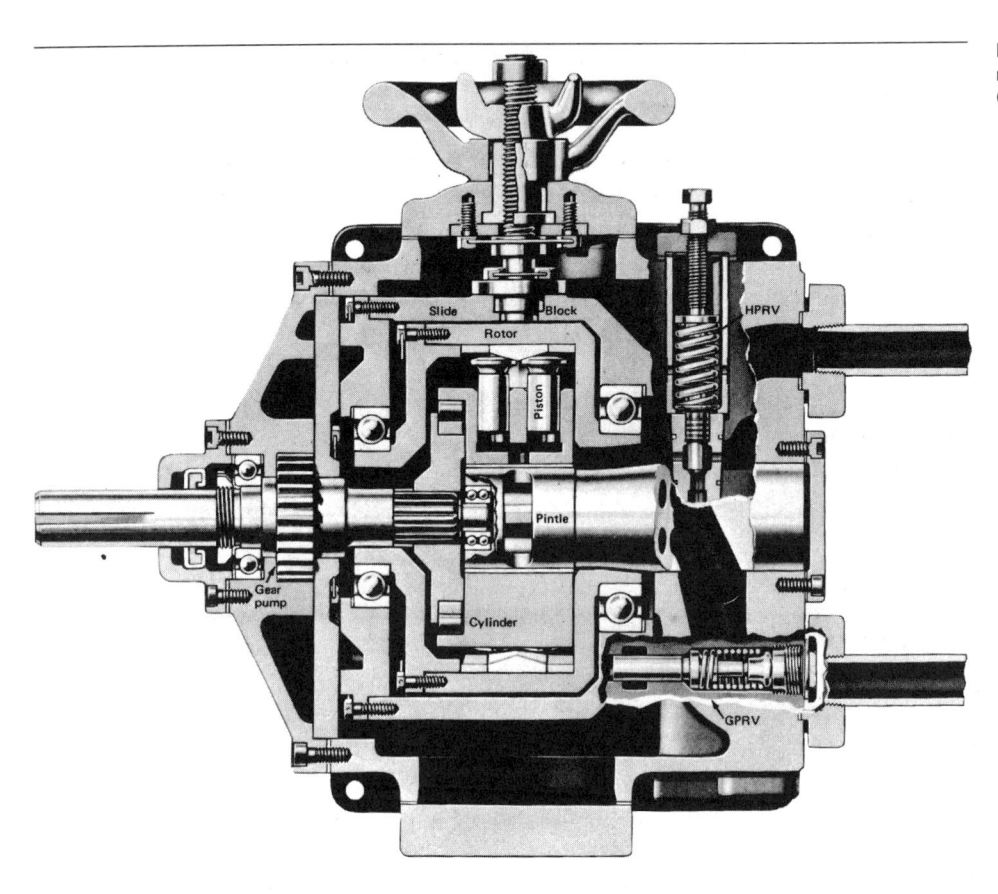

Figure 4.27 Cutaway view of a radial-piston pump. (Courtesy Oilgear)

Radial-piston pumps are available in both fixed- and variable-delivery models. The eccentricity of the cylinder block (which governs the rate of delivery) in variable-delivery, radial-piston pumps can be controlled by several methods. A hand-operated wheel that shifts a slide block (Figure 4.27) is used to increase or decrease the eccentricity of the rotor with respect to the cylinder block. Automatic control of displacement to accommodate changing circuit requirements is also provided by means of a hydraulically actuated slide block.

Advantages of piston pumps Piston pumps are capable of operating under high pressures. Many operate at pressures of 3000 psi, several at 5000 psi, and some pumps used in aircraft applications operate at 10,000 psi. Piston pumps are generally driven at high speeds, up to 6000 rpm. These pumps produce a seemingly nonpulsating hydraulic flow when driven by a suitable source of rotary power. They are positive-displacement pumps; that is, the delivery rate at a given drive speed is constant and virtually unaffected by variations in system pressure. They readily lend themselves to variable displacement. Some models have controls that permit reversible flow during pump operation.

Piston pumps are generally compact for their horsepower output. Of the three types of pump used in hydraulic circuit applications, piston pumps are the most efficient, since they are very closely fitted units. Volumetric efficiencies of 90 to 95% are not uncommon. Their mechanical efficiency approximates 90%, with an overall efficiency slightly lower.

Structurally, the only place for leakage to occur is at the piston. No side loads are imparted to the piston, so that pump life expectancy is normally several years. These pumps usually cannot be serviced in the field and are initially more expensive than other types.

Air compressors

The unlimited supply of air and its ease of compression make compressed air the most widely used fluid medium for pneumatic systems. On a visit to an industrial plant or an office, you are almost certain to observe a number of operations carried on by compressed air. The faster and more automatic the operations, the greater the likelihood that they are done by compressed air.

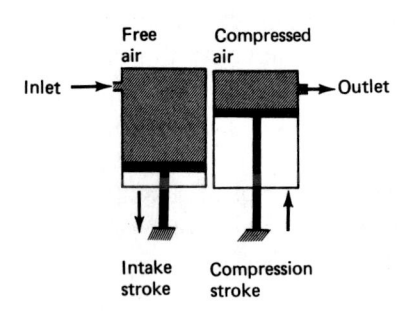

Figure 4.28 Positive-displacement compressor.

Air is a fluid just as hydraulic oil is. Air compressors are essentially energy input devices that convert mechanical energy into fluid energy. In pneumatic systems, compressors perform functions similar to those performed by pumps in a hydraulic system. These devices also require prime movers such as electric motors or engines.

As in the case of hydraulic pumps, a compressor works on the principle of increasing air pressure by confining it into a progressively diminishing volume, as illustrated in Figure 4.28. In general, compressors are classified by the type of compressing element, as positive-displacement or dynamic-displacement. Compressor *displacement* refers to the quantity of air that passes through a compressor in a single revolution or stroke. In the positive-displacement compressor, the air is confined within an enclosed space where it is compressed by decreasing its volume (Figure 4.28). In dynamic air compressors, the air is accelerated by rotor blades, causing some increase in pressure and a large increase in velocity. Subsequently, the velocity is reduced by diffusers, causing additional increase in pressure. Compressor *delivery* is the volume of air discharged by a compressor in a given time, generally expressed in standard cubic feet per minute (SCFM).[3]

Compressors may be further classified according to the specific design of the element used to create flow of air. The types commonly used to create air pressures for pneumatic power transmission systems are rotary or reciprocating.

Rotary

Positive-displacement rotary air compressors include the sliding-vane, lobed-rotor, liquid-piston, helical-screw, and the reciprocating-piston types.

Sliding-vane compressor The sliding-vane rotary air compressor, illustrated in Figure 4.29, consists of a cylindrical, slotted, eccentrically located rotor with sliding vanes, which is free to turn within

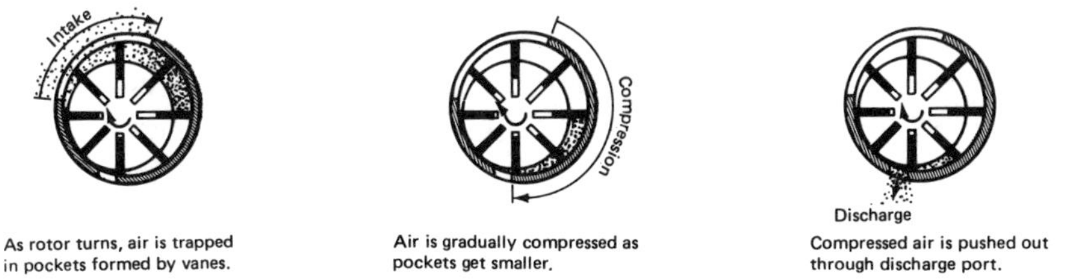

Figure 4.29 (a) Sliding-vane, rotary, positive-displacement compressor. (b) Rotary-vane, positive-displacement compressor. [(a) and (b) Courtesy Gast Manufacturing]

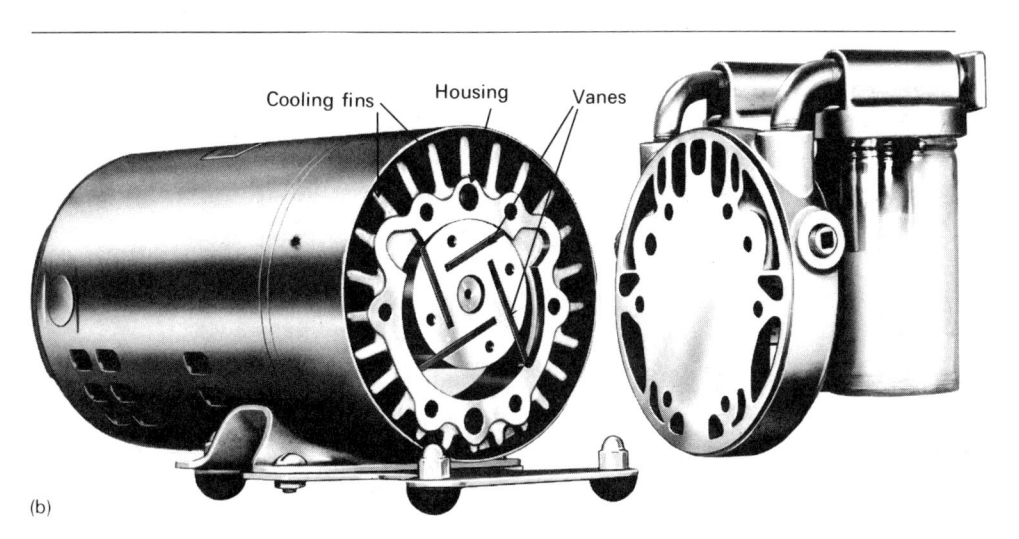

As rotor turns, air is trapped in pockets formed by vanes.
(a)

Air is gradually compressed as pockets get smaller.

Compressed air is pushed out through discharge port.

(b)

[3] SCFM is 1 ft³ of air per minute at the standard conditions of 68 °F, 14.7 psi, and 36% relative humidity.

a casing. Because the rotor is offset in the casing, the vanes moving past the inlet increase the space cavity and lower the pressure in the inlet chamber. This lower pressure permits atmospheric pressure (free air) to create flow into the inlet port. As the vanes move across an ever-smaller volume of cavity, the trapped air is compressed and discharged to the outlet port. Rotary sliding-vane compressors are used to obtain pressures up to 50 or more psig in single-stage designs and up to 150 psig in two stages. This type of low-pressure, low-volume compressor is used for instrument and laboratory equipment purposes.

Lobed-rotor compressor Two-stage, lobed-rotor compressors have two rotating elements that revolve in opposite directions within the chamber (Figure 4.30). There is no internal contact pressure between the impellers so that lubrication is not needed on these surfaces. The impellers are kept in proper rotative positions by a pair of gears external to the chamber. As the lobe elements rotate and disengage on the inlet side, the pressure decreases in the inlet chamber, permitting free air to enter at atmospheric pressure. As the lobe elements engage within the outlet portion of the chamber, the volume decreases, causing the trapped air to become pressurized and ultimately discharged through the outlet port. This type of compressor functions generally at lower pressures. A two-stage lobe-rotor compressor is used for pressures of approximately 30 psig. The lobed-rotor, like other impeller-type, compressors have capacities that vary from a few cubic feet per minute to several hundred cubic feet per minute.

Liquid-piston compressor The liquid-piston rotary compressor (Figure 4.31) has a ring of liquid, usually water, serving as the compressing medium, which moves in an elliptical path around a casing as the impeller turns. In effect, blades on the rotor form a series of buckets carrying the liquid around the casing. The liquid is forced outward against the housing and the resulting change in volumetric cavity causes the pumping action. In actual practice, the liquid following the contour of the casing surges back into the impellers near the outlet, causing the trapped air to be compressed and ultimately discharged. This type of compressor develops up to 75 psig in one stage.

Helical-screw compressor The rotary helical-screw compressor is made possible by improved materials and machining tolerances. The two helically grooved rotors (screws) mesh to compress air once they are in motion. The design of the screws enables them to move air from the inlet to the outlet port. Compression is achieved by rolling

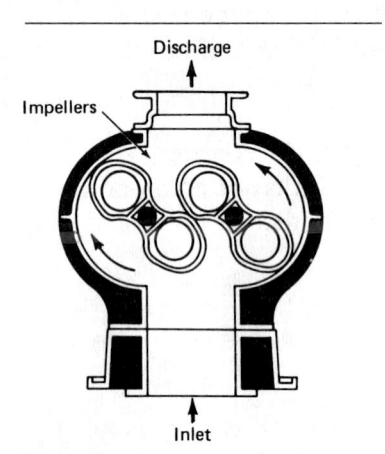

Figure 4.30 Two-lobed rotary compressor.

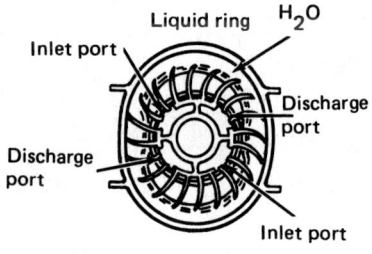

Figure 4.31 Liquid-piston rotary compressor.

the trapped air into a progressively smaller volume as the screws advance. This type of compressor, like the screw pump, has a low noise level, low loss of efficiency, and does not require cooling. Single-stage screw compressors are manufactured to operate with capacities of 100 cfm and pressures of 120 psi, which makes them very desirable for many pneumatic applications (Figure 4.32).

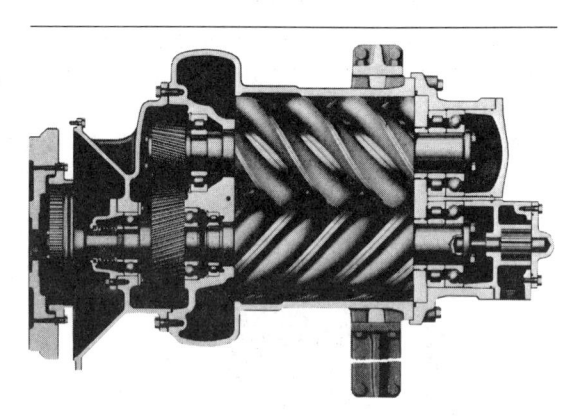

Figure 4.32 Single-stage screw rotary compressor. (Courtesy Kellogg-American)

Reciprocating

The reciprocating-piston compressor, the most widely used positive-displacement compressor, operates on the same principle as a piston bicycle pump. Reciprocating-piston compressors may be single or double acting by design and have one or more cylinders located in different configurations such as horizontal, vertical, or angled (Figure 4.33).

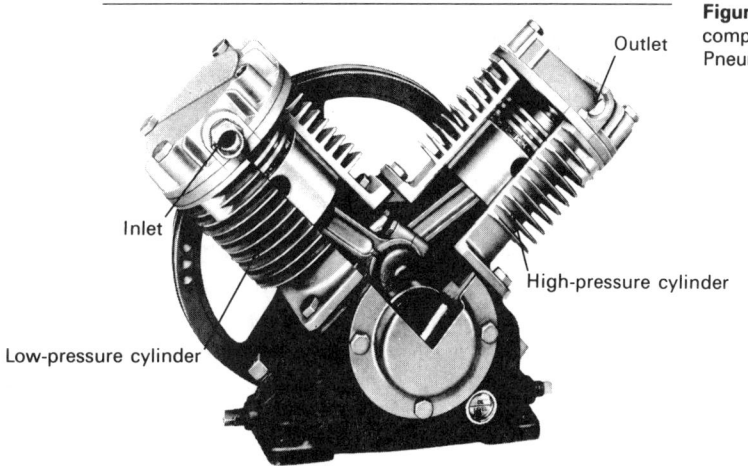

Figure 4.33 Reciprocating-piston compressor. (Courtesy Champion Pneumatic Machinery Company)

When valving and compression occur on only one side of the piston, the compressor is described as a *single-acting* piston compressor (Figure 4.34). *Double-acting* reciprocating-piston compressors have valves and compression occurring on both sides of the piston. This arrangement provides two compression cycles for each stroke of the piston. In a two-stage piston compressor, air enters the inlet of the first stage past the check valve while other air is being compressed by the piston and forced to the second stage through a cooler. Cooling between stages of multistage machines reduces the volume to be handled by the next stage and also reduces the work requirement. Moisture is precipitated during cooling, causing a further reduction in work. Compressed air from the first stage, reduced in volume by cooling, is further compressed by the second-stage piston.

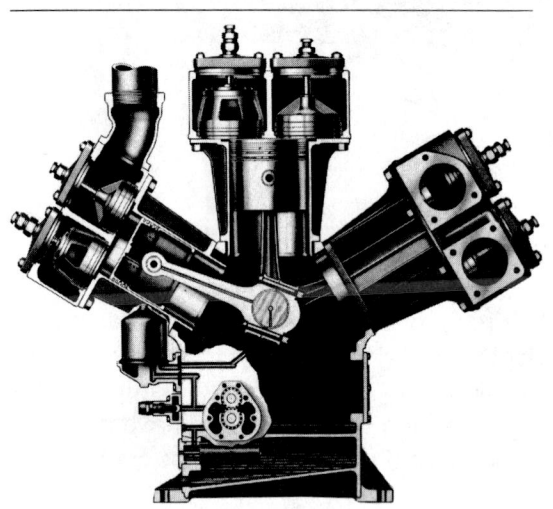

Figure 4.34 Two-stage, single-acting compressor. (Courtesy Ingersoll-Rand Company)

The vertical V-piston compressor has one low-pressure cylinder (right) and one high-pressure cylinder (left), which are arranged in a V-shaped configuration (Figure 4.34). Air is compressed in the right cylinder to an intermediate pressure. This air is then cooled in an intercooler and compressed in the left piston to a higher pressure.

As the air flows into the air-storage tank it cools further, causing some more moisture to precipitate. Over a period of time this moisture will accumulate in the storage tank and, unless it is drained periodically, will cause unnecessary deterioration of fluid power components. Aftercoolers, air dryers, and water separators are used to keep compressed air dry for optimum performance in pneumatic circuits.

It is always desirable to place compressors in areas where clean free air is available and heat can be dissipated most readily. It may be helpful to know that a compressor will compress 7.8 ft³ of free air to a volume of 1 ft³ at 100 psig.

Efficiency of energy input devices

The energy output of a machine is always less than the energy input. In converting one form of energy into another such as mechanical energy to fluid power, there are energy losses. The mechanical efficiency (e) of a machine is defined as

$$e = \frac{\text{work delivered by machine}}{\text{work put into the machine}}$$

or, in symbols,

$$e = \frac{G_o}{G_i} \times 100\%$$

If 100 units of energy were applied to a pump and 62 units of energy were available from the machine, its efficiency would be

$$e = \frac{62}{100} \times 100\%$$

$$= 62\%$$

Because power is a function of work, it is possible to state in symbols

$$e = \frac{P_o}{P_i} \times 100\%$$

where P_o = power output and P_i = power input

There are three categories of pump efficiency parameters: volumetric efficiency, mechanical efficiency, and overall efficiency.

Volumetric

Volumetric efficiency (e_v) is the ratio between the actual output at a given pressure and the theoretical output determined by the geometric displacement (Figure 4.35). It is expressed in percentage units.

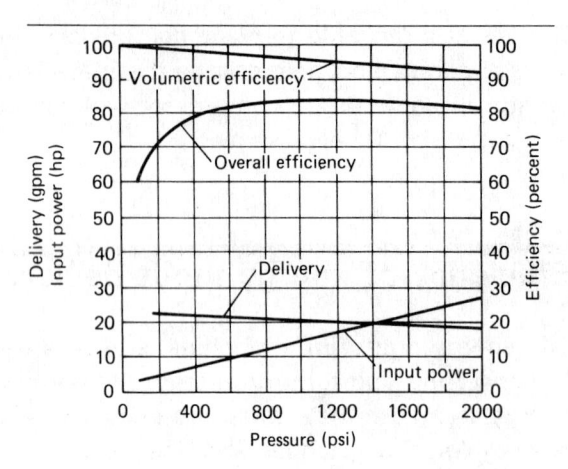

Figure 4.35 Typical pump-performance curves.

$$e_v\% = \frac{\text{actual volume output (gpm)}}{\text{theoretical volume output (gpm)}} \times 100$$

or, in symbols,

$$e_v\% = \frac{V_a}{V_t} \times 100$$

where V_a = actual volume (gpm) and V_t = theoretical volume (gpm)

For example, a pump, designed to deliver 10 gpm at a given pressure and speed, that under actual operating conditions delivered 8 gpm would have a volumetric efficiency

$$e_v\% = \frac{8}{10} \times 100$$

$$= 80\%$$

Volumetric efficiency will always be less than 100% because of leakage. Depending on the performance quality and design of individual pumps, gear pumps have a volumetric efficiency within the range of 80 to 90%; for vane pumps, the range is 80 to 92%; and for piston pumps, which have the highest volumetric efficiency, it is 90 to 98%.

Mechanical

Mechanical efficiency is the ratio between the theoretical input horsepower required to pump the geometric displacement and the actual input horsepower. It is also expressed in percentage units ($e_m\%$).

$$e_m\% = \frac{\text{theoretical input horsepower}}{\text{actual input horsepower}} \times 100$$

When the horsepower inputs are not given, mechanical efficiency can be computed in terms of torque inputs.

$$e_m\% = \frac{\text{theoretical torque required to drive pump (lb-in.)}}{\text{actual torque supplied to pump (lb-in.)}} \times 100$$

Theoretical input torque is

$$\text{Torque (lb-in.)} = \frac{\text{displacement (in}^3/\text{rev.)} \times \text{psi}}{2\pi}$$

Actual input torque is determined by laboratory tests.

$$\text{Actual torque (lb-in.)} = \frac{\text{brake hp} \times 63{,}025}{\text{rpm}}$$

$$\text{Brake hp (bhp)} = \frac{\text{hydraulic output (hp)} \times 100}{\text{pump volumetric efficiency}} \times 100$$

▶ **Example** What is the mechanical efficiency of a hydraulic pump with a displacement of 0.4 in³/rev., operating at 2000 psi, if the actual speed of the pump shaft is 1200 rpm and the brake horsepower is 2.7?

$$e_m\% = \frac{T_t}{T_a} \times 100$$

T_t = theoretical torque (lb-in.)

T_a = actual torque (lb-in.)

$$T_t = \frac{\text{displacement (in}^3/\text{rev.)} \times \text{psi}}{2\pi}$$

$$= \frac{0.4 \times 2000}{6.28}$$

$$= 127 \text{ lb-in.}$$

$$T_a = \frac{\text{bhp} \times 63{,}025}{\text{rpm}}$$

$$= \frac{2.7 \times 63{,}025}{1200}$$

$$= 142 \text{ lb-in.}$$

$$e_m\% = \frac{127 \text{ lb-in.}}{142 \text{ lb-in.}} \times 100$$

$$= 89\%$$

Overall

The overall efficiency is the product of volumetric and mechanical efficiency.

Overall efficiency (e_o%) = volumetric efficiency
$\times$ mechanical efficiency $\times$ 100

▶ **Example** Determine the overall efficiency of a pump that has a volumetric efficiency of 80% and a mechanical efficiency of 90%.

$$e_o = e_v \times e_m \times 100$$

$$e_o = \frac{80}{100} \times \frac{90}{100} \times 100$$

$$= 72\%$$

Compressor performance

An understanding of how to rate the performance of compressors is critical in the design or selection of an appropriate unit. Purchasers of large air compressors need to know what factors govern their performance. They should know the actual capacity of a compressor and the energy or power it requires per unit of air actually delivered. Before the performance of compressors is discussed, it may be helpful to define several terms used to identify specific measurements of efficiency.

Piston displacement refers to the net volume actually displaced by the compressor piston at the rated machine speed, generally given in cubic feet per minute (cfm).

Free air refers to air at normal atmospheric conditions. Because the altitude, barometric reading, and temperature vary at different localities and change with time; it does not mean air under identical conditions.

Actual capacity refers to the quantity of air being compressed and delivered to the discharge port at the rated speed of the machine and under rated pressure conditions. Actual capacity is expressed in cubic feet per minute at the first-stage intake pressure and temperature conditions.

Volumetric efficiency refers to the ratio of actual capacity to piston displacement and is expressed as a percentage.

Compression efficiency refers to the ratio of the theoretical horsepower required to compress a definite amount of gas. The theoretical power may be calculated according to the isothermal base or the adiabatic base and is correspondingly expressed as a percentage.

Mechanical efficiency refers to the ratio of the indicated horsepower in the compressing cylinder to the brake horsepower delivered to the shaft. It is also expressed as a percentage.

Overall efficiency is the product of the compression efficiency and the mechanical efficiency and is expressed as a percentage.

Clearance refers to the maximum cylinder volume on a working side of a piston minus the piston displacement volume per stroke. Clearance is expressed as a percentage of the displaced volume.

Comparing efficiency

In theory, air may be compressed *adiabatically* (without the addition or removal of heat) or *isothermally* (at constant temperature). Because air is normally cooled to room temperature in the receiver tank, heat from the compressed air is lost to the atmosphere in the adiabatic process. The isothermal process requires less work and is therefore preferred; it is achieved by cooling the air as it is compressed. Figure 4.36 shows the relationship of pressure and volume between two

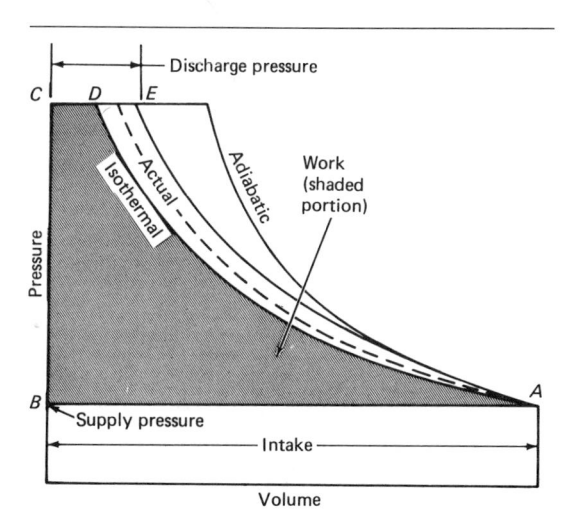

Figure 4.36 Adiabatic and isothermal compression diagram.

theoretical standards and the approximate actual compression. At 100 psig, 35% more power is required for the theoretical adiabatic process than for the theoretical isothermal process.

Work used by the air compressor may be expressed by the equation:

G = area under (pV)

where

G = work (ft-lb)
p = pressure (lb/ft^2)
V = volume (ft^3)

The total work is represented by the area to the left of the compression curves AE (adiabatic) or AD (isothermal). It is apparent that less work is involved in the isothermal compression process. The work area pV, under either the intake line BA or the discharge line EC or DC, may be regarded as flow work. The pV quantity is regarded as the flow work of a given mass of gas entering and leaving the compressor. The area to the left of the compression curves, expressed as G = area under (pV), represents the work of the double-acting compressor.

Determining capacity

Compressor capacity is the amount of air or gas measured at intake conditions that is compressed and actually delivered through the discharge valves in 1 min. (Recall that piston displacement is the volume exhausted by the piston in 1 min.)

▶ **Example** Find the displacement of an air compressor, given the following information: piston diameter, 12 in.; piston stroke, 11 in.; speed, 300 rpm; diameter of piston rod, 2 in.; the net area of the piston is equal to the piston's cross-sectional area minus half the cross-sectional area of the piston rod. (Remember that two strokes constitute one revolution in a reciprocating piston.)

Area of 12-in. piston = 113.1 in^2

Area of 2-in. piston rod = 3.14 in^2

133 Compressor performance

$$\text{Net area of piston} = 113.1 - 1.57 \text{ in}^2$$

$$= \frac{111.53 \text{ in}^2}{144 \text{ in}^2/\text{ft}^2}$$

$$= 0.7745 \text{ ft}^2$$

$$\text{Piston displacement} = \text{net piston area (ft}^2) \times \text{stroke (ft)} \times 2 \times \text{rpm}$$

$$= 0.7745 \times \frac{11}{12} \times 2 \times 300$$

$$= 426 \text{ ft}^3/\text{min}$$

The volumetric efficiency (e_v) of a compressor is the ratio of the volume of gas or air actually compressed to the theoretical compressor capacity. To determine the volumetric efficiency it is necessary to understand the terms *nominal displacement* and *theoretical capacity*. The nominal displacement of a compressor is a function of the bore of the cylinder, the stroke of the piston, and the speed of the machine in revolutions per minute. It is the space (volume) displaced by the piston, making no allowance for re-expansion or clearance.

▶ **Example** A double-acting piston compressor, with a 12-in. stroke and a piston diameter of 9.6 in., will have a nominal displacement of approximately 1 ft³/min. The nominal displacement of a compressor is computed by the following equation:

$$V_D \text{ (nominal)} = 0.000909 \, D^2LN$$
$$= 0.000909 \, (9.6 \text{ in.})^2 \times 12 \text{ in.} \times 1 \text{ rpm}$$
$$= 1.00528 \text{ ft}^3/\text{min}$$

or, for all practical purposes, 1 ft³/min.

In determining the nominal displacement of a compressor, we are *not* concerned with temperatures and pressures but with volumes only.

The volumetric displacement is obtained by multiplying the nominal displacement by the volumetric efficiency. If the compressor in the preceding example operates at 81% volumetric efficiency, the volumetric displacement will be

1.0 ft³ × 0.81 = 0.81 ft³/min.

The volumetric displacement of any double-acting compressor may be expressed as

$$V_D = \frac{2\pi}{4 \times 1728} D^2 L N e_v$$

$$= 0.000909\, D^2 L N e_v$$

where

V_D = volumetric displacement of compressor (ft³/min)

D = diameter of low-pressure cylinder (in.)

L = stroke of low-pressure cylinder (in.)

N = speed of compressor (rpm)

e_v = volumetric efficiency of low-pressure cylinder (as decimal fraction)

The capacity of a compressor refers to the actual amount of air compressed by the machine under standard conditions. For example, if the intake pressure is 14.7 psi (absolute) and the intake temperature is 60 °F, the capacity and the volumetric displacement will be practically identical. Such a condition would be likely to exist only if the air compressor were operating near sea level at a regulated temperature of 60 °F.

Since the intakes of many compressors do not operate at one atmosphere of pressure (14.7 psia), it is apparent that the term "capacity" refers to the volume of air compressed and is expressed in standard cubic feet per minute (SCFM). If the intake on the 9.6-in. diameter cylinder compressor in the example earlier was operating at 60 °F and 59 psia, the capacity would be

$$\frac{59}{14.7} \times 0.80 = 3.20 \text{ SCFM} \quad \text{(for 80\% volumetric efficiency)}$$

If a compressor running at a given speed has a nominal displacement of 100 ft³/min, with an intake at 20 psig, the theoretical capacity would be 2.35 (derived from a pressure multiplier table[4]) × 100 = 235 SCFM. If the machine actually compressed only 188 SCFM, the volumetric efficiency of the machine would be

[4] Available in the *National Compressed Air Institute Handbook*.

135 Compressor performance

$$\text{Volumetric efficiency } (e_v) = \frac{\text{actual intake volume of air compressed}}{\text{compressor theoretical displacement}}$$

$$= \frac{188}{235} \times 100\%$$

$$= 80\%$$

Clearance Volumetric efficiency is also a function of the clearance of the low-pressure cylinder and the ratio of compression. *Clearance* is defined as the volume remaining in the cylinder at the extreme position of the piston (dead center), divided by the displacement of the cylinder. Since not all the compressed gas is expelled beyond the discharge valve, some of this compressed air in the clearance volume expands again as the piston is retracted, delaying the opening of the intake valves, and consequently reduces the amount of air that is taken into the cylinder on the following stroke. The intake valve will open after the pressure in the cylinder has been reduced slightly below the intake pressure. The effects of clearance tend to reduce the capacity of a compressor.

Indicator cards Figure 4.37 shows an *indicator card*, which is a pressure volume record, for an air compressor taking in air along the base line *DA*. Compression progresses from *A* to *B*. The discharge

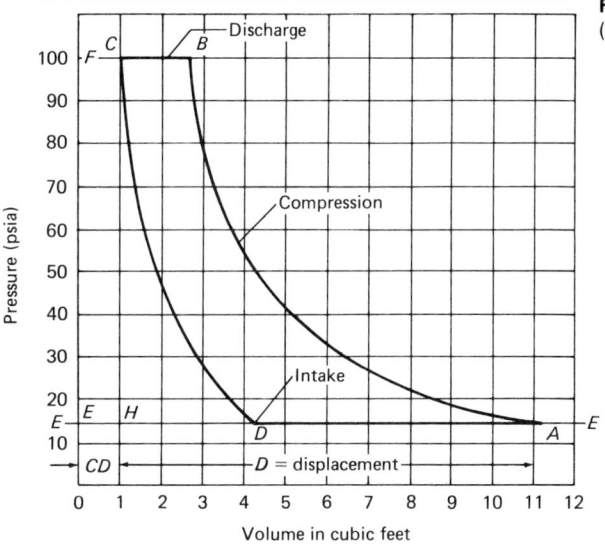

Figure 4.37 Indicator card (pressure-volume record).

valve opens at *B* and compressed air is discharged from *B* to *C*. At the end of the stroke, at Point *C*, a clearance volume of 1 ft³ of air at 100 psia still remains in the cylinder. As the piston retracts, this remaining air expands along line *CD* until it reaches the intake pressure line *DA* at Point *D*. At this point the intake valve opens and the air intake begins. Air is pushed into the intake along the line *DA*. Graphically, the volumetric efficiency is the ratio of area *DABC* to area *EABF*.

Compression efficiency Compression efficiency is the ratio of the theoretical power required to compress the amount of air actually delivered to the actual power developed in the compressing cylinder, as shown by indicator diagrams. If the theoretical power is based on isothermal compression, the efficiency is called *isothermal compression efficiency*.

▶ **Example** Suppose it is found by actual measurement that a compressor is delivering 1000 ft³ of free air (at a given intake temperature and pressure), compressed to 100 psig, in 1 min and is requiring 172 indicated horsepower (ihp) in the air cylinders, while the barometer remains at 30 in. What is the isothermal compression efficiency?

The fundamental relationship of the indicator card may be expressed as

$$\text{ihp} = \frac{pLAN}{33{,}000}$$

where

ihp = indicated horsepower actually developed in the compressor cylinder

p = mean effective pressure (MEP) throughout the stroke (lb/in²)

L = stroke (ft)

A = cross-sectional area of cylinder (in²)   (If the compressor is double-acting, multiply the area of the bore by 2 and subtract the area of the piston rod.)

N = speed of machine (rpm)

The indicated horsepower (ihp) per 100 ft³/min is

$$\frac{172}{10} = 17.2 \text{ ihp}/(100 \text{ ft}^3/\text{min})$$

Mechanical efficiency is the ratio of the air indicated horsepower (ihp) to the brake horsepower. The mechanical efficiency of well-designed air compressors ranges between 90 and 95%.

▶ **Example** A power-driven air compressor is selected, of the following size and displacement: diameter of low-pressure cylinder, 18 in.; diameter of high-pressure cylinder, 11 in.; stroke, 14 in.; speed, 257 rpm; discharge pressure, 100 psig; piston displacement, 1052 in^3. In an actual test, the brake horsepower (bhp) was measured as 167. The indicator diagram recorded an indicated horsepower (ihp) of 157. The mechanical efficiency would be

$$e_m = \frac{ihp}{bhp} \times 100$$

$$e_m = \frac{157}{167} \times 100 = 94\%$$

The overall efficiency of an air compressor is the product of the compression efficiency and the mechanical efficiency. Overall efficiency expresses a ratio between the power theoretically required to compress the air actually delivered and the real power input.

▶ **Example** Given a compression efficiency of 72% and a mechanical efficiency of 94%, determine the overall efficiency.

$$e_o = e_c \times e_m \times 100$$

$$= \frac{72}{100} \times \frac{94}{100} \times 100$$

$$= 67.7\%$$

Maintenance

While much of the theoretical knowledge provided in this chapter is helpful in the design and selection of industrial compressors, it is equally important to see that such units are properly installed and maintained. Compressed air systems are found in numerous factories, material-processing and packaging plants, large garages, assembly plants, laboratories, and other similar facilities. Air pressure is usually 80 to 125 psig for power requirements and is reduced to 15 to 35 psig for pneumatic instrumentation controls. Air consumption varies over a very wide range from less than 100 cfm to more than 20,000 cfm; however, the vast majority of plants consume 1000 to 6000 cfm. Frequently, compressors providing 1000 cfm or less are considered small-capacity units, while units above this range are considered large-capacity machines. Compressors should be located in clean accessible areas for inspection and maintenance and should be able to draw clean air from the outside of the compressor intake. It should also be possible to dissipate heat from the compressor area.

Summary

Energy input devices—pumps and compressors—require a prime mover such as an electric motor or an engine to drive them. Pumps and compressors convert mechanical force and motion into fluid energy. Hydraulic pumps create fluid flow, they do not pump pressure. Pressure is the result of resistance to flow encountered by a confined fluid.

Pumps are classified as either positive- or nonpositive-displacement units, while compressors are classified as positive- or dynamic-displacement units. These input devices can also be classified by reference to the construction design of the pumping or compressing element—centrifugal, reciprocating, and rotary. The most common rotary elements are the vane, the gear, and the piston.

Both pumps and compressors can be multistage to create higher pressure or they can be placed in parallel to combine flow and thereby increase volumetric output. Pumps and compressors, particularly piston-type units, have a high overall efficiency. A clean environment with proper conditioning of fluids will prolong their life. Contaminants, on the other hand, can create considerable maintenance costs with these precision-machined input devices. Since pumps and compressors constitute the greatest investment in any fluid power system and their efficient functioning is critical to the rest of such a system, they should be operated within the manufacturer's specifications regarding pressure, revolutions per minute, volumetric capacity, and any other imposed and limiting conditions.

Review questions and problems

1 What function does a hydraulic pump serve in a fluid power system?
2 What function does an air compressor serve in a fluid power system?

3 What is the principle of operation of (a) a pump? and (b) a compressor?
4 What causes fluid (either oil or air) to flow into a pump or compressor?
5 How is pressure created within a fluid power system?
6 Define the terms "displacement" and "delivery."
7 Why must positive-displacement pumps always be protected by a relief valve?
8 What is meant by a balanced-design hydraulic pump?
9 What common characteristic do a screw-type pump and a compressor share?
10 Why do vane pumps not lose volumetric efficiency with use, as do gear pumps?
11 What is the advantage of two-stage pumps and compressors?
12 Why is cooling required with reciprocating compressors?
13 Why must receiver tanks be drained periodically?
14 Determine the volumetric efficiency of a pump rated at 10 gpm that, when operated at specified rpm and psi, delivered only 8 gpm.
15 Determine the mechanical efficiency of a hydraulic pump with a displacement of 0.4 in^3/rev., operating at 1000 psi and an rpm of 1200, and a brake horsepower of 2.5.

Selected readings

- Atland, George. *Practical Hydraulics.* Vickers Division of Sperry Rand Corporation, Troy, Michigan, 1968.
- Basal, P. R., Jr. (ed.) *Mobile Hydraulics Manual* M-2990-S. Sperry Rand Corporation, Troy, Michigan, 1967.
- Fitch, Ernest C. Jr. *Fluid Power and Control Systems.* McGraw-Hill, New York, 1966.
- *Fluid Power.* Bureau of Naval Personnel Navy Training Course, NAVPERS 16193-A, U.S. Government Printing Office, Washington, D.C., 1966.
- *Fundamentals of Service Hydraulics.* Deere and Company, Moline, Illinois, 1967.
- Glenn, Ronald E., and James E. Blinn. *Mobile Hydraulic Testing.* American Technical Society, Chicago, 1970.
- Henke, Russell W. *Closing the Loop.* Huebner, Cleveland, 1966.
- ———. *Introduction To Fluid Mechanics.* Addison-Wesley, Reading, Massachusetts, 1966.
- *Hydraulic Power Transmission.* Engineering Bulletin no. HP-221S. Standard Oil, American Oil Company, Chicago, 1966.

- McNickle, L. S., Jr. *Simplified Hydraulics.* McGraw-Hill, New York, 1966.
- Oster, Jon. *Basic Applied Fluid Power: Hydraulics.* McGraw-Hill, New York, 1969.
- Pease, Dudley A. *Basic Fluid Power.* Prentice-Hall, Englewood Cliffs, New Jersey, 1967.
- Stewart, Harry L., and John M. Storer. *Fluid Power.* Howard W. Sams, Indianapolis, Indiana, 1968.
- *Compressed Air and Gas Handbook,* 3rd ed. Compressed Air and Gas Institute, New York, 1961.

Chapter 5
Energy modulation devices

This chapter describes the fundamental control functions performed by valves in a fluid power circuit. Essentially, these are pressure control, direction control, and rate of fluid flow control. Control valves are named according to the functions they serve in a circuit. For each of the three types (pressure-control valves, directional-control valves, and flow-control valves), construction, operation, and selection for a particular application are discussed. The most sophisticated and complex valve is the servo control valve, which is used in closed-loop circuits to control an actuator by means of the feedback principle.

Key terms

- **Counterbalance valve** A valve whose primary function is to restrict fluid flow from a primary to a secondary port and to maintain a pressure level sufficient to balance a load held by an actuator.
- **Directional-flow valve** A valve whose primary function is to direct or prevent flow through selected passages.
- **Flow-control valve** A valve whose primary function is to control flow rate.
- **Flow-control, pressure-compensated valve** A flow-control valve that controls the rate of flow independent of system pressure.
- **Flow-control pressure-temperature compensated valve** A pressure-compensated, flow-control valve that controls the rate of flow independent of fluid temperature.
- **Normally closed valve** A valve that blocks flow between ports.
- **Normally open valve** A valve that does not restrict flow.
- **Orifice** A restriction, the length of which is relatively small with respect to its cross-sectional area.
- **Pressure-control valve** A valve whose primary function is to control pressure.
- **Pressure-control relief valve** A valve whose primary function is to provide pressure limitation in a circuit.
- **Pressure-control sequence valve** A valve whose primary function is to direct flow in a predetermined sequence.

- **Pressure-control unloading valve** A pressure valve whose primary function is to permit a pump or compressor to operate at a minimum load.
- **Pressure-reducing valve** A valve whose primary function is to meter the flow into a branch outlet to maintain pressure at a preset value.
- **Seating action** A valve design in which flow is stopped by a seated obstruction in the flow path. The element may be a ball, a diaphragm, a disk, or a spool.
- **Valve** A device that controls fluid flow, direction, pressure, or flow rate.
- **Valve actuator** The valve part through which force is applied to move or position a flow-directing element.
- **Valve position** The point at which flow-directing elements provide a specific flow condition in a valve.

Energy modulation devices (valves) are integral components of a fluid power system; they control pressure, direction, and rate of fluid flow. Their fundamental control functions in a hydraulic circuit are: (1) restricting or directing the flow of fluid within the circuit; and (2) modifying the energy or pressure level of the fluid stream by regulating either flow or pressure (Figure 5.1). Valves in a circuit

Figure 5.1 Energy modulation devices in a hydraulic system.

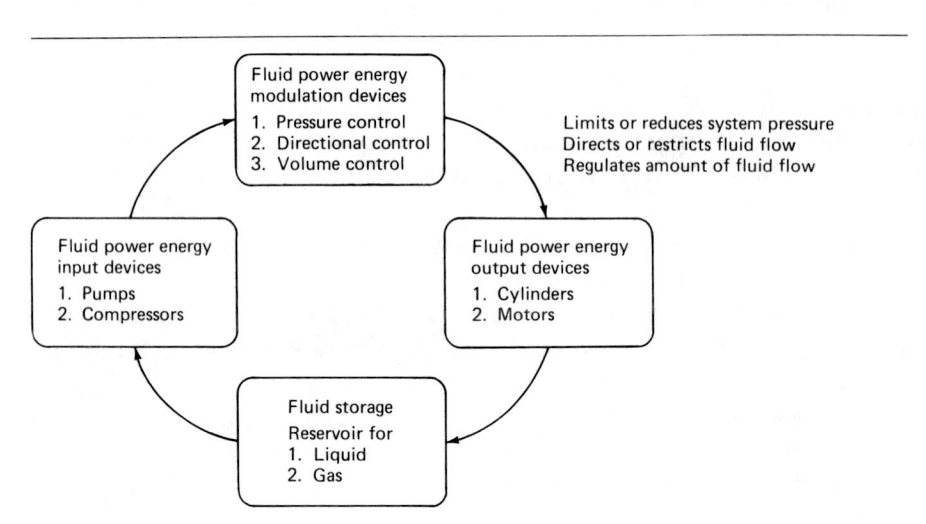

that regulate pressure or create required pressure conditions are referred to as *pressure-control valves*. Valves that direct, divert, combine, or deactivate flows in a circuit are called *directional-control valves*. And valves that regulate the amount of fluid flow within a circuit or a branch of a circuit are referred to as *volume-control valves*. Ideally, valves should not affect the total energy transfer in a circuit. In practice, however, since they are mechanical devices and operate in a dynamic condition, heat energy losses are encountered in performing the three major modulation functions. Valves are usually named according to the functions they serve within a circuit. They may also be named according to their construction, which can vary from a simple ball and seat to a multielement, spool-type valve coupled with electric controls.

Pressure-control valves

In a fluid power circuit, the power required is a function of both pressure and flow. In designing a circuit to perform a required job within a given time, the logical order would be to determine (1) the system pressure, (2) the actuator size, and (3) the flow. The upper limit of the force exerted by a linear fluid power cylinder (or the torque produced by a rotary fluid power motor) is a function of the maximum system pressure and area, or displacement. Precise control of system pressure enables actuators to apply precisely controlled forces and to provide protection against undue stresses to the machinery.

The selection of pressure-control valves is determined by two important conditions—operational circuit requirements and environmental requirements. Operational circuit requirements include the size of pressure valves as determined by the requirements of pressure level, flow capacity, and the sequence of operations. The pressure-control valve must be able to operate within the range of minimum and maximum circuit pressure requirements. Environmental conditions such as heat, vibration, shock, and contamination will affect the performance of a pressure-control valve.

Pressure-control valves are either normally closed or normally open. The normally closed type (Figure 5.2) blocks flow between two ports until a predetermined pressure is reached, which unseats the ball or spool and permits fluid flow. All pressure valves, except pressure-reducing valves, are normally closed.

Simple-relief

The function of a relief valve is to make certain that the system pressure does not rise above a preset value. Relief valves can be con-

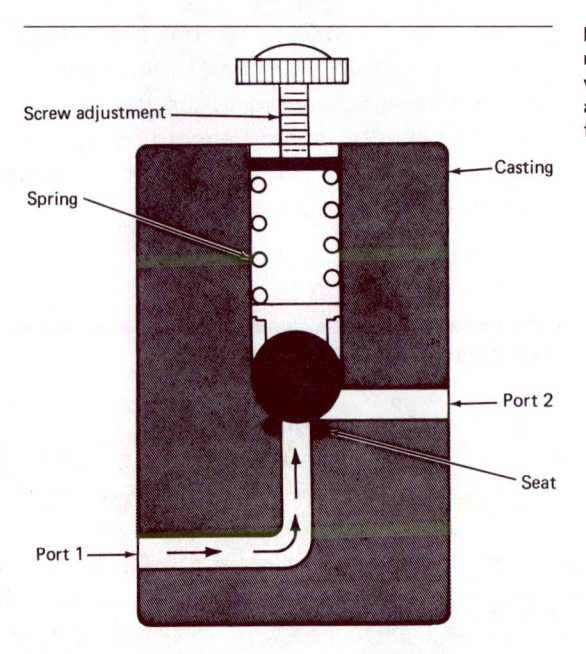

Figure 5.2 Basic operation of a normally closed pressure-control valve. A spring holds the ball against a seat, allowing no flow from Port 1 to Port 2.

sidered safety valves and must be large enough to handle the entire pump output volume flow. Normally, the relief valve is closed [Figure 5.3(a)] until the pressure level exceeds the preset value. When it reaches that value (the *cracking* pressure), it unseats the ball or poppet, allowing some fluid to flow [Figure 5.3(b)].

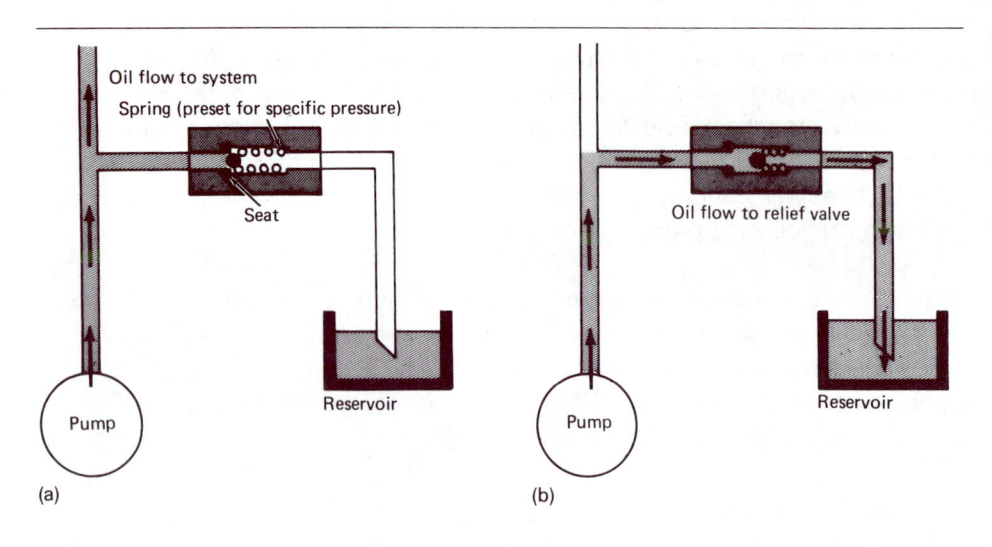

Figure 5.3 A relief valve. Normally, the relief valve is closed (a) until the pressure level exceeds the preset value (b), at which point it unseats the ball or poppet and allows some fluid to flow.

In order for the full pump volume to flow through the relief valve, the fluid pressure must exceed the mechanical pressure sufficiently to open the valve for full delivery back to the reservoir. This is known as the *full-flow pressure* of the relief valve. When the line pressure drops, the valve closes quietly and smoothly. The simple-relief valve has a direct spring-loaded ball, poppet, or spool to restrict fluid flow.

Simple-relief valves respond directly to the higher resistance of the mechanical spring setting and produce a chattering effect during the intervals of beginning and terminating discharge to the reservoir. These disadvantages are overcome by the design and operating characteristics of a compound-relief valve.

Compound-relief

This is a two-stage pressure-control valve. In the closed position [Figure 5.4(a)], oil at the system pressure flows through the primary inlet port, around the main spool, and out the primary outlet port. Primary ports may be used interchangeably, but the secondary port must be connected to return flow to the reservoir.

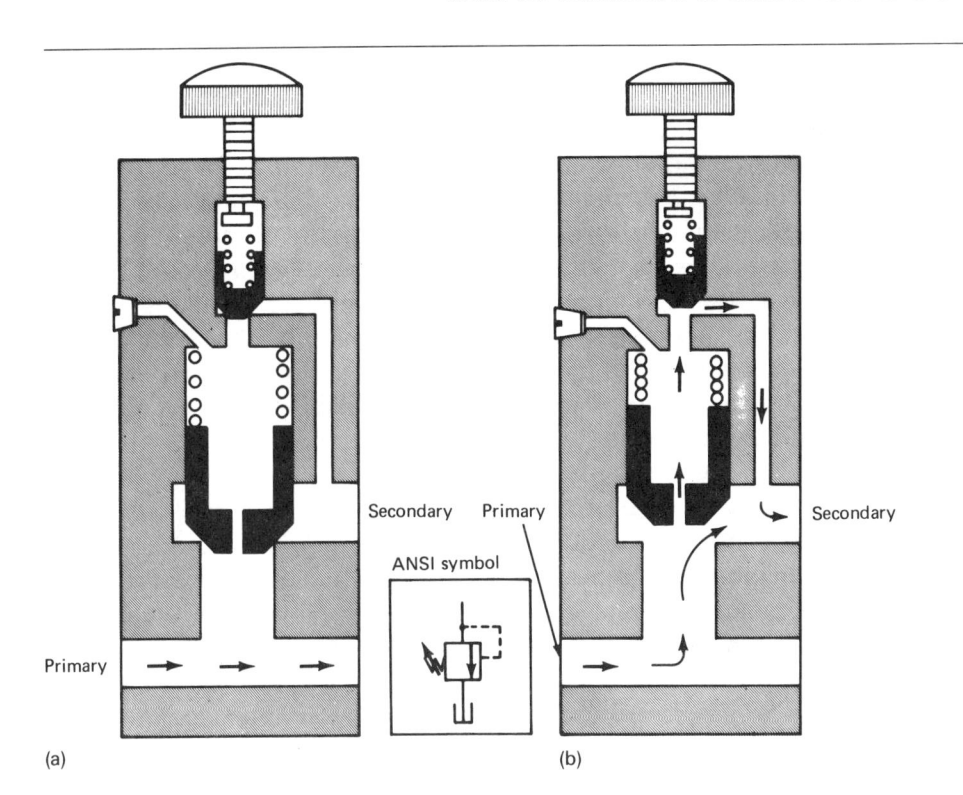

Figure 5.4 Compound-relief valve (pilot operated). (a) Normal position. (b) Activated condition.

The operation of a pilot-relief valve is a two-stage process. By means of a restricted passage, oil flows at a controlled rate from the inlet to the control chamber. This passage may be in the piston or the housing. System pressure operates on both sides of the piston, causing it to be pressure balanced. Because the main spool is pressure balanced, any increase in system pressure operates on both sides of the piston and the spool remains seated with the assistance of the mechanical spring bias. When the system pressure exceeds the pilot-relief valve setting, the mechanical spring is compressed, unseating the pilot valve and permitting pressurized fluid to return to the reservoir [Figure 5.4(b)]. As soon as this happens, pressure in the control chamber is limited by the escape of fluid past the pilot piston and through the cored (restricted) passage in the housing back to the reservoir. Since the orifice is not large enough to take the full flow of the pump at that pressure, the pressure in the primary chamber tends to increase. When the system pressure increases to the level at which it is sufficient to overcome the pressure in the control chamber and the spring bias of the main spool, the main spool is forced off its seat. This allows fluid from the pump to return through the secondary port back to the reservoir.

When the system pressure drops, the main spool again becomes pressure balanced and assumes its normally closed position. In the balanced-piston relief valve, the pressure in the control chamber above the piston is limited by the setting of the pilot-relief spring. This can be used to great advantage when it is desirable to have the pump divert its flow back to the reservoir at near-zero pressure. By simply connecting the control chamber to one side of a shutoff valve and connecting the valve to the reservoir, it is possible to have the control chamber at atmospheric pressure when the shutoff valve is open, thus "venting" the relief valve. In this situation the maximum pressure to which the pump is exposed is the value of the spring bias of the balanced piston. When the shutoff valve is closed, the control chamber is again limited by the pilot-relief valve setting.

If, instead of a shutoff valve, another simple-relief valve is used, it is possible to have the simple-relief valve control the limit of pressure in the control chamber, up to the value of the pilot-relief valve setting of the balanced piston-relief valve. This technique is known as *remote control of the main relief valve.*

Pilot-operated relief valves are used more frequently in applications requiring more exact pressure levels. These valves can have infinite pressure-level adjustments within a given range. A spring-tensioning device is used to set the valve slightly in excess of the system working pressure. For example, if 1000 psi is needed to perform a given operation, the relief valve is set at some value in excess of 1000 psi. An accepted practice is to set the relief valve 10% higher than system requirements. Excessive pressures impose unnecessary strain on components, provide less safety, and increase cost of operations. If orifices and passages are clogged with contaminants, the valve will be inoperative.

Unloading

An unloading valve is a normally closed valve [see Figure 5.5(a)], whose primary purpose is to permit a pump to operate at minimal load. An external signal is always provided for an unloading valve. This valve works on the principle that pump delivery is diverted through the secondary port back to the reservoir when sufficient pilot pressure [Figure 5.5(b)] is applied to move the spool against the spring force. The spool is kept open by the pilot pressure until the

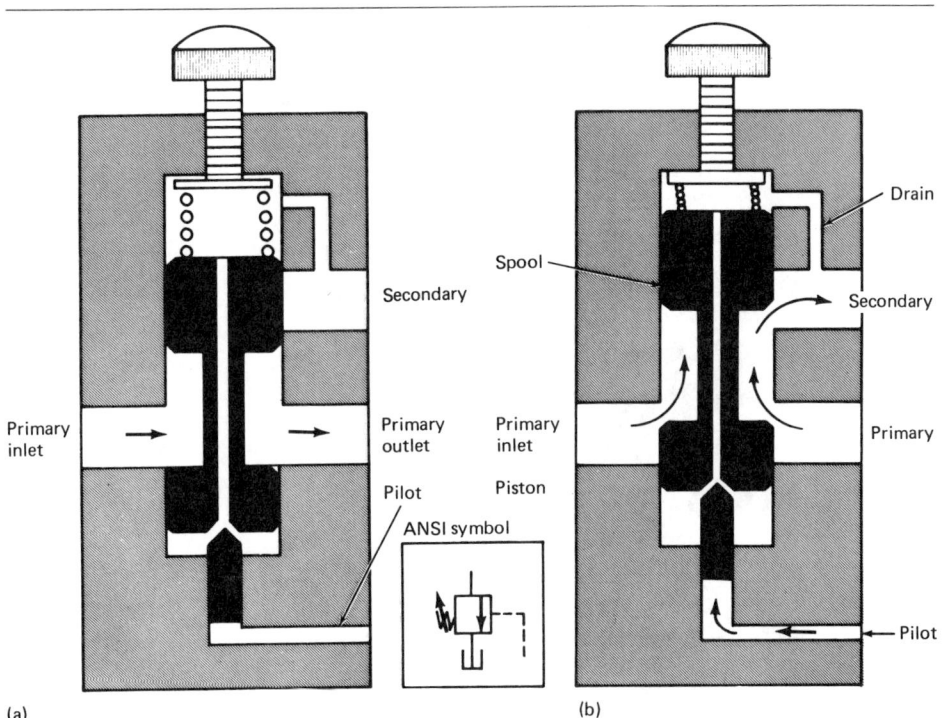

Figure 5.5 Unloading valve (pilot-operated). (a) Normal position. (b) Activated condition.

pilot sensing pressure is less than the spring bias. Spring cavities in an unloading valve usually are internally drained [Figure 5.5(b)]. The unloading valve is useful in fluid power systems that have two or more fixed-delivery pumps to control the amount of flow at any given time. In applications such as rapid approach, slow drill feed, and rapid return, unloading valves are particularly efficient.

Many hydraulic systems are designed to provide high volume at low pressure during the advance phase, then low volume at high pressure during the work phase of the motion cycle. Unloading valves conserve horsepower by unloading the high-volume pump during the work phase.

In air compressors, the primary function of unloading valves is to permit a compressor to start under minimal load. This is accomplished by a normally closed valve that is designed to vent to atmosphere.

Sequence

Sequence valves are also pressure-control devices and are normally closed [Figure 5.6(a)]. A sequence valve's primary function is to direct flow in a predetermined sequence based on a given pressure rise.

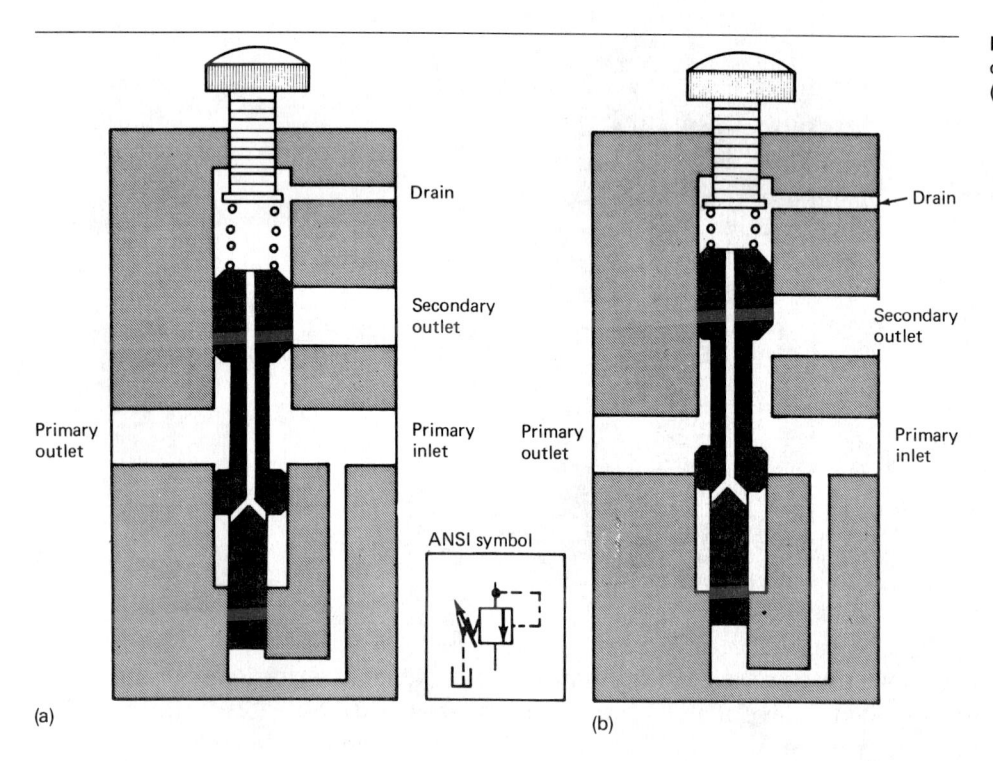

Figure 5.6 Sequence valve (pilot-operated). (a) Normal position. (b) Activated condition.

The piston is normally held in the closed position by an adjustable spring. As the fluid pressure reaches the sequence-valve pressure setting, the signal from a direct or external pilot causes the spool to shift upward and open the secondary port. Fluid is then directed to the next actuator [Figure 5.6(b)]. When the circuit requirements are such that reverse flow through a sequence valve is necessary, a check valve must be used in conjunction with the sequence valve. Some manufacturers include an integral check valve within the sequence valve. It is important to remember that a sequence valve must be externally drained. If the oil is not drained but is trapped in the spring chamber, it will help the mechanical spring to hold the spool closed, ultimately preventing the spool from shifting. If the circuit calls for sequencing by position, as well as by pressure rise, externally piloted sequence valves may be used. The balanced-piston sequence valve, coupled with an external pilot, will shift at about 20 psi differential across the piston.

Counterbalance

The primary function of a counterbalance valve is to restrict fluid flow from a primary to a secondary port [Figure 5.7(a)] and to maintain a pressure level sufficient to balance a load being held by a cylinder or motor. Like other pressure-control valves, the counterbalance valve operates on the principle that fluid is trapped under pressure until pilot pressure, either direct or remote, overcomes the spring force setting in the valve and the spool moves up to throttle return flow through the valve to the reservoir [Figure 5.7(b)].

An integral check valve allows free flow around the piston when the directional valve is shifted to raise the load. Counterbalance valves are normally closed and internally drained. These valves are used in such applications as loaders, lift trucks, vertical presses, and similar situations where loads are positioned and held momentarily. Essentially, a counterbalance is used as a braking valve for decelerating large inertial forces.

A remotely controlled counterbalance valve can be used to impose resistance on hydraulically sustained loads. This valve controls the lowering of a piston or rotation of a motor by sensing the pressure from the high-pressure side of a circuit. If the pressure drops suddenly because of a downward movement of a piston, the remote

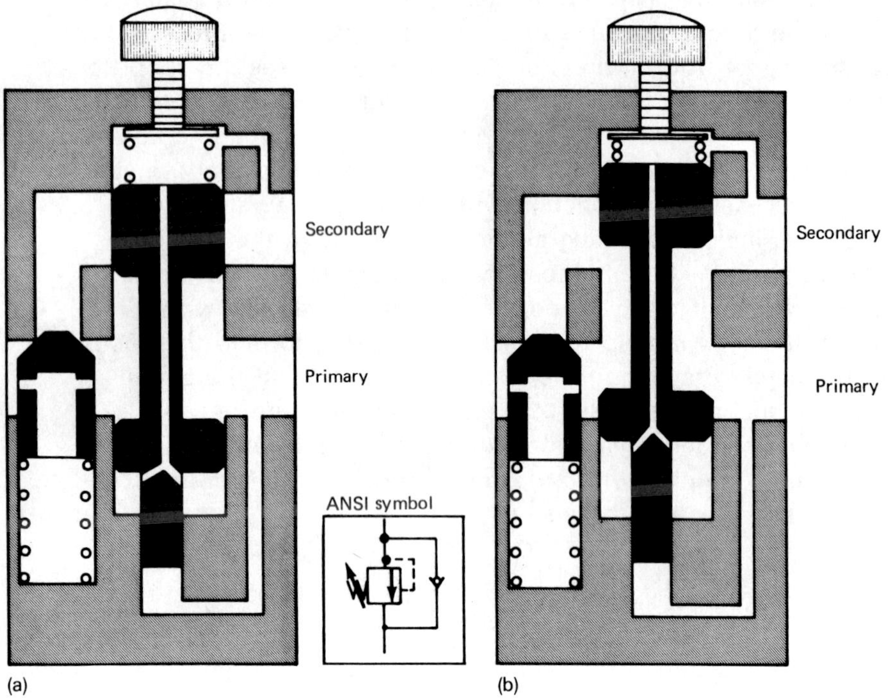

Figure 5.7 Counterbalance valve (pilot-operated). (a) Normal position. (b) Activated condition.

sensing pressure drops, causing the spool to shift downward and restrict flow to the secondary port.

Pressure-reducing

The purpose of a pressure-reducing valve is to reduce and maintain a lower pressure in a branch circuit than is available in the main line. Pressure-reducing valves are used for clamping and other similar applications.

A pressure-reducing valve consists primarily of a pressure-balanced spool valve that meters the flow into a branch outlet to maintain pressure at a preset value. In a two-stage, pressure-reducing valve, the pressure setting is provided by an adjustable, simple-relief valve.

A pressure-reducing valve is a normally open valve [Figure 5.8(a)]. Its operation is based on the principle that pilot pressure from the controlled-pressure side opposes an adjustable bias spring normally holding the valve open. When the two forces are equal, the pressure downstream is controlled at the valve setting. When the branch-circuit

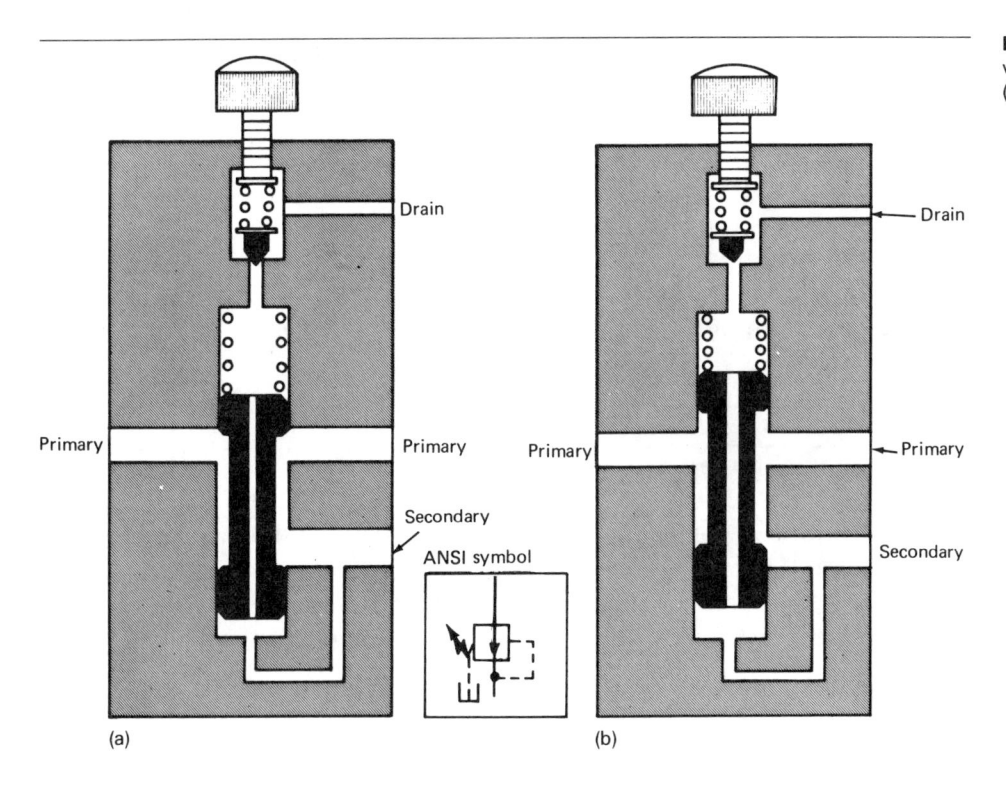

Figure 5.8 Pressure-reducing valve. (a) Normal position. (b) Activated condition.

pressure drops below the relief-valve setting, the main spring forces the spool valve to open, allowing full pressure flow. As the branch-circuit pressure builds up, above the setting of the relief valve, the relief-valve poppet opens, thereby limiting the pressure in the spring chamber above the spool and causing it to close [Figure 5.8(b)]. As the spool moves upward, it restricts the outlet passage and creates a pressure drop between the inlet and outlet ports. Pressure at the outlet is limited to the sum of the equivalent forces of the relief spring and the main spring. In normal operation, the outlet port never closes entirely, for it must allow sufficient flow to meet any work requirements on the low-pressure side of the valve plus enough flow through the spool passage to maintain the pressure drop needed to hold the spool at the control position. The oil flowing through the spool passage must be externally drained to the reservoir.

Directional-control valves

Directional-control valves primarily provide a means of controlling when and where fluid is delivered in the circuit to perform various functions. Directional-control valves start, stop, accelerate, decelerate, and control the direction of motion of actuators. They may be used to

- activate a circuit,
- deactivate or isolate a circuit,
- reverse the direction of flow within a circuit,
- divert return flow from an output device,
- combine flow from two or more branches, or
- separate flow from one branch to two or more branches.

Internal control elements

Direction-control valves can be classified according to the structural internal control element—sliding spool, rotary spool, rotary plate, poppet, and ball. The constructional design of the elements makes certain classes particularly suitable for specific circuit applications and conditions.

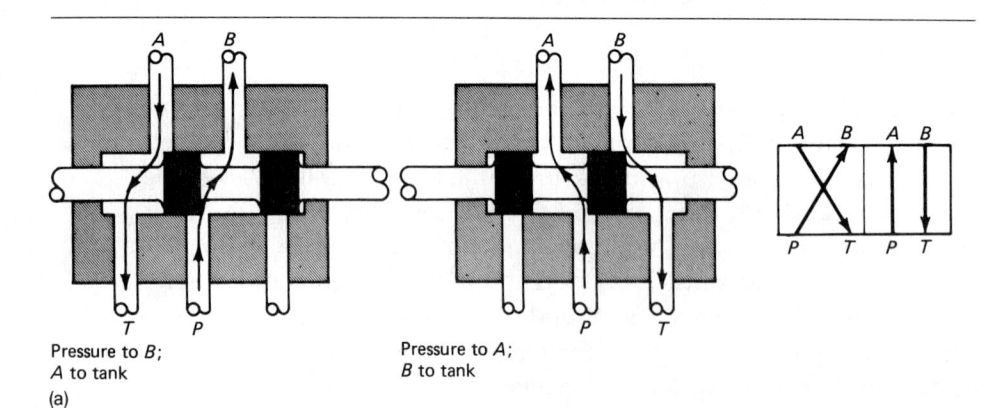

Figure 5.9 (a) Diagram of a spool four-way directional-control valve. (b) Photograph of a sliding-spool directional-control valve. [(b) Courtesy Rivett, Division of Applied Power Industries]

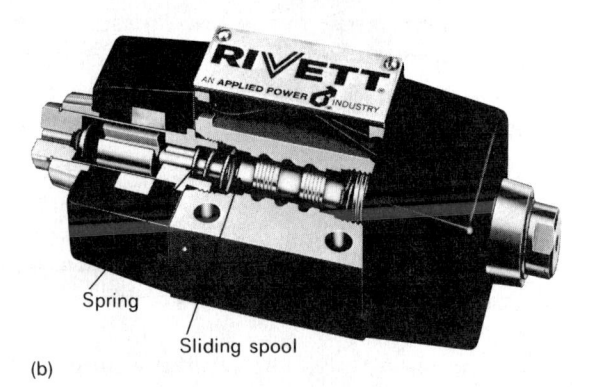

The sliding-spool valve (Figure 5.9) has a spool fitted to a valve body. Moving the spool linearly varies the direction of fluid flow. The rotary-spool valve (Figure 5.10) has a spool fitted to the valve body, which is rotated to change the direction of flow. (In the illustrations, the following abbreviations are used: *A* and *B*, cylinder ports; *P*, pump; and *T*, tank.) The rotary-plate valve has a plate that provides a shearing action as it rotates, opening and closing ports (Figure 5.11).

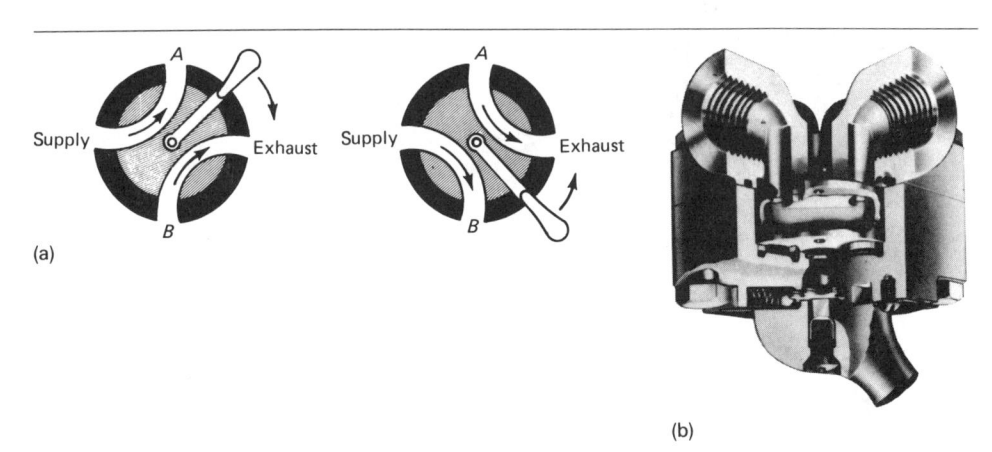

Figure 5.10 Rotary-spool valve. [(b) Courtesy Teledyne Republic Manufacturing]

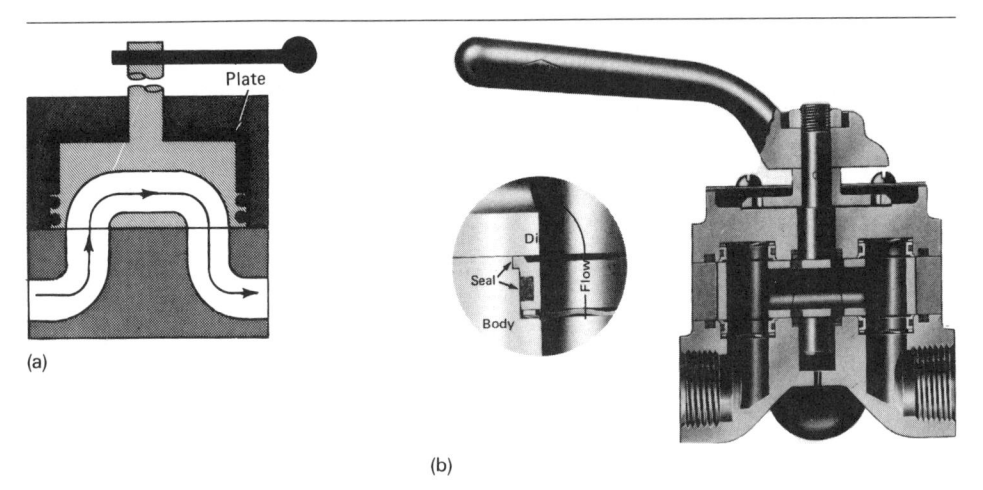

Figure 5.11 Rotary-plate valve. [(b) Courtesy Teledyne Republic Manufacturing]

The poppet valve has a poppet fitted to a seat (Figure 5.12). This type of valve has good sealing characteristics. Any increase in system pressure above the poppet tends to decrease the leakage between the poppet and the seat. It also increases the force holding the poppet against the seat, ensuring a tighter seal. The ball-type directional-control valve is similar to a poppet valve in that it also provides a good seal. These valves are generally used for high-pressure circuits.

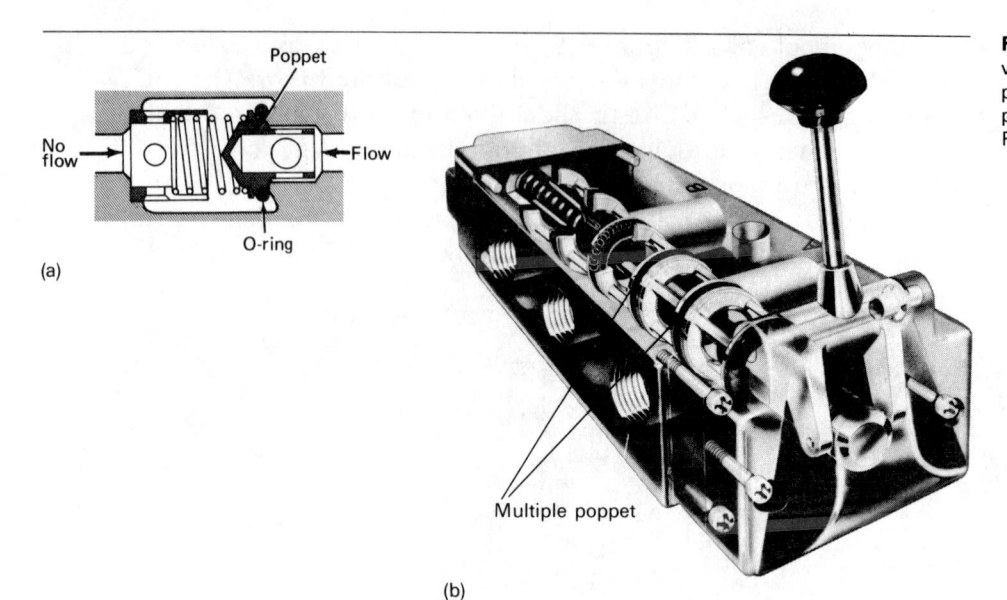

Figure 5.12 (a) Poppet check valve. (b) Manually operated five-port, four-way, directional-control poppet valve. [(b) Courtesy Skinner Precision Industries]

Directional-control valves can be classified by: (1) ways, (2) position, (3) center, (4) actuation, and (5) spring condition (Figure 5.13).

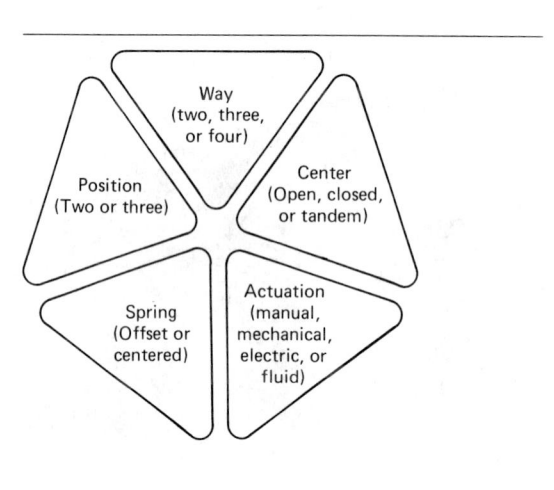

Figure 5.13 Characteristics of directional-control valves.

Ways

Directional-control valves are commonly described by the number of working ports. The *one-way valve,* or check valve, is of poppet or ball construction and allows flow in only one direction. The check valve has the outlet as a working port because its primary function is to restrict flow directed from the outlet to the inlet port. The check valve permits flow in only one direction and prevents flow in the opposite direction (Figure 5.14). For example, a check valve in the

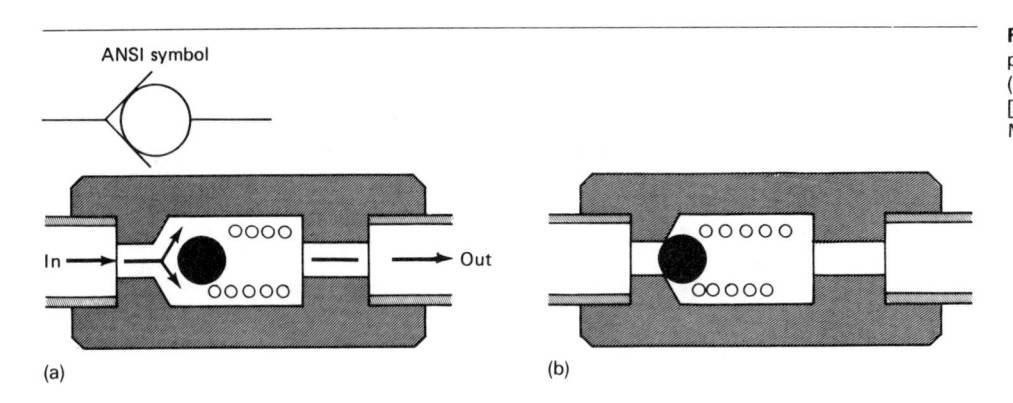

Figure 5.14 Check valve. (a) Valve permits free flow in one direction. (b) No flow through valve. [(c) Courtesy Teledyne Republic Manufacturing]

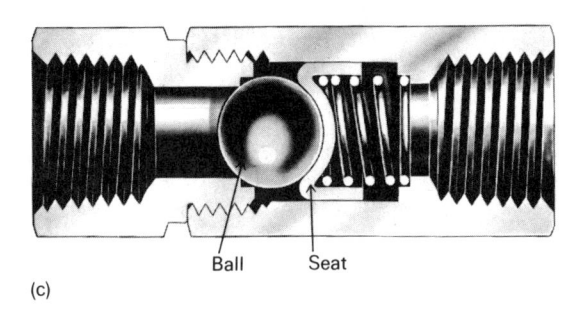

pressure line of the low-pressure, variable-volume pump protects it from any back flow from the high-pressure, fixed-displacement pump. The oil returns over the relief valve if the system exceeds the relief-valve setting.

Two-way valves have two working ports, an "in" and an "out" port [Figure 5.15(a)]. They are used either to open [Figure 5.15(b)] or to close a path for flow in a single line. In essence, a two-way valve is an off-on valve. Two-way valves are used frequently in hydraulic circuit applications and are seldom used in pneumatic circuits.

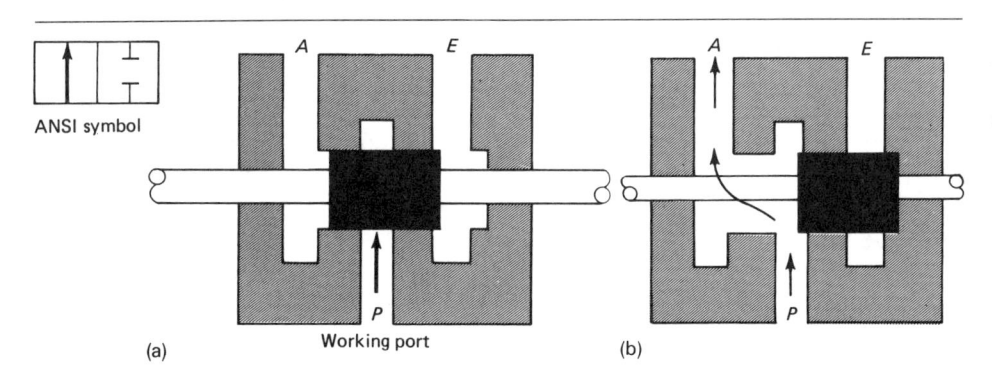

Figure 5.15 Two-position, two-way, sliding-spool valve. (a) Working ports are blocked. (b) In position 2, working ports are open.

A *three-way valve* (Figure 5.16) has three working ports. It can have one inlet and two outlets, or two inlets and one outlet. If it has two inlets and one outlet, it is used as a selector to furnish two pressures to a common outlet. If it has two outlets and one inlet, it can also be used as a selector to determine which outlet will receive the fluid. In

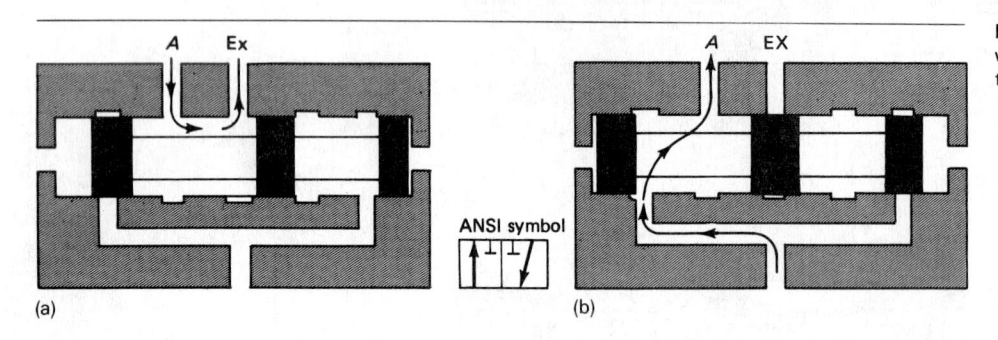

Figure 5.16 Two-position, three-way, sliding-spool valve. (a) Position 1. (b) Position 2.

pneumatic circuits, a three-way valve can be used to alternately pressurize or exhaust air. A three-way valve can be used in pneumatics for special applications. For example, Figure 5.17 illustrates a quick-exhaust, or shuttle, valve, commonly referred to as an *OR* valve. It is less confusing, however, to characterize valves by their functions rather than by their applications.

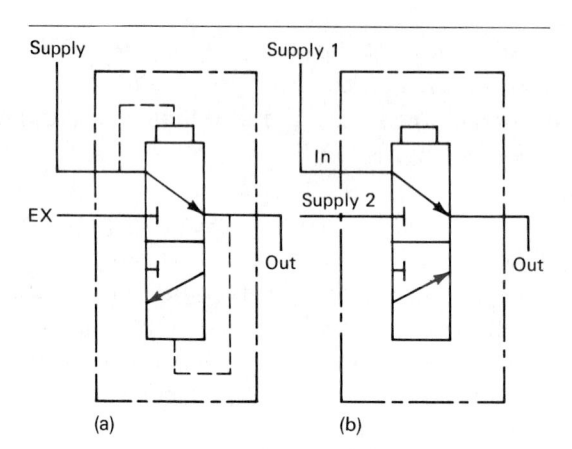

Figure 5.17 Special applications of two-position, three-way, directional-control valve. (a) Quick-exhaust valve. (b) Shuttle valve (often called an *OR* device).

Four-way valves have four connections to the circuit: one line connects the valve to the pump or compressor; another line exhausts the fluid from the valve; the other two lines connect to either end of the actuator. The designations *P* for pressure, *T* for tank, and *A* and *B* for ports on either end of the cylinder are used on four-way valves (Figure 5.18).

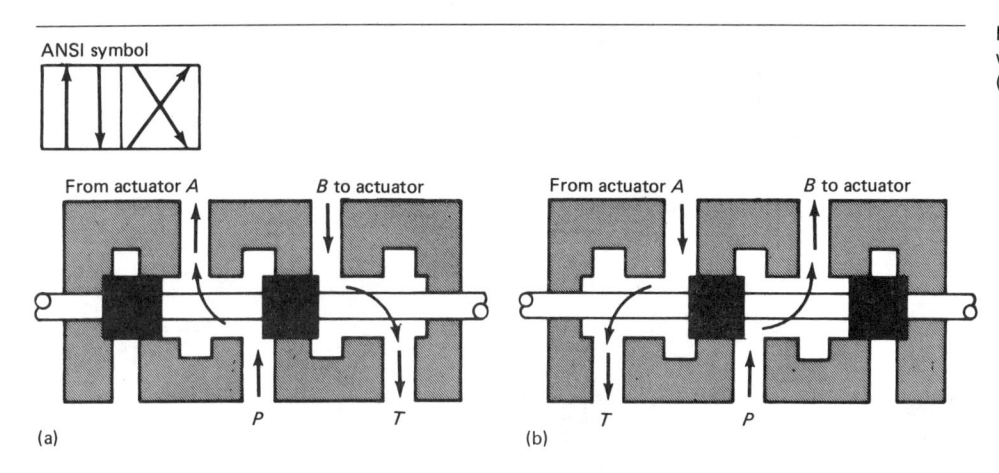

Figure 5.18 Two-position, four-way, directional-control valve. (a) Position 1. (b) Position 2.

Position

Position determines the number of alternative flow conditions the valve can provide. For example, Figure 5.15(a) shows a no-flow condition in that position. This would be a nonfunctional valve if it were limited to this one position. However, with two positions available [Figure 5.15(b)], it is possible to have either a no-flow or a flow condition, whichever control function is required.

Two-position, two-way valves are generally used in circuit applications where shutoff valves are required in a branch or main circuit.

With *two-position, three-way* valves (Figure 5.16), flow into the valve is blocked by the spool in Position 1, while the fluid from the cylinder is being exhausted. In Position 2, flow is allowed through the valve, while the exhaust line is blocked by the spool. Two-position, three-way valves are used extensively where cylinders are actuated by fluid under pressure and returned mechanically. In pneumatic control circuits, two-position, three-way air valves are used as limit valves, push buttons, and control devices.

Two-position, four-way valves (Figure 5.18) allow flow from the energy-input device through the valve to one side of the actuator in Position 1, while fluid is exhausted from the other side of the actuator through the valve. In Position 2, the flow is reversed. Two-position, four-way directional valves are used in applications where actuators are fully extended or retracted under pressure.

In *three position, four-way* directional valves (Figure 5.19), the two extreme positions are identical to those of a two-position, four-way valve. In the center position, however, the flow is blocked in all ports. These valves are used in applications that do not require full extension or retraction under pressure, particularly where circuit applications call for a third option.

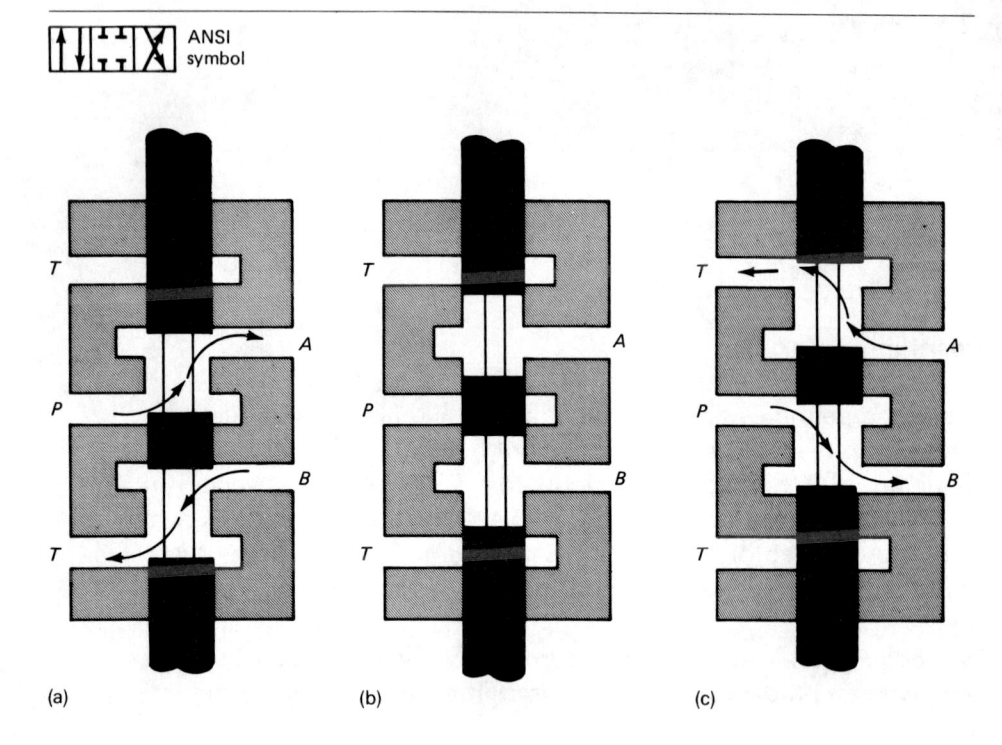

Figure 5.19 Three-position, four-way, directional-control valve. (a) Position 1. (b) Position 2. (c) Position 3.

Center

There are several variations of the flow pattern of the center position. These are made possible by the configuration of the spool or the passages of the valve body. Symbolically, these center configurations are represented as shown in Figure 5.20.

Actuation

Actuation refers to the various methods of moving the valve element from one position to another (Figures 5.21 and 5.22). Valves can be actuated in four basic ways: manually, mechanically, electrically, or

159 Directional-control valves

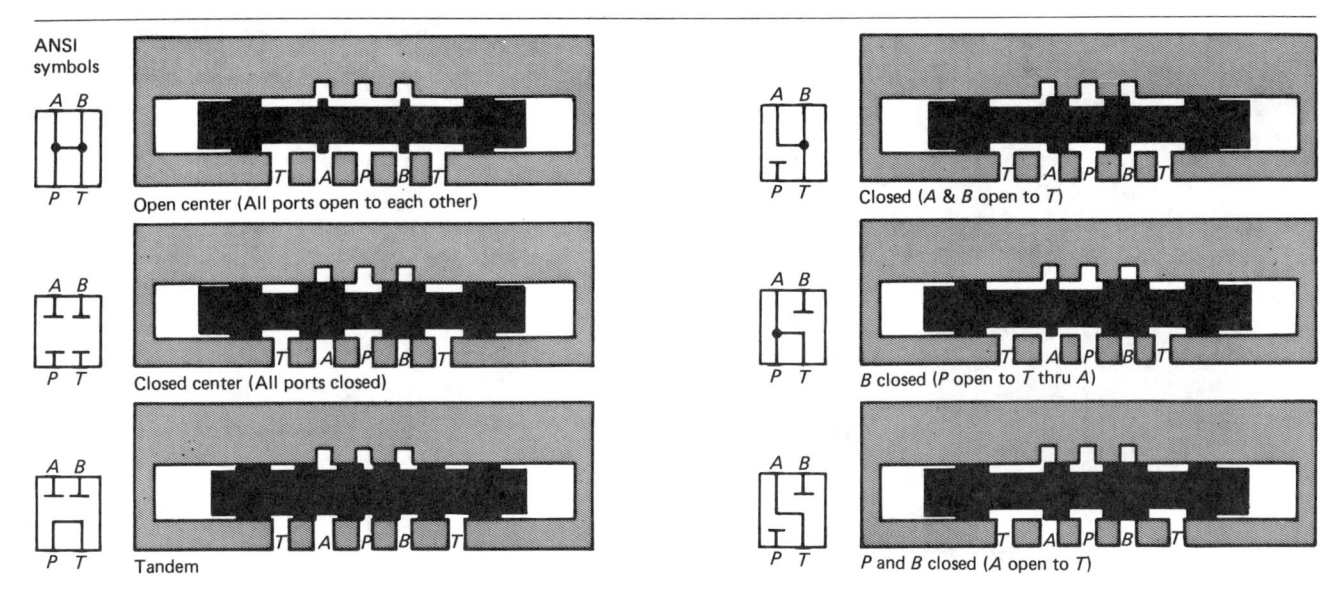

Figure 5.20 Center configurations of directional-control valves.

- Open center (All ports open to each other)
- Closed center (All ports closed)
- Tandem
- Closed (A & B open to T)
- B closed (P open to T thru A)
- P and B closed (A open to T)

with fluid. Under different circumstances, any one method or combination of these may be used with a given valve. Manual actuators include hand and foot devices for operating a valve. Electric, or solenoid, actuators use electromagnetic devices to cause linear motion of a shaft that moves the valve element from one position to another. Solenoids are generally actuated by some external signal

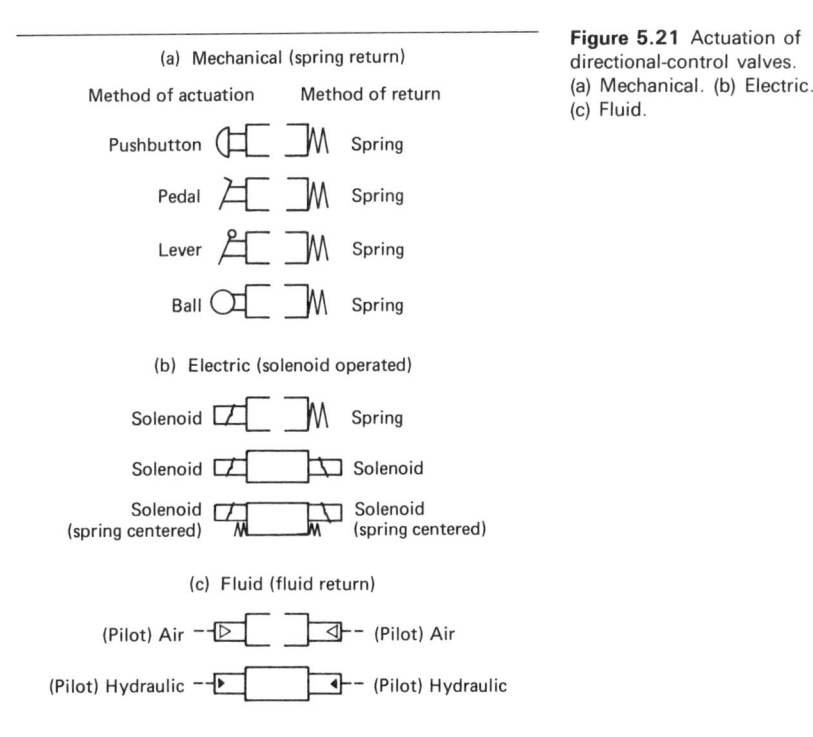

Figure 5.21 Actuation of directional-control valves. (a) Mechanical. (b) Electric. (c) Fluid.

(a) Mechanical (spring return)

Method of actuation	Method of return
Pushbutton	Spring
Pedal	Spring
Lever	Spring
Ball	Spring

(b) Electric (solenoid operated)

- Solenoid — Spring
- Solenoid — Solenoid
- Solenoid (spring centered) — Solenoid (spring centered)

(c) Fluid (fluid return)

- (Pilot) Air — (Pilot) Air
- (Pilot) Hydraulic — (Pilot) Hydraulic

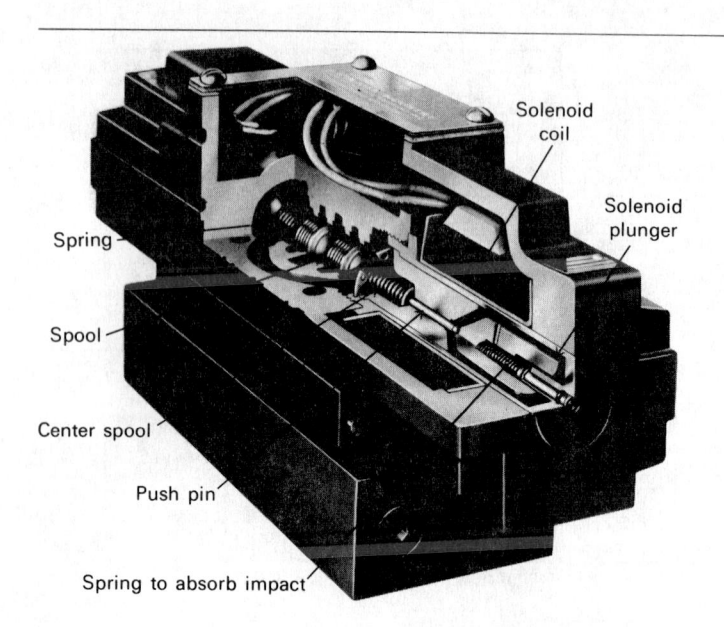

Figure 5.22 Solenoid-actuated, spring-centered, directional-control valve. (Courtesy Webster Electric Company, Inc.)

such as a manual push button or some position- or pressure-sensing device. Fluid actuators used in directional-control valves can be hydraulic or pneumatic. Directional-control valves may be actuated by a combination of electric and fluid devices. This combination is generally used where large flows are controlled by the directional-control valve.

In a large-capacity directional-control valve, the force needed to shift the spool directly would require a very large solenoid. If a smaller spool directional-control valve (master) actuated by a solenoid is added, it can direct pilot pressure to shift the large spool. The solenoid-controlled, pilot-operated valve serves as a "master," by directing flow to either end of the pilot-operated main spool [Figure 5.23(a)]. The master valve is mounted piggyback on the larger ("slave") valve [Figure 5.23(b)]. Air valves can also be remotely controlled [Figure 5.23(c)] and actuated by pilot pressure.

Spring condition

Spring conditions refer to the mechanical action on the spool of the valve when springs are included in a valve design. When a spool is held in one extreme position by the force of a spring, the spool is spring offset [Figure 5.23(c)]. It is normally held in this position

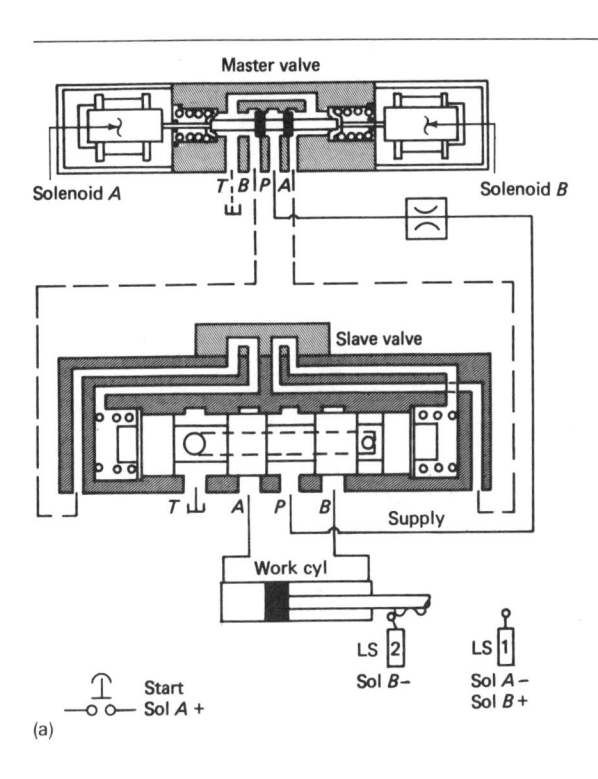

Figure 5.23 (a) Solenoid-actuated, directional-control, three-position, four-way, pilot-operated, spring-centered slave valve. (b) Piggyback-mounted directional-control valves on a manifold block. (Courtesy Almo Manifold and Tool Company) (c) Solenoid-controlled, pilot-operated, spring-returned, four-way, five-port, directional-control air valve. (Courtesy Ross Operating Valve Company)

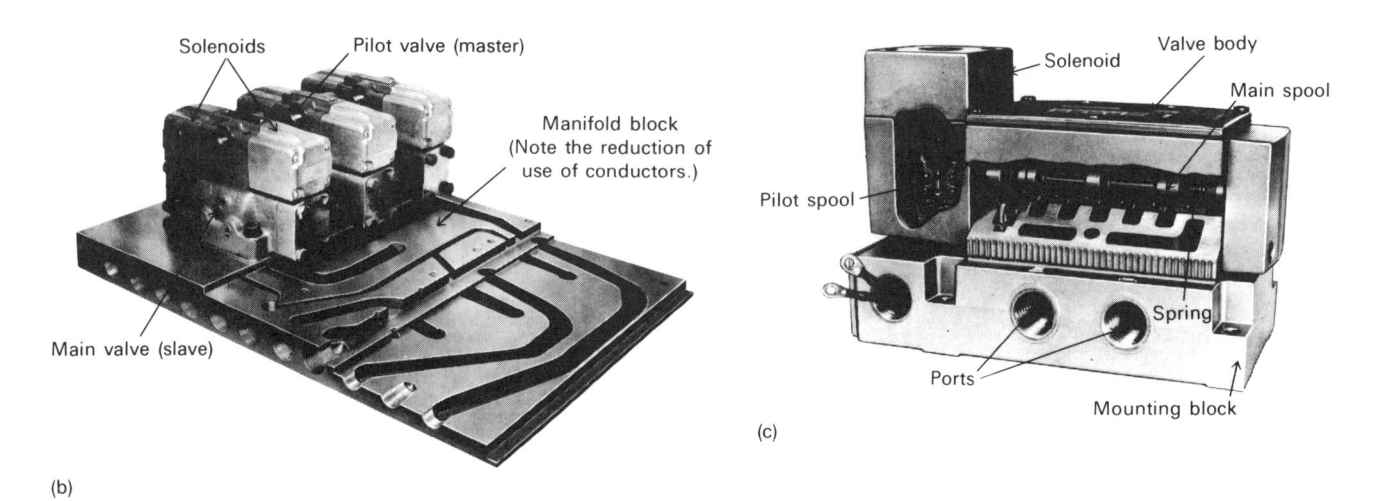

until it is shifted to its alternate position by the actuating device overcoming the spring force. When the spool is held in the center position by a spring force, it is described as "spring-centered." This condition may be accomplished by one spring that is compressed in either direction. Normally, these directional valves are manually actuated. Spring-centered spools with two springs are used more commonly and can be actuated by fluid, electric, or manual devices.

Servo

As more critical and precise control was called for in automation, space vehicles, and other sophisticated machine applications, there was a demand for more sophisticated directional controls. A servo system is a combination of elements for the control of a source of power. The output of the system (or some function of the output) is fed back for comparison with the input and the difference between these quantities is used in controlling the power (Figure 5.24). The variables controlled in such a system are: (1) position of a shaft, (2) speed of a motor, and (3) acceleration of a load (inertia).

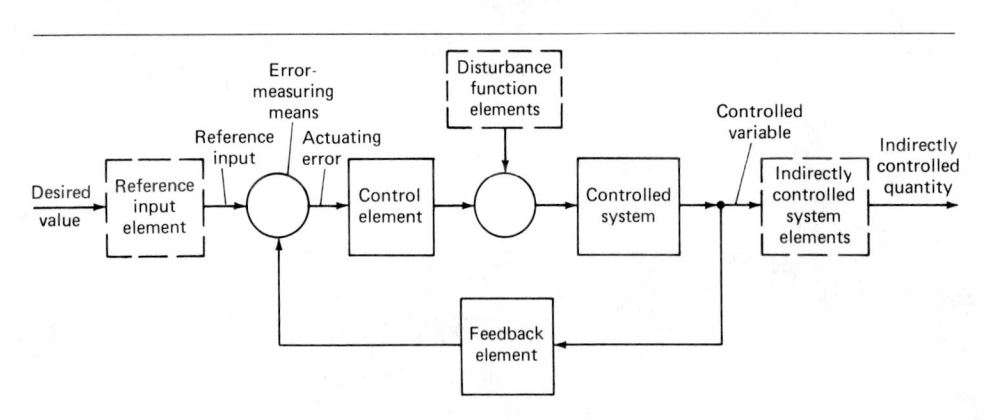

Figure 5.24 Simplified block diagram of feedback control system.

Feedback loop

The primary function of an electrohydraulic servo valve is to control hydraulic output more accurately by regulating velocity, acceleration, or position in response to an electric input signal. The amount and the direction of flow are related to the polarity and magnitude of the electric signal. These valves can be used to direct and meter flow to and from a hydraulic actuator or to control the flow of a variable-displacement pump. Servo control valves are used in closed-loop systems in which the controller is continuously aware of the condition of a cylinder or fluid motor that is actuated by means of feedback. The feedback principle, operating in a fluid system, compares the output with the reference signal and makes adjustments to reduce the difference. An input electric signal to the servo valve must be converted into mechanical motion to actuate the control. The components that make up a basic closed-loop electrohydraulic servo system are shown in the block diagram (Figure 5.25).

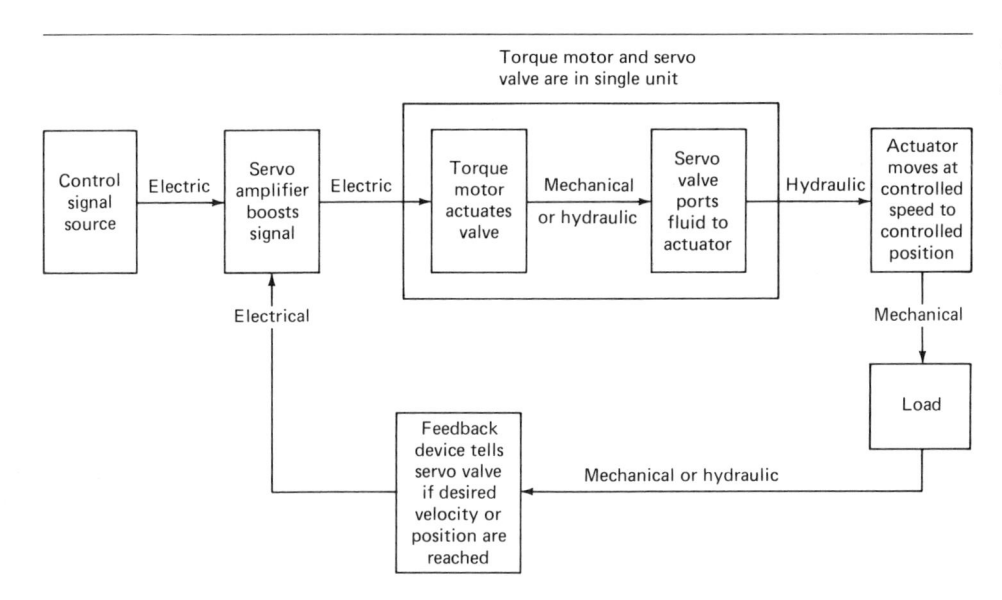

Figure 5.25 Block diagram of electrohydraulic servo valve system.

Servo valves control the flow of liquid to or from a load actuator in proportion to the input current signal to the force motor (Figure 5.26). A very small amount of oil flows continuously out of the pressure cavity P through the 90-micron screen and through the flexible pipe connected to the force motor coil and out of the pro-

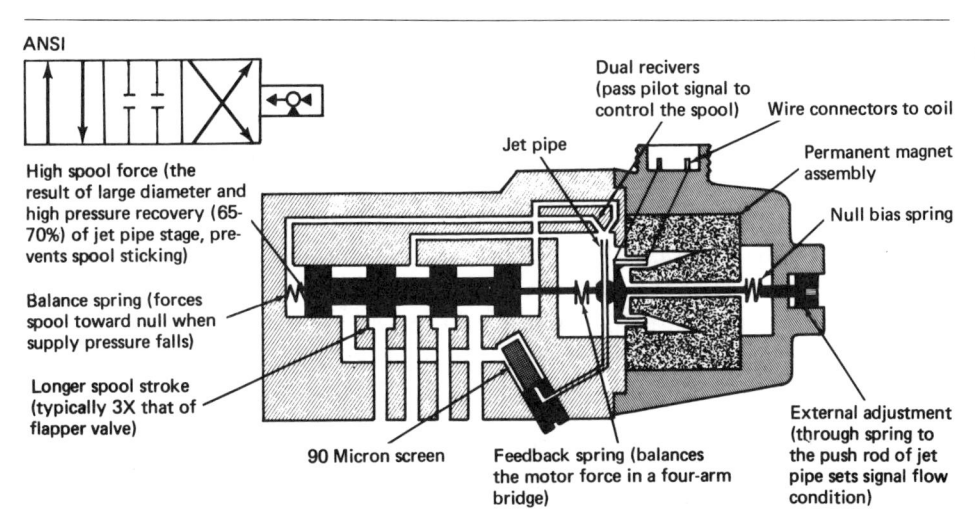

Figure 5.26 Hydraulic servo valve construction and operation. (Courtesy Abex Corporation, Denison Division)

jector jet. The coil armature is supported and aligned at both ends by frictionless flexure springs of beryllium copper. The oil flowing from the jet impinges on two receiver pipes that are connected to each end of the spool bore. At null, equal-pressure forces are developed on each end of the spool, holding it in a fixed position.

When a current signal is applied to the force motor, it flexes the jet pipe through a small angle. If the coil flexes the jet pipe to the right, more oil will impinge on the right-hand receiver pipe, increasing the pressure there and in the right-hand cavity of the spool bore. Conversely, less pressure will be developed in the left-hand pipe and spool-bore cavity. Thus a differential force is exerted on the spool, which causes it to move to the left.

As the spool moves, the counteracting force transmitted to the jet pipe by the feedback spring is reduced relative to the force of the null spring on the right-hand end of the cap. The spool continues to move to the left until the force created by the feedback spring is equal to the force developed by the torque motor and the null spring. When this occurs, the jet pipe is again statically positioned over the two receiver pipes, and no differential force is acting on the spool, which remains at its new position. Thus the second-stage spool has assumed a position proportional in direction and magnitude to the input current signal in the force motor. If the signal current polarity is reversed, the spool will move to the other side.

Electrohydraulic servo controls are playing an increasingly significant role in meeting demands for accuracy and response speed through closed-loop servo systems. These controls are finding widespread applications on high-performance machine tools, steel mill machinery, construction equipment, test equipment, and space vehicle controls.

Flow-control valves

Flow-control valves are used in fluid power circuits to control the rate of fluid flow from one part of the circuit to another. Their function is to regulate the amount of fluid passing through the valve, thereby controlling the speed of the actuator (Figure 5.27). Valves that control the volume of fluid flow usually accomplish this through a metering orifice. These valves may vary in construction and design from a simple needle valve to a sophisticated pressure-temperature compensated, variable flow-control valve.

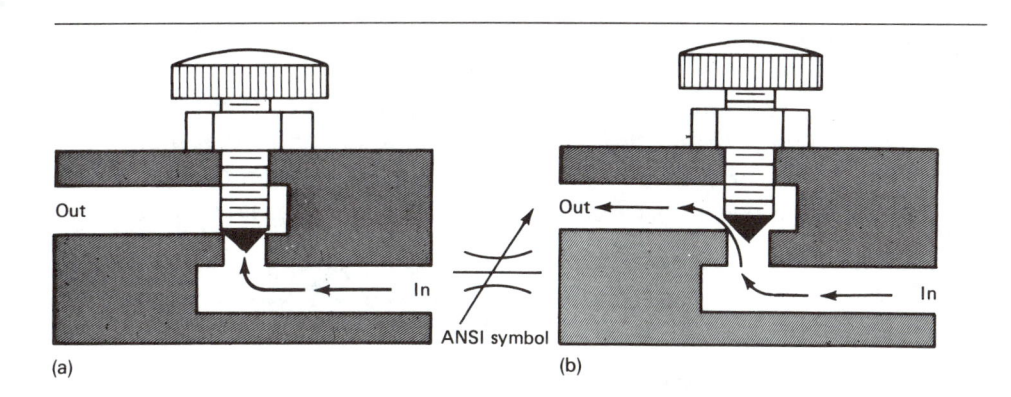

Figure 5.27 Flow-control valves regulate the amount of fluid passing through the valve to control the speed of the actuator.
(a) Needle valve is closed.
(b) Needle valve is partly open.

Noncompensated

Simple needle valves function by restricting or throttling flow (Figure 5.28) and are termed "noncompensated." Flow and pressure drop across an orifice are related. For example, in a simple orifice, as flow increases, pressure drop increases in direct proportion to the square of the flow change. The reverse is also true. Because of this operating phenomenon, noncompensated flow-control valves cannot be used where precise flow requirements are critical.

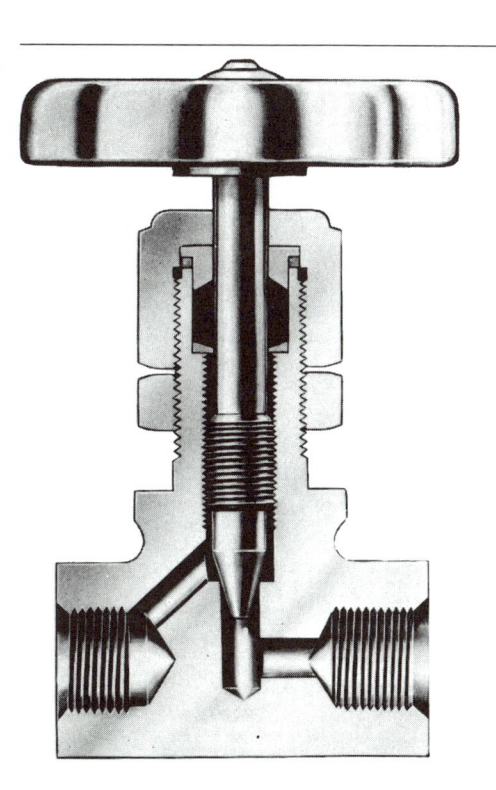

Figure 5.28 Simple needle flow-control valve. (Courtesy Teledyne Republic Manufacturing)

Compensated

A fixed-volume, pressure-compensated, flow-control valve (Figure 5.29) maintains a constant flow regardless of variations in the inlet flow to the valve. For example, if the inlet flow rate rises, the pressure-compensated valve partly closes to reduce the outlet flow. The

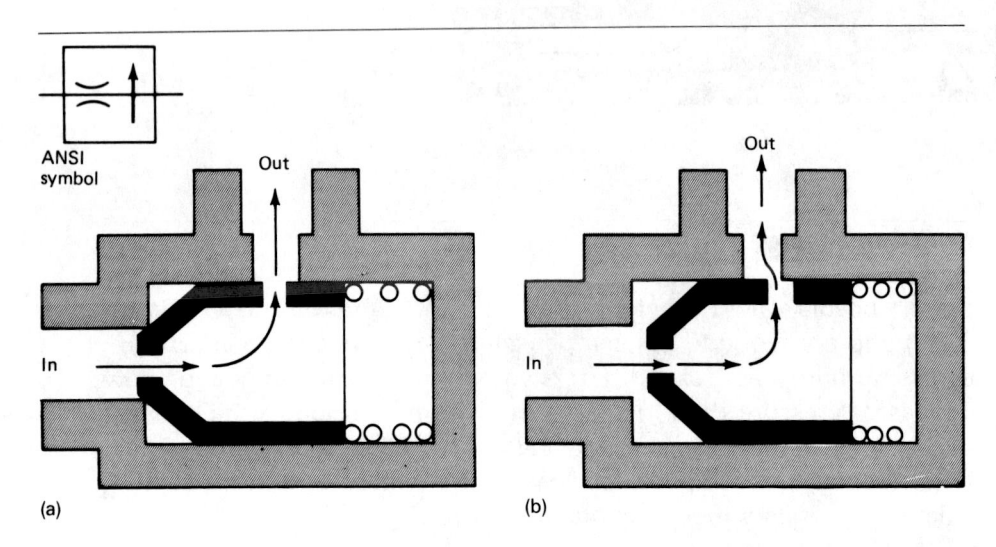

Figure 5.29 Fixed-volume, pressure-compensated, flow-control valve. (a) Position 1. (b) Position 2.

result is that the total volume of fluid through the valve always remains fixed. Where a constant flow regardless of pressure changes is critical, a fixed-volume, pressure-compensated valve is used.

Variable-volume, pressure-compensated, flow-control valves (Figure 5.30) use adjustable volume-control devices. In numerous fluid power applications, cylinder or fluid motor speeds must be adjustable. Several devices can be used for adjusting the orifice area. The most common ones are tapered slots or metering spools. A variable-flow, pressure-compensated, flow-control valve maintains constant flow with varying inlet and outlet pressures.

Temperature-compensated valves are also used in hydraulic applications. Oil temperature changes do alter the flow rate through fixed-orifice, flow-control valves, both compensated and noncompensated. As the viscosity of a fluid decreases, the flow rate through the orifice increases. To compensate for this change in flow rate, some flow-control valves use sharp-edged adjustable metering orifices that reduce flow variations due to temperature and viscosity changes. Another method uses dissimilar metals that expand at different rates with increasing temperature.

167 Flow-control valves

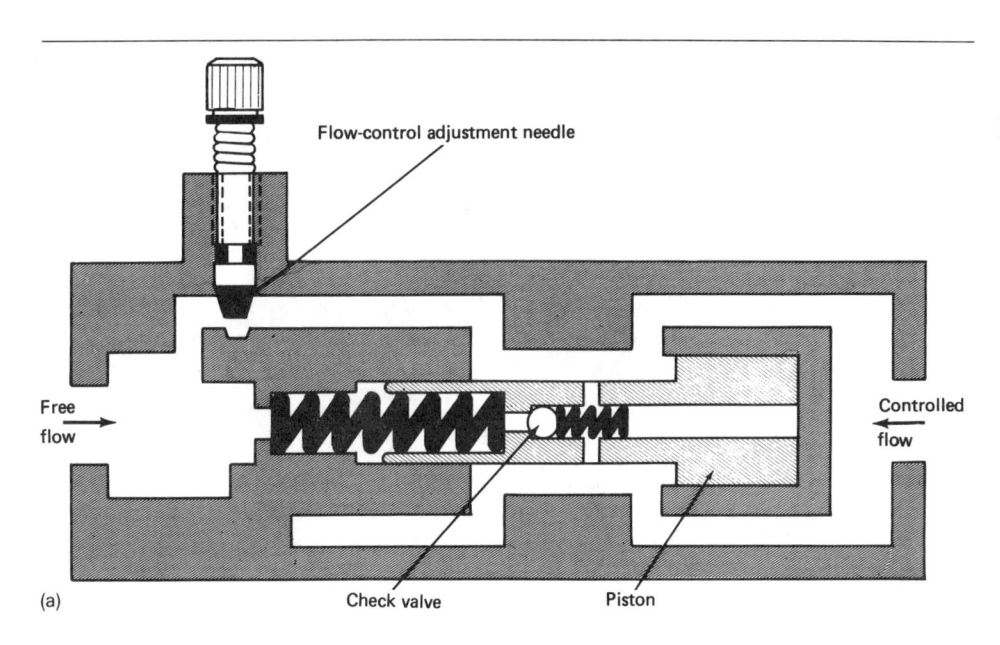

Figure 5.30 (a) Variable-volume, pressure-compensated, flow-control valve. (b) Industrial flow-control valve. [(b) Courtesy Sperry/Vickers]

Deceleration flow-control valves are normally open, cam-actuated, two-way valves. Their function is to decelerate a heavy load by gradually reducing the fluid flow rate by means of a tapered spool rather than a sharp-edged spool (Figure 5.31).

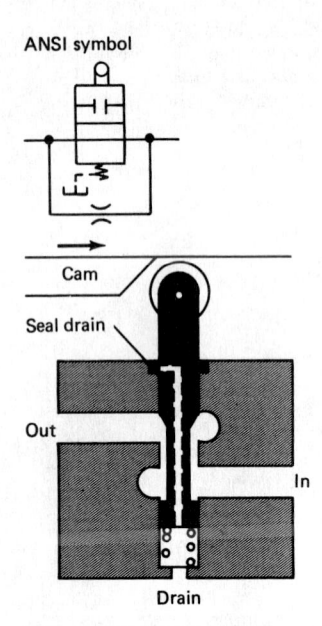

Figure 5.31 Cam-operated, deceleration, flow-control valve.

Summary

Fluid power valves are devices that control fluid flow as to pressure, direction, or flow rate. These are the three major modulation functions of a valve.

The function of a pressure-control valve is to control pressure within a circuit or branch of a circuit. The pressure-relief control valve has a primary function of providing pressure limitation in a circuit, thereby protecting components against excessive system or subsystem pressure.

A sequence valve has the primary function of directing flow in a predetermined sequence in reaction to pressure differences. Two or more actuators can be sequenced with appropriate installation and setting of sequence valves. The primary function of a pressure-control unloading valve is to permit a pump or compressor to operate at a minimum load. For example, if a large quantity of fluid is necessary initially and not necessary once the load is in motion, an unloading valve will divert fluid flow back to the reservoir.

A counterbalance valve has as its primary function the restriction of fluid flow from primary to secondary ports and the maintenance of a pressure level sufficient to balance a load held by an actuator. It is a pressure-sensitive valve and serves the same function as a mechanical counterbalance weight.

Directional-control valves may direct, divert, combine, stop, and reverse flows in a circuit. Structurally, these valves are classified as sliding-spool, rotary-spool, rotary-plate, and ball types. Their structural design renders certain classes of directional-control valves particularly suited for specific applications. Directional-control valves can also be described in terms of ways, position, center, actuation, and spring condition. Being able to recognize and determine the function of a specific type of valve, first schematically and then in component form, will assist a designer in connecting an appropriate set of components to construct a given circuit. The number of working valve ports indicates the number of ways. A pneumatic control valve may have five ports that result from having two exhaust or two pressure ports. This is still classified as a four-way valve. The position refers to the possible locations of the control element. A manually controlled valve may indicate whether it is a two- or three-position valve.

Servo valves are used in closed-loop circuits, in which case the valve is continuously aware of the fluid actuator by means of feedback. The valve is constantly responding to this feedback, which compares the output with the reference signal and makes adjustments to reduce the difference.

Flow-control valves are used in fluid power circuits to control the rate of fluid flow (liquid or air) from one part of a circuit to another. Their function is to regulate the amount of fluid flowing (gpm or cfm) through the valve, thereby controlling the speed of an actuator. Fluid-control valves usually accomplish the metering of flow through a metering orifice. Some are pressure and/or temperature compensated.

Review questions and problems

1 What are the three primary functions of energy-modulation devices in a fluid power system?
2 List three functions performed by pressure-control valves in a fluid power circuit.
3 What three factors, in logical order, must be known to design a circuit to perform a specified job?
4 What is meant by a *normally closed pressure-control valve?*
5 What is the function of a relief valve?
6 Explain how a compound-relief valve operates.
7 Explain how an unloading valve operates.
8 Why must a sequence valve be drained externally?
9 Name three possible applications of a counterbalance valve.
10 What does a counterbalance valve do in a circuit?
11 Is a counterbalance valve (a) normally closed or normally open? (b) drained internally or drained externally?
12 What is the purpose of a pressure-reducing valve?
13 List six possible functions that directional-control valves can serve.
14 Draw ANSI symbols for a four-way, three-position directional-control valve including: (1) open center, (2) closed center, and (3) tandem center.
15 By what four methods can a directional-control valve be actuated?

16 What is meant by a spring-centered directional-control valve?
17 Explain what is meant by the feedback principle.
18 How is flow control accomplished in many of the flow-control valves?
19 What specific type of flow-control valve should be used where a constant flow is critical regardless of pressure changes in the circuit?
20 Where are deceleration flow-control valves used?

Selected readings

- Fitch, Ernest C., Jr. *Fluid Power and Control Systems.* McGraw-Hill, New York, 1966.
- Glenn, Ronald E., and James E. Blinn. *Mobile Hydraulic Testing.* American Technical Society, Chicago, 1970.
- Hedges, Charles S. *Industrial Fluid Power,* Vol. I. Nomack Machine Supply Co., Dallas, Texas, 1965.
- *Industrial Hydraulics* Manual 935100–A, Vickers Division of Sperry Rand Corporation, Troy, Michigan, 1970.
- McNickle, L. S., Jr. *Simplified Hydraulics.* McGraw-Hill, New York, 1966.
- Newton, Donald G. *Fluid Power for Technicians.* Prentice-Hall, Englewood Cliffs, New Jersey, 1971.
- Oster, Jon. *Basic Applied Fluid Power: Hydraulics.* McGraw-Hill, New York, 1969.
- Pease, Dudley A. *Basic Fluid Power.* Prentice-Hall, Englewood Cliffs, New Jersey, 1967.
- Stewart, Harry L. *Hydraulic and Pneumatic Power for Production,* 3rd ed. Industrial Press, New York, 1970.
- Stewart, Harry L., and John M. Storer. *Fluid Power.* Howard W. Sams, Indianapolis, Indiana, 1968.

Chapter 6
Energy output devices

This chapter deals with energy output devices—cylinders and motors. It includes the principles of their operation and construction and explains their speeds. Classification of cylinders according to body styles and of motors according to converting elements is discussed in considerable detail. Applications, advantages, and safety of cylinders and motors are also covered.

Key terms

- **Actuator** A device that converts fluid energy into mechanical force or motion.
- **Cap** A cylinder end closure that completely covers the bore area.
- **Centerline mounting** A mounting that permits connection of a cylinder on a plane in line with the piston rod centerline.
- **Cushion** A device that provides controlled resistance to motion.
- **Cylinder** A device that converts fluid energy into linear mechanical force or motion. It usually consists of a movable element such as a piston, plunger, or ram operating within a cylindrical bore.
- **Double-acting cylinder** A cylinder in which fluid force can be applied to the movable element in either direction.
- **Fixed mounting** A mounting that provides rigid connection between the cylinder and the mating element wherein the piston rod reciprocates in a fixed line.
- **Head** A cylinder end closure that covers the differential area between the bore area and the piston rod area (sometimes referred to as the "rod end").
- **Motor** A device that converts fluid energy into mechanical force and motion. It usually provides rotary mechanical motion.
- **Pivot mounting** A mounting that permits a cylinder to change alignment in a plane.
- **Sealing device** A device that prevents or controls the escape of a fluid or the entry of a foreign material.
- **Single-acting cylinder** A cylinder in which fluid force can be applied to the movable element in only one direction.

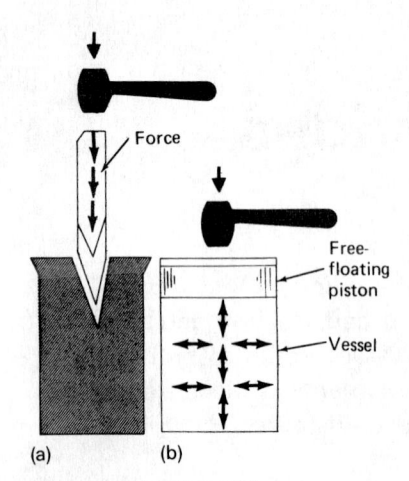

Figure 6.1 (a) Lines of force are determined by the direction of impact in solids. (b) Lines of force are transmitted equally in all directions and at right angles to the containing surfaces of fluids.

The purpose of energy-output devices is to convert one form of energy, force, or power into another form of energy, force, or power and to transmit that force to perform useful work. Each of the three types of substances (solids, liquids, and gases) is capable of transmitting forces to perform useful work. Each substance, because of its physical structure and characteristics, has advantages and disadvantages in transmitting forces.

When we strike the end of a solid bar, the main force of the impact is carried straight down to the other end [Figure 6.1(a)]. The direction of the blow almost entirely determines the direction of the transmitted force. When a force is applied to the end of a column of confined liquid or gas, however, it is transmitted not only straight down through the other end but, according to Pascal's law, equally and undiminished in every direction throughout the column [Figure 6.1(b)].

Liquids and gases must be completely confined to transmit forces without complete loss. An actuator is a device that converts fluid power (pressurized fluid) into mechanical force and motion. The output motions may be linear, rotary, or combinations of the two (Figure 6.2).

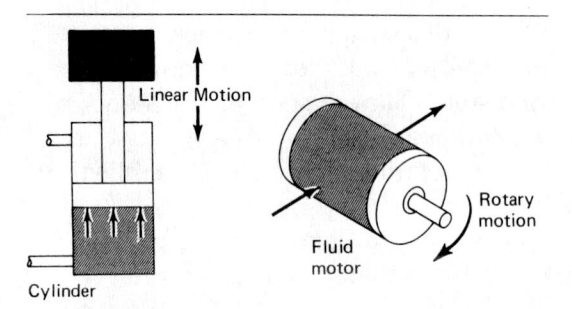

Figure 6.2 Fluid actuators produce linear or rotary motions.

Principles of actuator operation

The fact that liquids are incompressible for most practical purposes (less than 0.05% at 1000 psi) permits them to transfer power in an efficient manner within a confined system. Figure 6.3 shows this principle in operation. A force of 50 lb applied to Piston *A* produces a pressure on the confined fluid. This pressure is constant throughout the enclosed fluid. The pressure developed at Piston *A* is transmitted

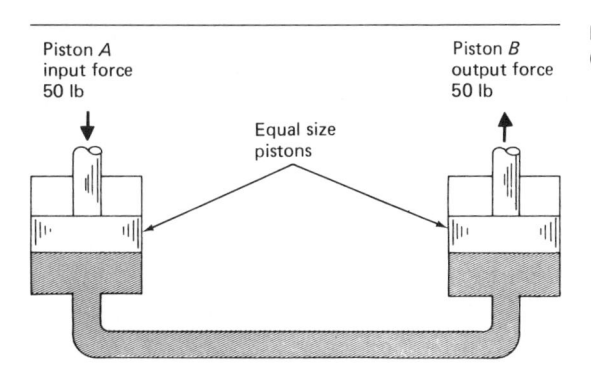

Figure 6.3 Transmission of a force (50 lb) from Piston A to Piston B.

to Piston B and, provided that the pistons have the same area, Piston B produces an output force equal to the force applied to Piston A (50 lb). In actual practice the force at Piston B would be slightly less because of heat and frictional losses.

For a given quantity of energy or force, a certain amount of work can be accomplished. A small piston (Figure 6.4) under a force of 500 lb will produce a pressure of 100 psi on a 5-in² piston surface. The

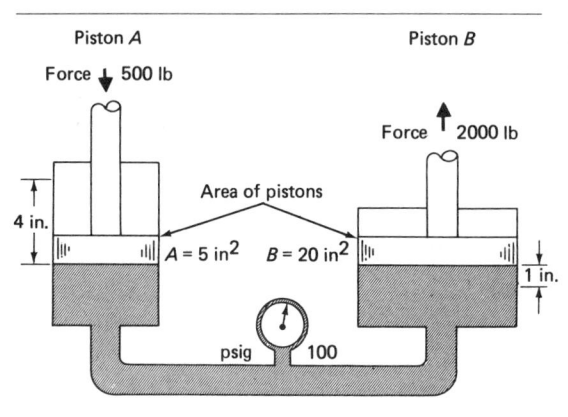

Figure 6.4 Multiplication of forces (Piston B moves considerably less than Piston A).

pressure per unit area exerted anywhere on a mass of liquid is transmitted in all directions. It acts with the same intensity on all surfaces at a right angle to those surfaces. Since the output piston has an area of 20 in², the output force is 2000 lb.

Force = pressure × area

Force = $100 \frac{lb}{in^2} \times 20 \; in^2$

$ = 2000 \; lb$

It becomes apparent that it is possible to multiply forces by using pistons of dissimilar size. The hydraulic jack is designed to take advantage of multiplication of forces. Note that, in Figure 6.4, while the input piston moved 4 in. down, the output piston moved only 1 in. upward. Although the forces have been multiplied four times, this gain was achieved at the sacrifice of distance of travel. The work done by each piston remains balanced. The total output of work cannot exceed the input of work. As with all systems, losses in various forms occur and the output is less by some percentage than the input.

$$\text{Work} = \text{force} \times \text{distance}$$

$$\text{Work of Piston } A = 500 \times 4$$
$$= 2000 \text{ in-lb}$$

$$\text{Work of Piston } B = 2000 \times 1$$
$$= 2000 \text{ in-lb}$$

The multiplication of forces is similar to the mechanical advantage of a lever, as shown in Figure 6.5. In Figures 6.4 and 6.5, while the output force is increased, the distance is reduced. This principle of mechanical advantage (gain in force and loss in distance) is fundamental to all power transmission systems involving a change of force.

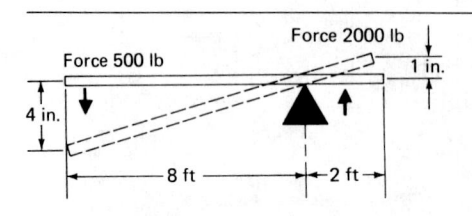

Figure 6.5 Mechanical advantage of a lever.

Fluid transmission devices have some advantages over mechanical devices such as levers, cams, pulleys, belts, or gears. Fluids take on the shape of their confining containers and thus can transmit forces around bends or corners. Fluids can be piped over greater distances and can be divided or combined for smaller or greater volumetric flow to regulate the speed of an actuator. The flexibility of fluid power distribution through conductors to the actuator provides the designer of equipment with greater geometric freedom and margin of safety. Fluid under pressure acts against the piston surface to develop a force that will overcome the resistance of the opposing load.

Actuating cylinder

A cylinder is a device that converts fluid (liquid or gas) energy into mechanical energy or motion. It usually consists of a movable element such as a piston and piston rod, or plunger, operating within a cylindrical housing [Figure 6.6(a)]. From the standpoint of performance design, there are two forms of cylinder, known as the *plunger* (*ram*) and the *piston*. Fluid under pressure need be applied to only one end of the ram-type cylinder to force the plunger outward. It is therefore referred to as a *single-acting cylinder* [Figure 6.6(b)]. The plunger retracts because of the weight of the load or from some other mechanical force such as a spring. A piston cylinder can also be a single-acting device. A single-acting cylinder is controlled by reversing a directional valve and permitting the flow from the pump and the cylinder to return to the reservoir. In some applications the flow from the cylinder is metered to control the speed of retraction.

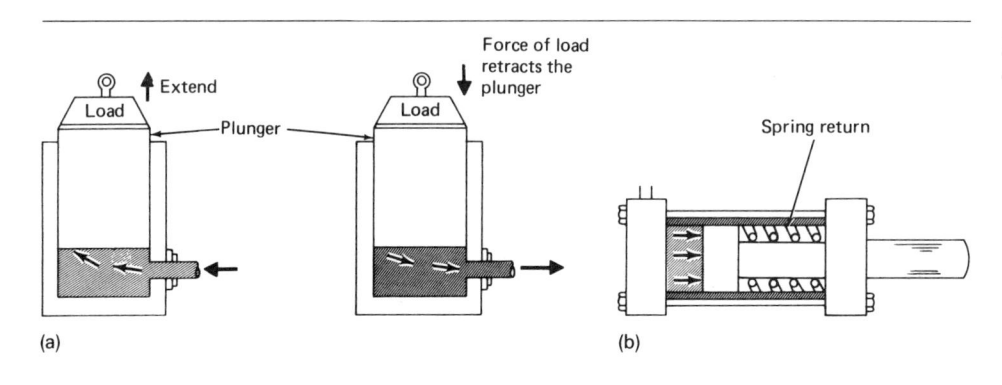

Figure 6.6 (a) Plunger, or ram, cylinder. (b) Single-acting cylinder.

A double-acting cylinder has ports that allow a fluid (liquid or gas) to enter the cylinder at either the cap end or the rod (head) end (Figure 6.7). Forcing fluid to flow from the pump to the cap end extends the rod while simultaneously discharging the fluid back to the reservoir in the case of hydraulics or to the atmosphere in the case of pneumatics. To retract the rod, the flow is reversed. More accurate control of both the extension and the retraction stroke is possible with a double-acting cylinder.

The double-acting cylinder in Figure 6.8(a) is a differential cylinder. That is, the cross-sectional area is greater on the cap end than on the rod end. The cross-sectional area that the rod occupies must be subtracted from the bore area to determine the effective area in contact with the pressurized fluid [Figure 6.8(b)]. For the same pres-

176 Energy output devices

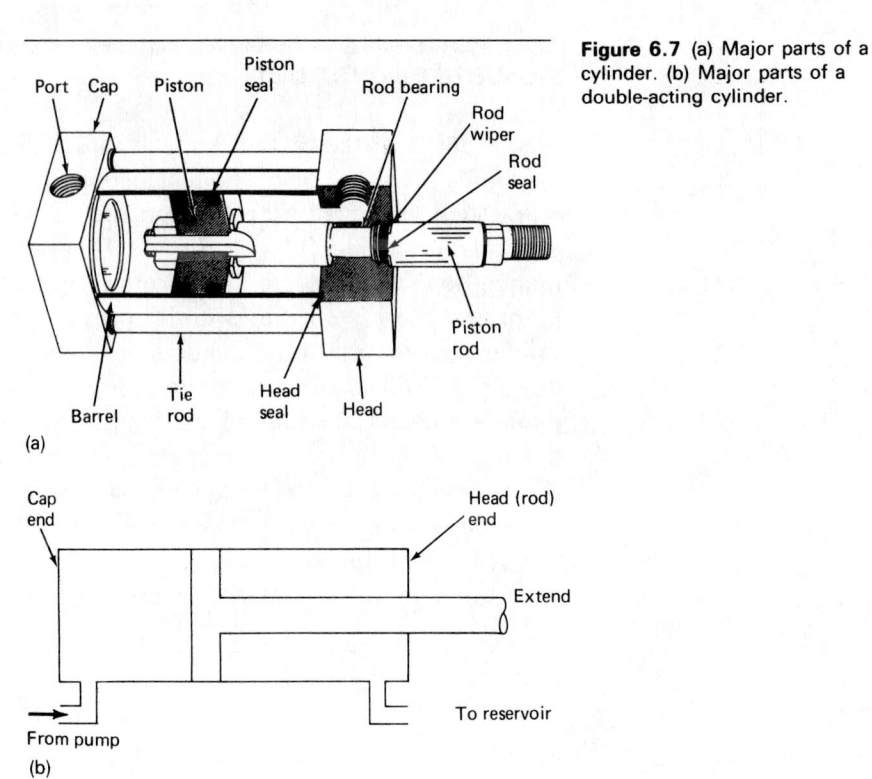

Figure 6.7 (a) Major parts of a cylinder. (b) Major parts of a double-acting cylinder.

sure, a greater force can be transmitted when fluid is delivered to the cap end. Speed is sacrificed to gain greater force. The rod moves more rapidly on the retraction stroke for a given delivery of fluid per minute since the rod displaces a portion of the cylinder volume.

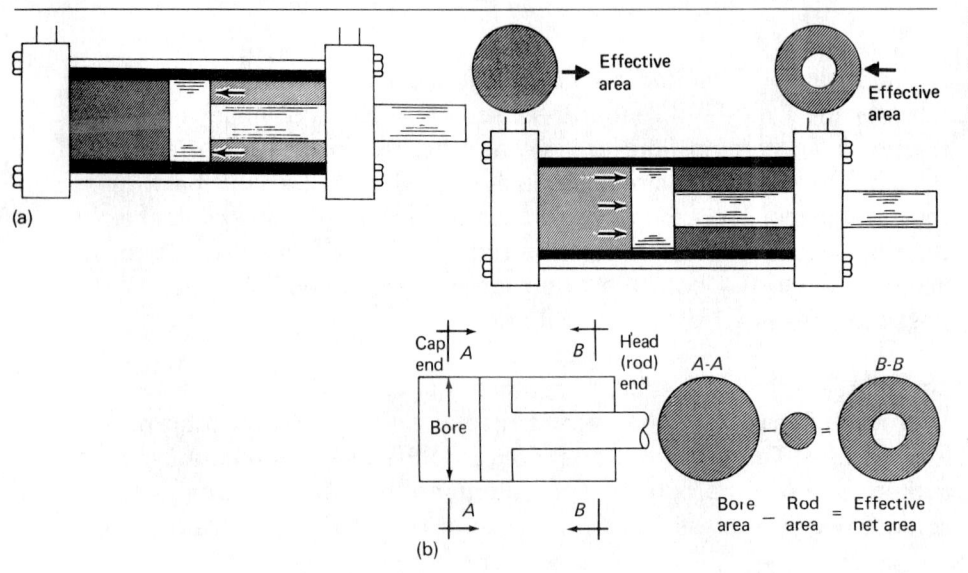

Figure 6.8 (a) Double-acting cylinder. The piston is moved in both directions by a pressurized fluid. (b) Effective net area is less than the bore area. (c) Double-end rod cylinder.

(c) Equal force and velocity in both directions

Figure 6.8 (cont'd)

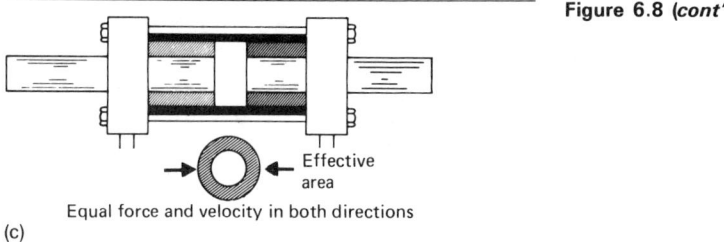

The double-end rod cylinder has a rod protruding from both ends [Figure 6.8(c)]. This permits equal force and velocity in both directions. Standard cylinders usually have a bore area:annular (rod) ratio of approximately 6:5. Heavy-duty rods may decrease that ratio to 2:1.

Velocity

The velocity of a cylinder as it extends or retracts depends on how fast a fluid is delivered to the cylinder for a given diameter or bore of cylinder. The velocity varies directly with the gallons per minute of liquid delivery (gpm) and inversely with the area of the cylinder (Figure 6.9). Mathematically, these relationships may be expressed as

$$v(\text{ft}/\text{min}) = \frac{\text{gal}/\text{min} \times 19.25}{A(\text{in}^2)}$$

Velocity of cylinder A is greater than cylinder B.

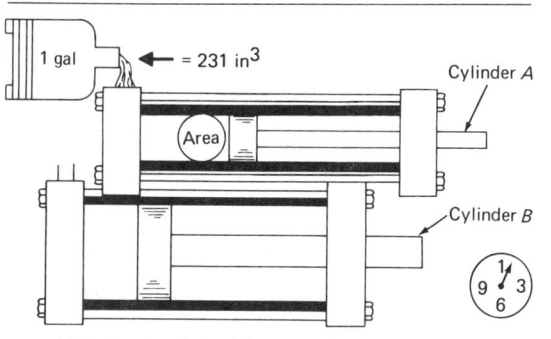

Figure 6.9 Velocity varies inversely with the area of the piston (expressed in feet per minute).

Since the velocity of a cylinder is generally expressed in feet per minute (ft/min), you can take the volume of 1 gal (231 in³) and divide this by the area of the cylinder

$$\frac{231 \text{ in}^3}{5 \text{ in}^2}$$

which will indicate the distance in inches 1 gpm will move that size of piston (Figure 6.10). The smaller the area of a piston, the greater the velocity. To convert the distance from inches to feet it is more convenient to divide 231 by 12 = 19.25 and thus use the factor of 19.25.

$$v(\text{ft/min}) = \frac{\text{gal/min} \times 19.25}{\text{area}}$$

$$v = \frac{\text{gpm} \times 19.25}{A}$$

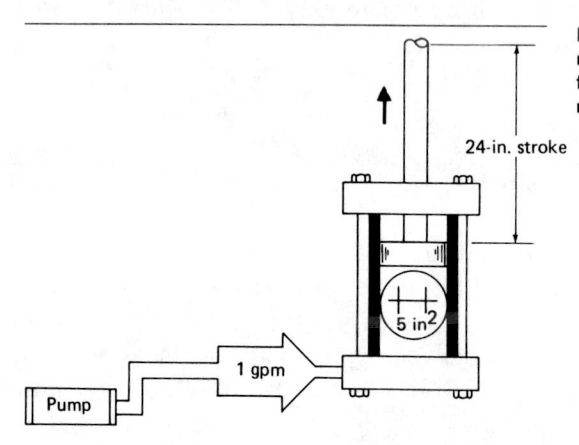

Figure 6.10 The volume of liquid required to move the piston one full stroke is called the displacement of the cylinder.

▶ **Example** If a pump could deliver 10 gpm of liquid to the cylinder in Figure 6.10, the velocity of the cylinder would be

$$v = \frac{10 \text{ gpm} \times 19.25}{5 \text{ in}^2} = \frac{38.5 \text{ ft}}{\text{min}}$$

The velocity in feet per second is obtained by dividing the factor 19.25 by 60 = 0.3208; therefore,

$$v(\text{ft/sec}) = \frac{\text{gpm} \times 0.3208}{A}$$

Frequently it is necessary to determine the size of the pump in order to produce a specific velocity in a given cylinder. The preceding equation can be stated in terms of required gpm pump (Figure 6.11).

$$\text{gpm} = \frac{v \text{ ft/sec} \times A \text{ in}^2}{0.3208}$$

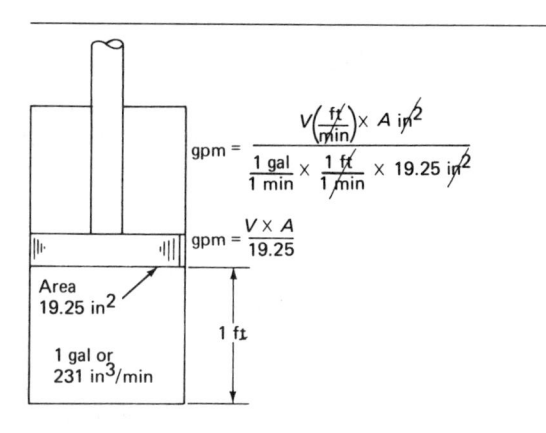

Figure 6.11 Derivation of the constant 19.25. One gallon (231 in³) per minute will move a piston with an area of 19.25 in² 1 ft in 1 min.

$$\text{gpm} = \frac{V\left(\frac{ft}{min}\right) \times A\ in^2}{\frac{1\ gal}{1\ min} \times \frac{1\ ft}{1\ min} \times 19.25\ in^2}$$

$$\text{gpm} = \frac{V \times A}{19.25}$$

To extend the piston in Figure 6.10 at a velocity of 2 ft/sec, determine the gpm pump required.

$$\text{gpm} = \frac{2 \times 5}{0.3208}$$

$$= 31.17$$

On the retraction stroke, if the piston rod has an area of 1 in², the annulus area of 4 in¹ for the same pump delivery would have a velocity of

$$v = \frac{31.17 \times 19.25}{4}$$

$$= 150.0\ \text{ft/min}$$

It is evident that a double-acting, differential cylinder will retract faster than it will extend [Figure 6.12(a)]. For differential cylinders,

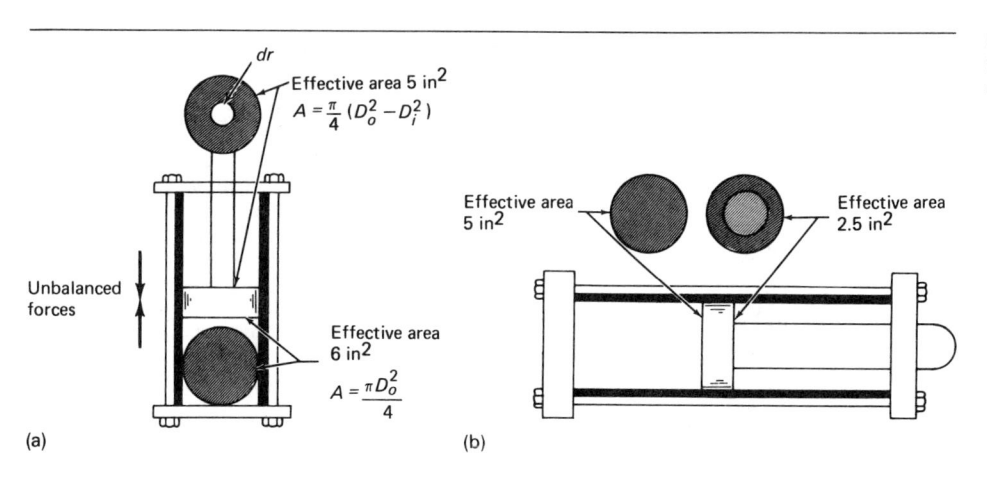

Figure 6.12 (a) Double-acting differential cylinder. (b) A 2:1 ratio cylinder.

the area ratios are determined by the size of rod used with a cylinder of a given bore size. The ratio of the bore and rod size depend on the construction, application, and length of stroke. The increased size of rods used in heavy-duty cylinder applications may reduce the ratio to about $2:1\frac{1}{2}$. Other cylinders used in regenerative circuits have a ratio of 2:1 [Figure 6.12(b)].

Most hydraulic systems are designed by starting with the work to be accomplished and sizing of cylinders; then other components are chosen to match the forces and motions of the cylinders.

Two additional special features have been developed in cylinder designs. One of these is the stepped-plunger, or piston, cylinder illustrated in Figure 6.13(a). A stepped piston allows a cylinder to provide a rapid starting stroke with a low force and a slower and more powerful working stroke. At first, the pressurized fluid pushes only against the smaller part of the piston, which moves rapidly until the work is contacted. Then the entire piston surface comes into contact with the pressurized fluid for the power stroke [Figure 6.13(b)].

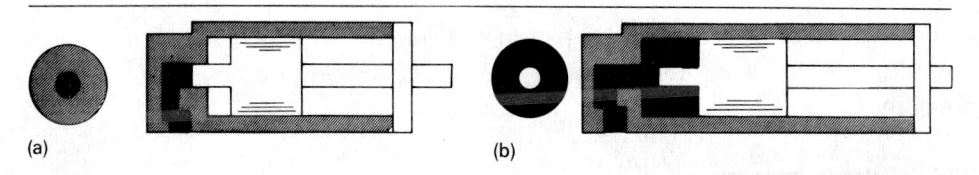

Figure 6.13 (a) Stepped-piston cylinder. (b) Power stroke.

The second special feature is the telescopic cylinder [Figure 6.14(a)]. As the name suggests, this consists of a nested group of cylinders. Telescopic cylinders are used on installations where the working stroke is long but the retracted length must be exceedingly short such as in the operation of a hydraulic hoist for a dump truck [Figure 6.14(b)]. One of the drawbacks of this design is that the area of each successive inner cylinder is smaller and smaller, thus changing the force exerted and speed proportionately. Telescopic cylinders are generally designed to be either single acting or double acting [Figure 6.14(c)].

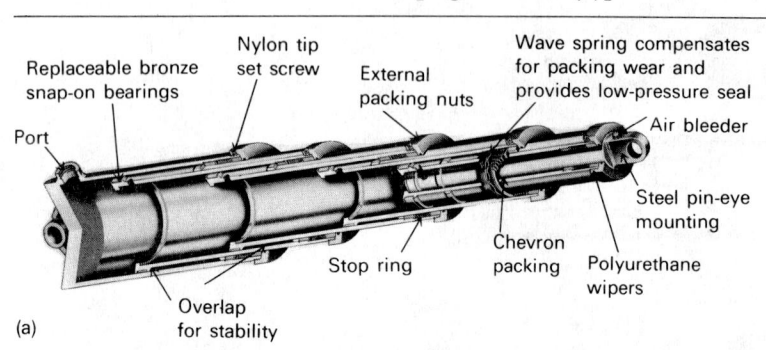

Figure 6.14 (a) Single-acting telescopic cylinder. (Courtesy Commercial Shearing) (b) Application of telescopic cylinders in the operation of a hydraulic hoist for a dump truck. (Courtesy Euclid) (c) Double-acting telescopic cylinder. (Courtesy Commercial Shearing)

Figure 6.14 (cont'd)

(b)

(c)

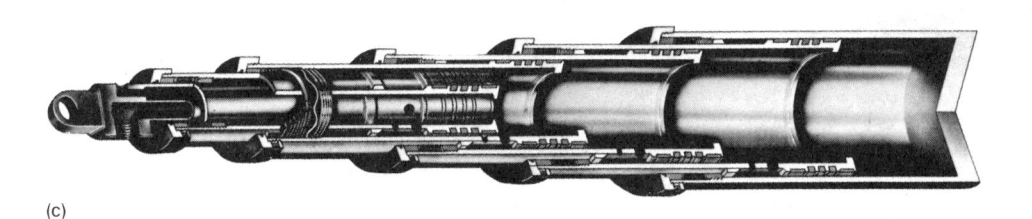

In both types of design—the single-acting and double-acting cylinders—it is apparent that the seals, ports, and machined interior surface of a cylinder housing provide a complete confining closure in which Pascal's law can be advantageously applied.

Construction

A cylinder is constructed basically of a barrel in the form of a tubular housing, a closely fitting piston with adequate seals, and a rod (Figure 6.15). The barrel is usually a seamless steel tubing, or cast, and the interior is machined and polished to be true and very smooth. Brass barrels are used for pneumatic cylinders. The steel piston rod is machined and highly polished. For some applications, the rod may also be hard-chrome-plated to resist pitting and scoring. It is supported by a bushing mounted in the end cap. Pistons are usually made of cast iron or steel and are fitted with either piston rings or flexible seals.

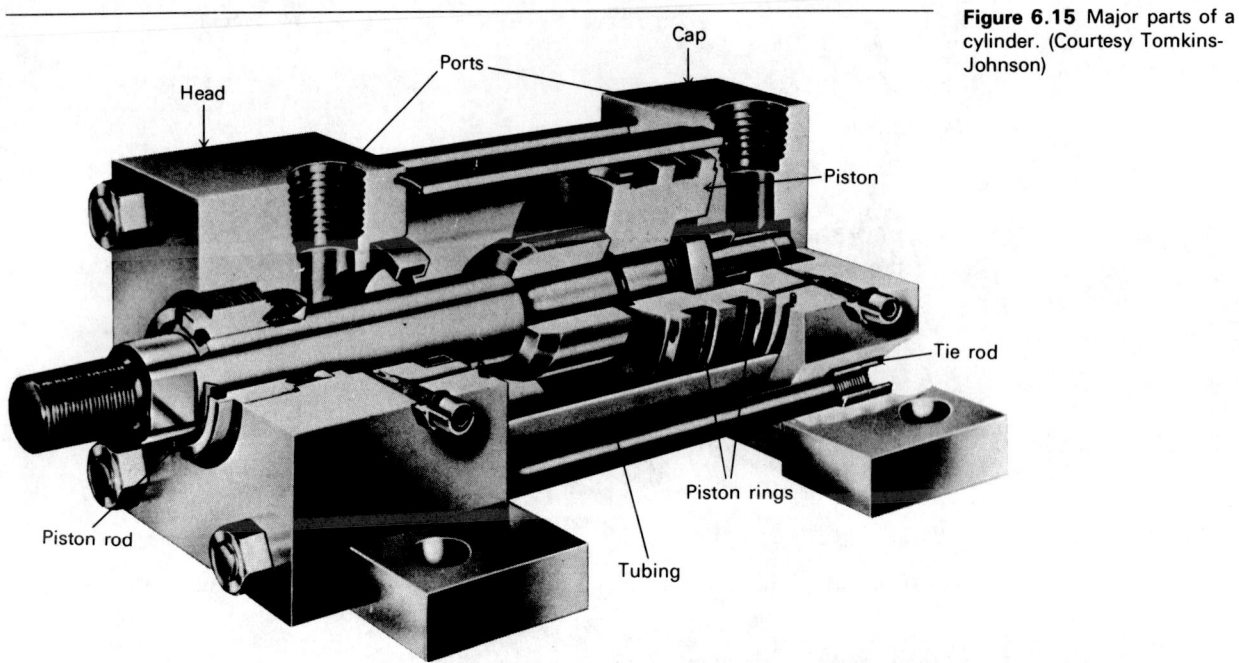

Figure 6.15 Major parts of a cylinder. (Courtesy Tomkins-Johnson)

Hydraulic cylinders are equipped with several other important parts. Seals and wipers are installed in the rod end cap to prevent foreign particles from entering the cylinder and causing external leakage (Figure 6.16). Leakage is undesirable both internally and externally.

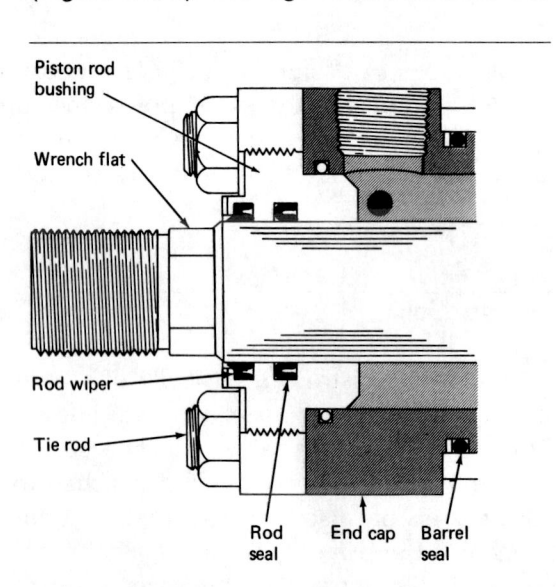

Figure 6.16 Cylinder seals and wipers. (Courtesy Vickers Division/Sperry Rand)

Another design feature of some cylinders is a *cushion* [Figure 6.17(a)], which is built in to decelerate the piston at the end of its stroke. This is very important in applications where the inertia forces

183 Actuating cylinder

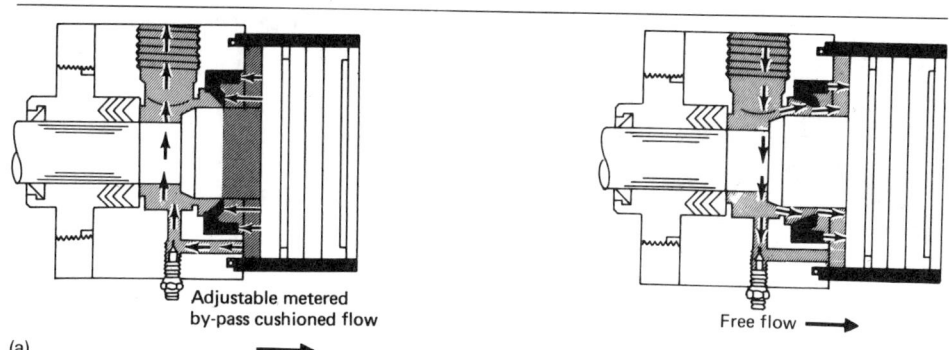

Figure 6.17 (a) Cushioning of air cylinders. (Courtesy Vickers Division/Sperry Rand) (b) Cushion adjustment. (Courtesy Pathon Manufacturing Company) (c) Cushioning of hydraulic cylinders. (Courtesy Pathon Manufacturing Company) (d) Adjustable cushioning of hydraulic cylinders.

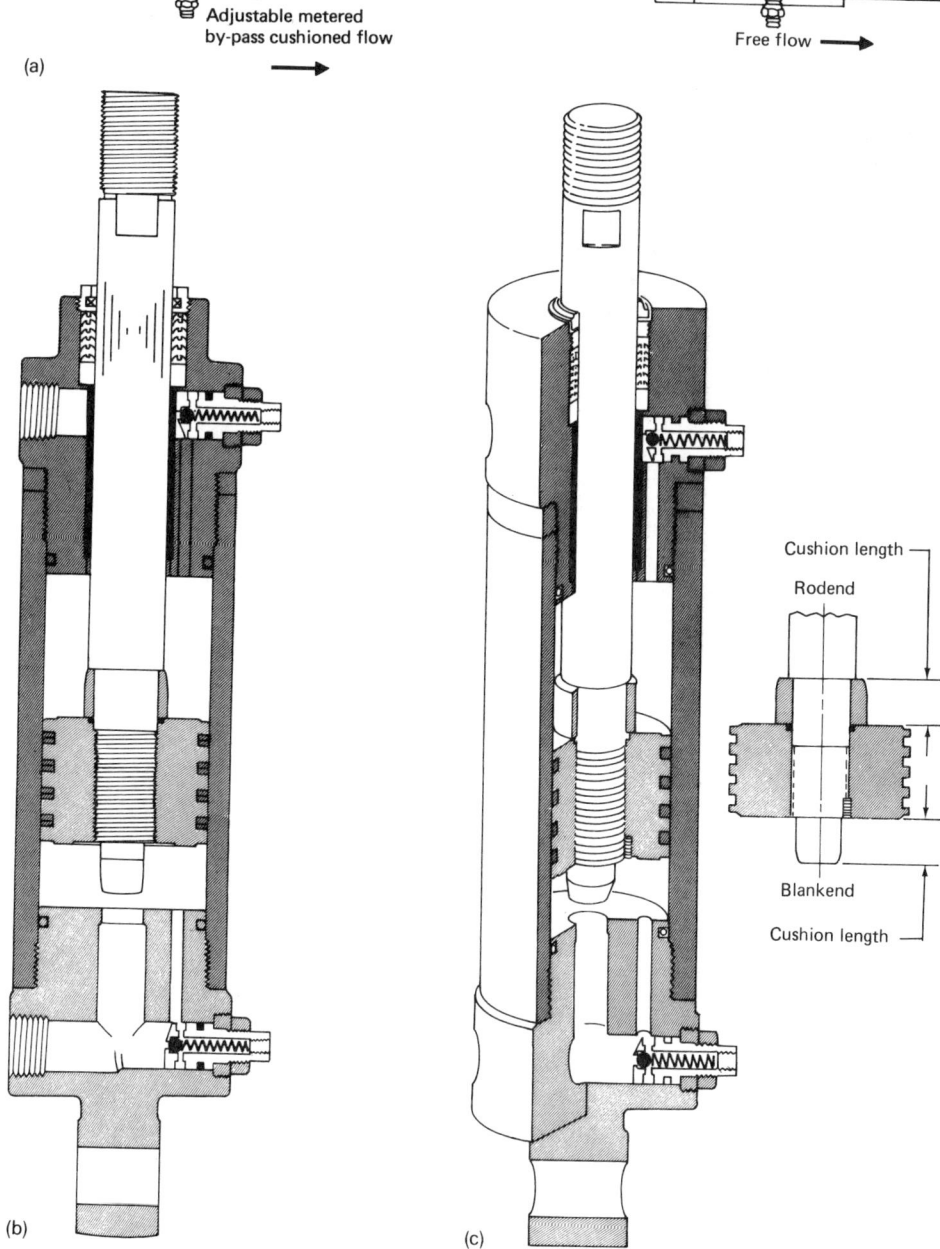

(a) Adjustable metered by-pass cushioned flow

Free flow

Cushion length
Rodend
Blankend
Cushion length

(b)

(c)

Figure 6.17 (*cont'd*)

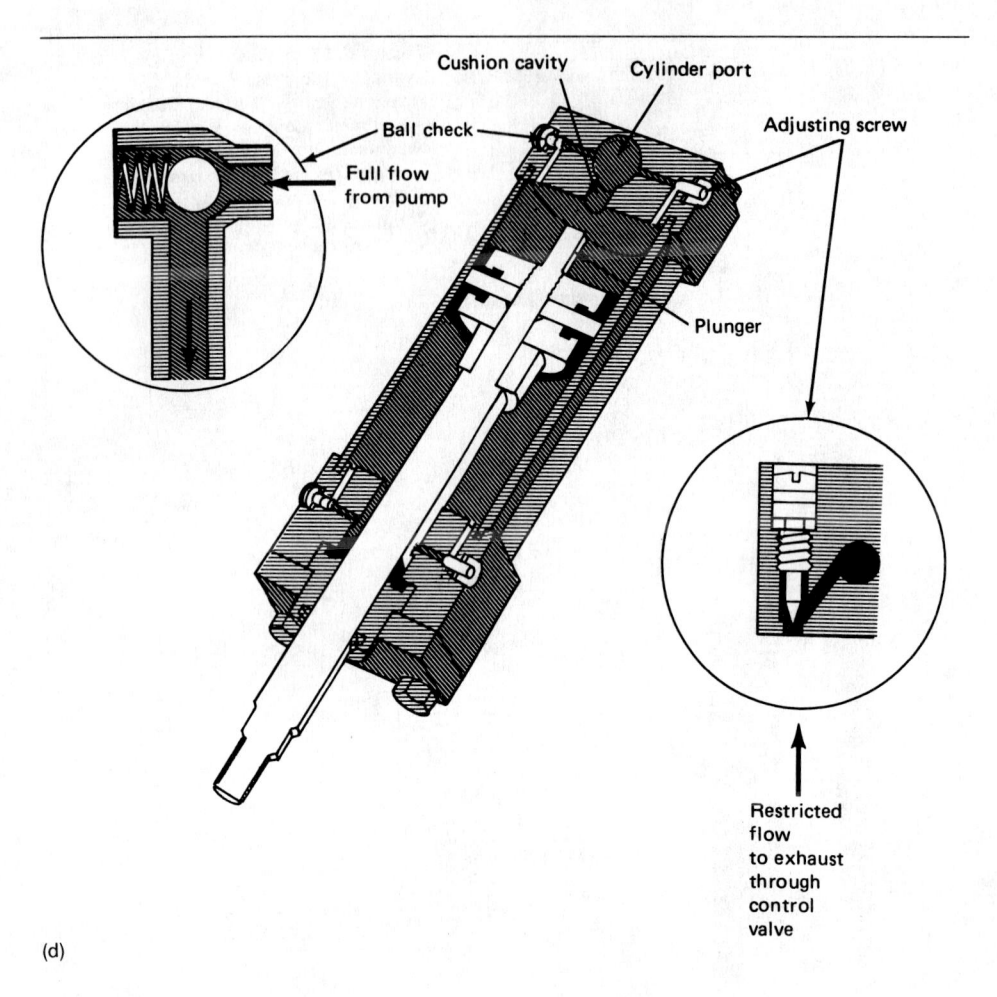

(d)

of heavily loaded and rapidly moving pistons must be arrested to protect the piston against impact damage. Several methods are used to decelerate (hydraulically) the piston at the end of the stroke [Figure 6.17(b)]. One method used frequently is a tapered plunger on the rod, which upon entering a counterbore in the cylinder head, closes the normal discharge port and forces the oil to leave through a small orifice in the end cap. An adjustable valve is provided that increases or decreases the orifice size to control the deceleration rate [Figures 6.17(c) and (d)].

Body style

Cylinders are manufactured in a wide variety of shapes, sizes, and structural designs. The tie-rod construction is perhaps the oldest and commonest type (Figure 6.18). Four or more tie rods hold the end caps against the tubular housing. High-tensile steel rods are used to absorb the internal energy stresses. For this reason the heavy-duty cylinders used in the automotive, mobile, and machine-tool industries are made by the tie-rod method of construction.

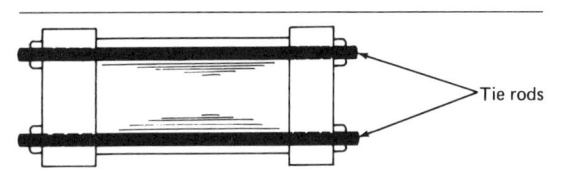

Figure 6.18 Tie-rod cylinder construction.

The threaded-construction cylinder has the end caps threaded to fit the internal thread of the tubular housing (Figure 6.19). This is a relatively new body style, made possible by improved machining capabilities and materials. In applications where space is an important consideration, such cylinders have become widely used. The food-processing and food-packaging industry makes extensive use of this cylinder body style.

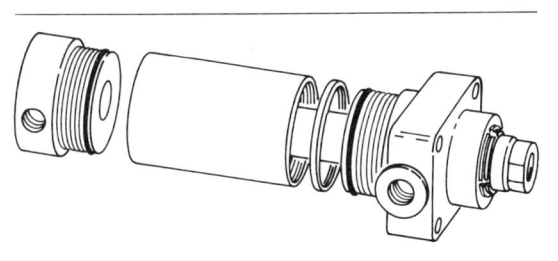

Figure 6.19 Threaded caps and barrel cylinder construction (screw-thread assembly).

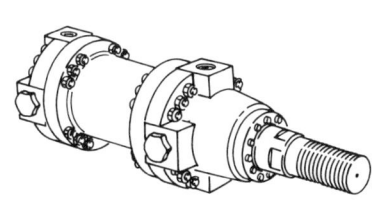

Figure 6.20 Mill construction

The mill-type cylinder has the end caps fastened to mating flanges on the tube with bolts or cap screws (Figure 6.20). These cylinders usually have a thick-wall tubular housing, providing great strength and heat resistance. Such cylinders are commonly used in the steel-making industry.

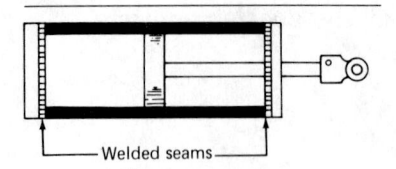

Figure 6.21 Mobile cylinder; caps and barrel welded together.

The mobile cylinder is generally constructed by a single machined casting or a two-piece permanent assembly with head cap and tubular housing welded together (Figure 6.21). These cylinders are less expensive to mass produce but are not repairable.

The standard cylinder identification code illustrated in Figure 6.22 facilitates the interchangeability of cylinder components. A coding system used by most manufacturers permits designers to obtain critical and necessary dimensions to facilitate circuit design planning.

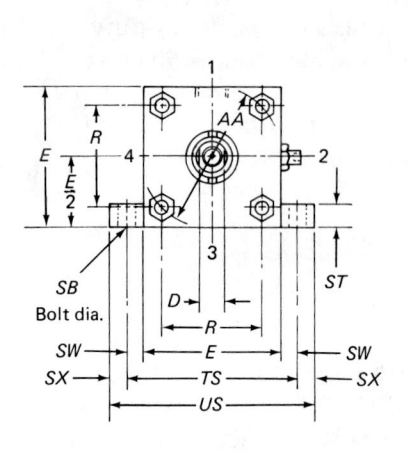

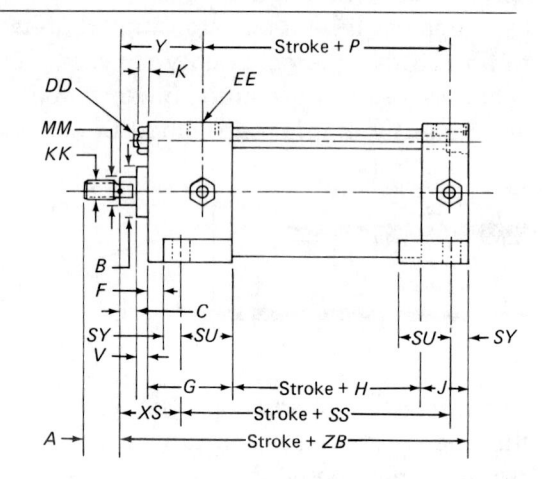

Figure 6.22 Standard cylinder identification code.

Mounting configurations

A cylinder can be attached to a load through mechanical linkages so as to multiply forces by sacrificing distance or to gain distance by reducing force. For example, a scissor-type lift utilizes the force capabilities of hydraulics to get considerable increase in distance while sacrificing force. Such machines can be seen at airports lifting cargo onto aircraft. The designer of a fluid power system decides what linkage to select based on the force, distance, geometry, space available, and motions desired for a particular application. Several cylinder leverage mountings are illustrated in Figure 6.23 to demonstrate the principle of mechanical advantage in raising a given load. While many applications require a straight lift or a straight push, many more complex systems involve other mechanical linkages to obtain rotary or positional advance motions. Cams, cogs, gear segments, and screw drives may be employed. The mathematical and geometrical solutions to sophisticated cylinder mountings are outside the scope of this text. However, such information is available in more advanced books.

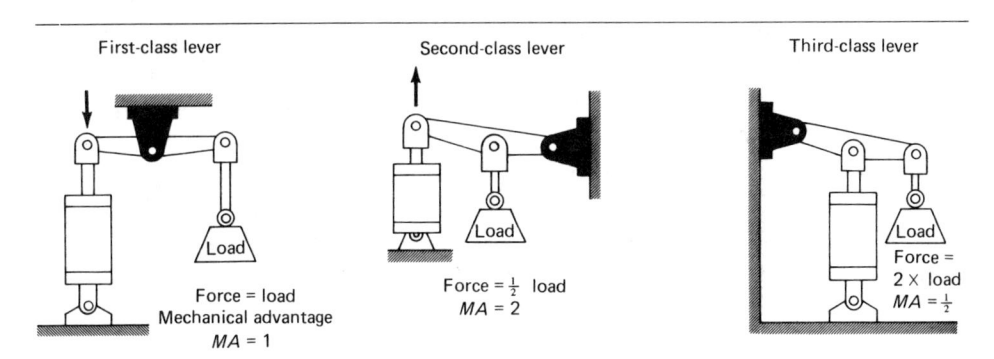

Figure 6.23 Cylinder application to levers.

First-class lever
Force = load
Mechanical advantage
MA = 1

Second-class lever
Force = ½ load
MA = 2

Third-class lever
Force = 2 × load
MA = ½

The types of mountings available on cylinders are numerous and varied (Figure 6.24). The most common are the rod, lug, flange, trunnion, and clevis mounting. Each of these has a variety of design features. Selection of a mounting style depends on a number of

Figure 6.24 Cylinder mounts.

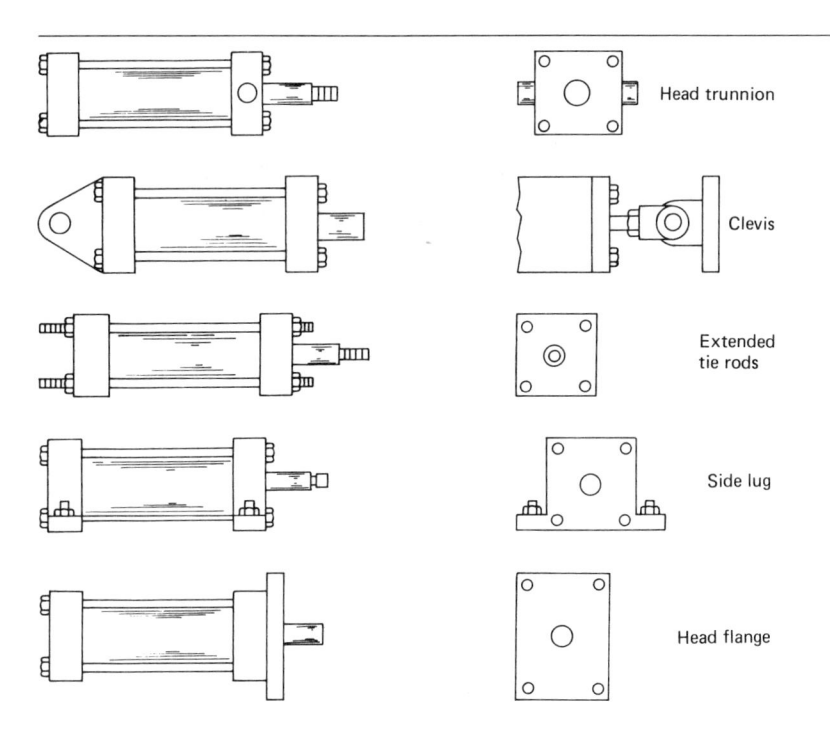

Head trunnion

Clevis

Extended tie rods

Side lug

Head flange

factors and machine-design principles. One of the most important factors in choosing cylinder mounting styles is whether the major force applied to the machine will result in tension or compression of the piston rod. Alignment, length of stroke, shock, acceleration, and other severe service conditions must also be considered in selecting mounting styles. Three mounting configurations will be discussed briefly in this section.

1 Fixed centerline mounts are the strongest, most rigid method of fastening a cylinder (Figure 6.25). They are used for applications where thrusts occur linearly along the centerline with the piston rod. Fixed centerline-mounted cylinders cannot tolerate misalignment because this would create compound stresses and cause binding and excessive friction of the piston rod. If there is no misalignment, the mounting bolts or tie rods are either in shear or tension with no compound stresses. For maximum strength, centerline lug-mounted

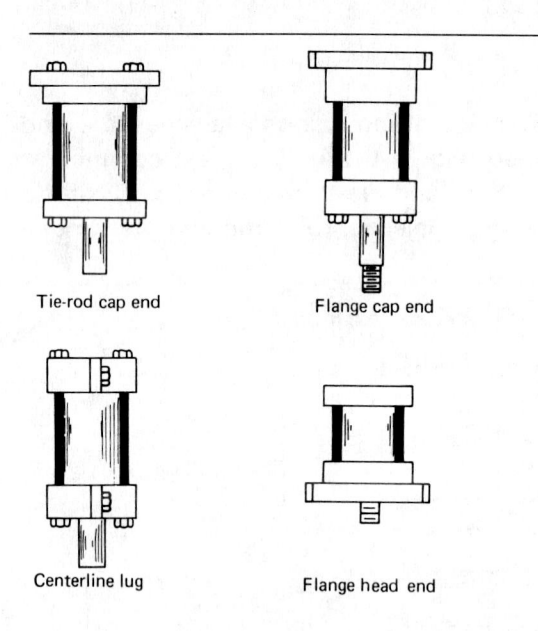

Figure 6.25 Fixed centerline mounts.

Tie-rod cap end

Flange cap end

Centerline lug

Flange head end

cylinders should have the mounting bolts placed in the cap end for thrust loads and in the head rod end for tension loads. If the centerline lug cylinder is located by means of dowel pins or keys, only one end should be pinned because the cylinder must expand when it is pressurized. Warping of the cylinder may occur due to shock loads and operating temperatures if diagonal corners are pinned. The choice of which end to pin depends on the directions of the major shock load.

2 Fixed noncenterline mounts are designed for applications where extremely large linear thrusts are encountered (Figure 6.26). Noncenterline mounts are the most often used type of fixed mount because of their convenience. Cylinders with fixed noncenterline mounts often have integral keys or pins to take up shear loads and provide accurate alignment, while only simple tension loads are exerted on mounting bolts. Keys or pins simplify installation and servicing.

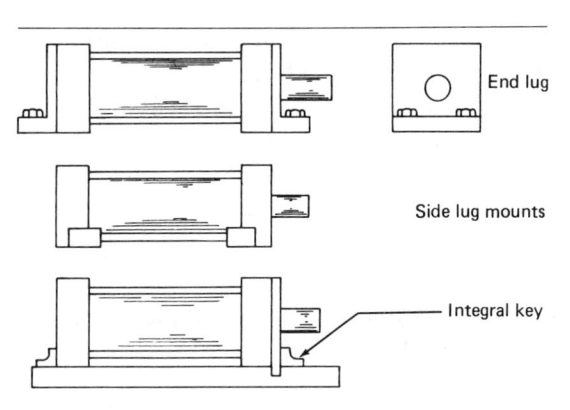

Figure 6.26 Fixed noncenterline mounts.

End lug

Side lug mounts

Integral key

3 Pivoted centerline mounts are used to accommodate thrusts occurring within a range of changing planes such as loads traveling in a curved path (Figure 6.27). Lifting the box of a dump truck is a typical example employing a pivoted centerline mount. Clevis, trunnion, and ball-joint mounts allow the thrust loads to be transferred along the cylinder centerlines.

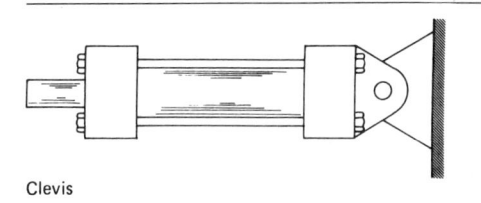

Figure 6.27 Pivoted centerline mounts.

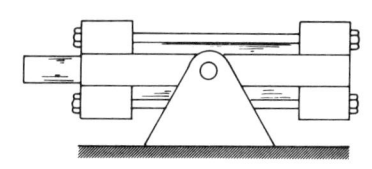

Clevis

Intermediate trunnion

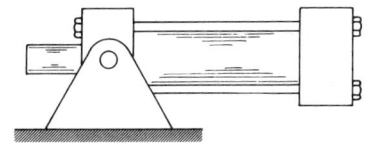

Head trunnion

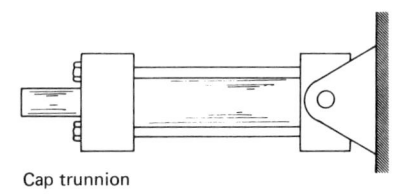

Cap trunnion

Proper alignment and secure fastening of cylinders are most critical to the performance life and efficiency of a cylinder. It is good practice to keep the ports of cylinders covered with plugs prior to their installation to prevent foreign matter from entering the cylinder. It is equally important that cylinders be handled carefully, particularly to avoid burrs on a high-quality piston rod finish.

Air cylinders

Air cylinders are very similar to hydraulic cylinders with regard to their basic operation. In a compressed air (pneumatic) cylinder, the air pressure is applied to one side of the piston, causing it to move, while the air is exhausted from the opposite side of the piston. Air cylinders are classified as single-acting or double-acting, depending on whether compressed air is applied to one or both sides of the piston. Cylinders are also classified as single-ended or double-ended, depending on whether the piston rod extends through one end or both ends of the cylinder.

In a single-acting, single-ended cylinder, the piston is extended by air pressure [Figure 6.28(a)]. The retraction stroke is accomplished by the mechanical force of a spring. A counterbalance weight could also accomplish the same results. In a double-acting air cylinder, the air pressure is applied to either side of the piston to move it, while the air is simultaneously exhausted back from the other side through a valve to atmosphere [Figure 6.28(b)].

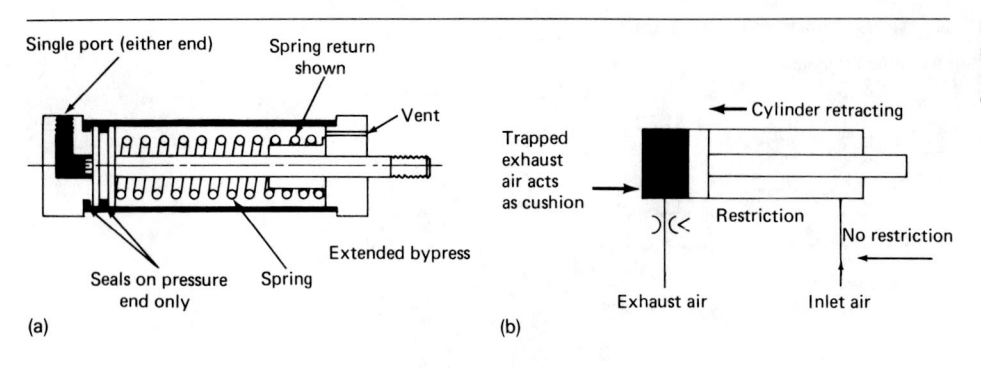

Figure 6.28 (a) Single-acting, single-ended air cylinder. (b) Double-acting, single-ended cylinder with cushion.

Cylinder cushions may be built into either or both ends of an air cylinder. Cushions eliminate the impact between the piston and the cylinder caps when the cylinder is in operation [Figure 6.29(a)]. Most

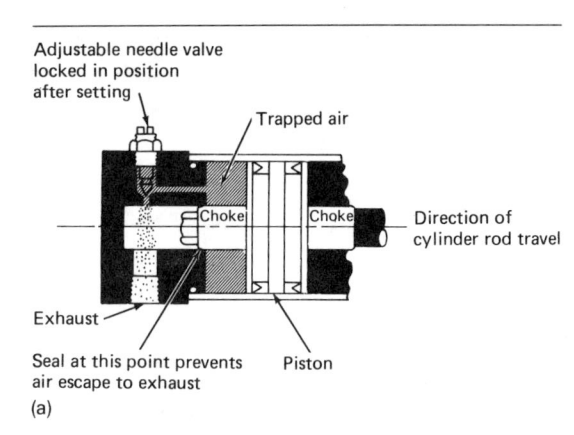

Figure 6.29 (a) Cylinder cushion regulates deceleration. (b) Internal construction parts of a double-acting air cylinder. [(b) Courtesy Allenair Corporation]

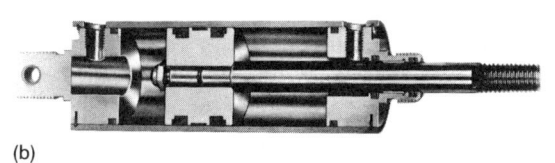

cushion designs trap air internally between the piston and the cap end and meter this exhausted air at a desired rate preset for conditions by means of an orifice and needle valve adjustment [Figure 6.29(b)].

Construction

Air cylinders are made in various shapes and sizes, and from a variety of materials. Because compressed air may vary only from 60 psi to 150 psi, air cylinders do not have to withstand such large pressures as hydraulic cylinders do. Air cylinders may be of die-cast construction as well as cylinder tube and cap designs. A number of materials are used for air cylinder tubes. Smaller air cylinders operating at lower pressures may be made from drawn brass or aluminum. These metals are not subject to corrosion as steel if moisture is not removed from the compressed air. Brass and aluminum are also better conductors of heat and help to dissipate heat in high-frequency cycle applications. Larger air cylinders have tubing made of castings of iron, aluminum, bronze, or steel. Adequate lubrication for such cylinders is more critical.

Air cylinders have many applications in a wide variety of industries, particularly automation. They can operate at high frequencies in reasonably high-temperature environments and provide for interfacing with electric, hydraulic, and mechanical controls. Air cylinders also have limitations—they are not practical for applications requiring accurate feeds or enormous forces.

Selection

Selection of a cylinder depends on numerous factors, most of which are unique to a particular application. There are, however, some basic considerations that apply to most situations.

Size The size of the cylinder is determined by the force required, the operating pressure, and the effective pressure area for each phase of the cycle.

Action The types of motion required for a specific application dictate whether a single-acting, double-acting, single- or double-ended cylinder should be used.

Mounting The mounting style must be selected according to equipment motion requirements and load strains and stresses.

Cushioning The cushion action of a cylinder is determined by the load, its speed, and cycle time, and the physical design of the cylinder.

Rod size The rod to piston ratio depends on the stability and rigidity, length of stroke, loads, cycle speeds, and other job specifications. Standard- and special-ratio rod to piston combinations are available.

Cylinder packing and seals The type of fluid must be compatible with the packing and seals. Tolerance of leakage, shock, temperature, and precision of operation dictate the nature and types of packing and seals.

Space limitation The machine design may provide physical limitations as to the size or mounting style of a cylinder.

Generally, designers of fluid power–assisted systems need to make a system analysis including the magnitude of the force, the direction of the force, the distance through which the force will act (stroke), the time in which the force will act, and the sequence of operations.

Fluid motors

A fluid motor, either hydraulic or pneumatic, may be used instead of a fluid cylinder to convert fluid energy into mechanical motion. Fluid motors are designed to provide rotary drives and may be connected directly or through gears and clutches to the actuated part.

Operation

A fluid motor works on the principle of consuming the fluid energy, put in motion by the pump, to drive the element attached to the shaft of the motor. In this manner, a motor is capable of converting fluid energy into a rotating force and motion. Work by the fluid motor is accomplished by the conversion of fluid pressure energy into torque on the shaft. The volume delivered by the pump determines shaft speeds.

A fluid motor and a pump produced by the same manufacturer may look alike in many respects and use interchangeable parts, but their function is opposite in terms of energy conversion (Figure 6.30).

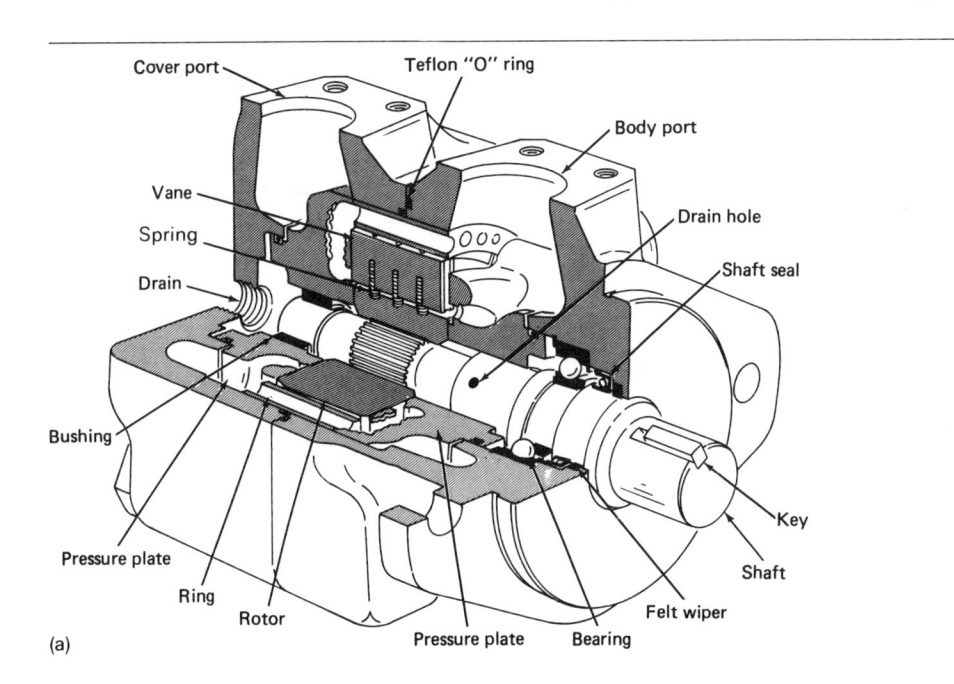

Figure 6.30 Comparison of pumps and motors. (a) Motor. (b) Pump.

Figure 6.30 (cont'd)

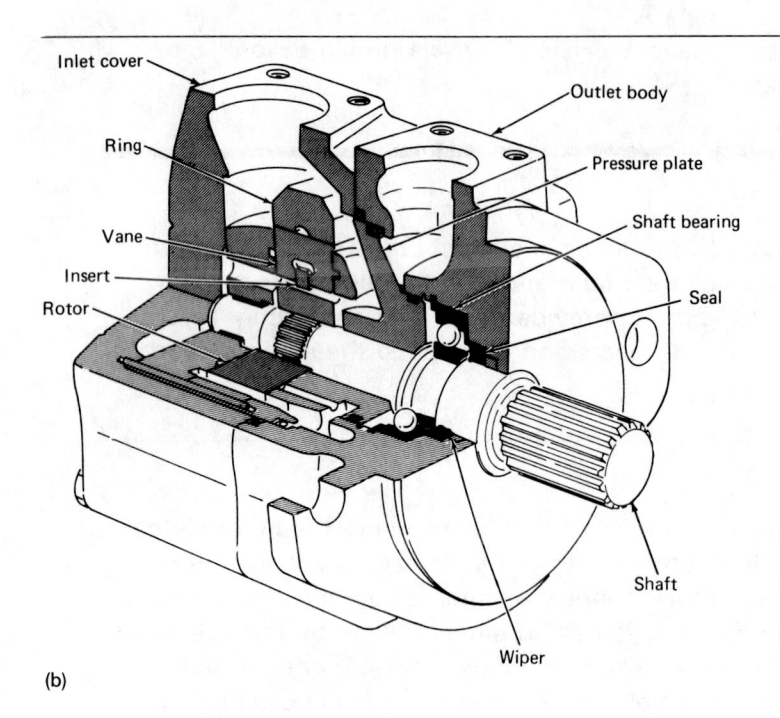

Fixed- and variable-displacement motors

There are two major classes of fluid motors, fixed displacement and variable displacement. Within the fixed-displacement class, there are gear, vane, and piston fluid motors. Piston designs are used in variable-displacement fluid motors.

The displacement of a hydraulic motor is designated in cubic inches per revolution of the motor shaft (Figure 6.31). This is the same as

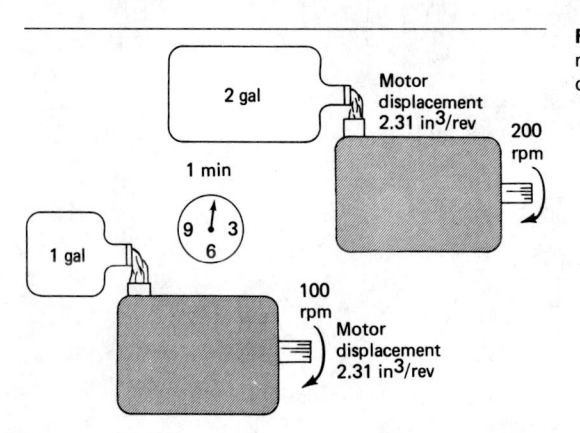

Figure 6.31 The speed of fluid motors is a function of motor displacement and pump delivery.

the rating used with hydraulic pumps. Fixed-displacement fluid motors displace a specific amount of fluid with each revolution. The speed of any fixed-displacement fluid motor depends on the displacement per revolution and the amount of fluid delivered to it by the pump or the pump and accumulator.

A variable-displacement piston fluid motor is constructed with an adjustment device that can vary the displacement per revolution. With such an adjustment, it is possible to control the speed of the fluid motor from zero to its maximum limit, with a constant delivery of oil from the pump.

Speed

If the displacement and the required speed of a fluid motor are known, it is possible to calculate the pump delivery requirements:

$$\text{gpm} = \frac{\text{speed (rpm)} \times \text{displacement (in}^3/\text{rev.)}}{231}$$

▶ **Example** A fluid motor with a displacement of 2.31 in³/rev., required to run at 1000 rpm, would require a pump delivery of

$$\text{gpm} = \frac{1000 \text{ rev/min} \times 2.31 \text{ in}^3/\text{rev.}}{231 \text{ in}^3/\text{gal}}$$

$$= 10 \text{ gpm}$$

If the displacement of the fluid motor and the pump delivery are known, it is possible to calculate the drive speed of the fluid motor.

$$\text{rpm} = \frac{\text{pump delivery (gpm)} \times 231}{\text{displacement (in}^3/\text{rev. of motor)}}$$

Given a pump with a delivery of 10 gpm and a motor with a 2.31 in³/rev. displacement, the resulting rpm would be

$$\text{rpm} = \frac{10 \text{ gal/min} \times 231 \text{ in}^3/\text{gal}}{2.31 \text{ in}^3/\text{rev.}}$$

$$= 1000 \text{ rpm}$$

From this example, it is apparent that increasing the displacement of a fluid motor decreases its speed; conversely, decreasing its displacement increases its speed. With the variable-displacement motor it is possible to regulate its speed. It is also possible to use a variable-displacement pump to regulate the gallons per minute delivered to the fluid motor as a means of regulating its speed.

Torque

The term *torque* is defined as the product of the force and the perpendicular distance from the axis to the line of action of the force. Motor torque is usually measured in pound-inches (Figure 6.32).

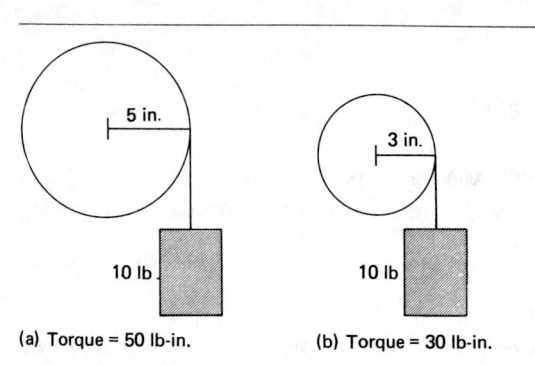

Figure 6.32 Torque is expressed in pound-inch units.

(a) Torque = 50 lb-in.

(b) Torque = 30 lb-in.

Fluid motors provide rotational power to perform work. Within a fluid motor, the turning force of its shaft is a measure of torque, which is equal to the force or load multiplied by the radius arm. This means that the torque generated by a fluid motor depends on the fluid pressure and the radial distance from the center of the motor shaft. In basic mechanics, you learned this relationship as follows:

$$T(\text{lb-in.}) = Fd$$

where

T = torque
F = force
d = distance of radius arm

From Pascal's law, you learned that $F = pA$; that is, force equals unit pressure times the exposed area. Substituting pA for F in the preceding equation gives the relationship $T = pAd$. If you know the area of Vane B, which is the unbalanced vane producing the torque, and the distance from the center of the attended vane to the center of the shaft, you can determine the torque produced (Figure 6.33).

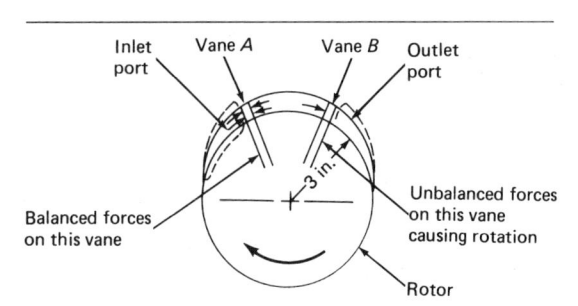

Figure 6.33 Pressure differences on motor vanes.

▶ **Example** A vane with an area of 1 in² that is 3 in. from the center of the shaft under 600 psi would generate a torque equivalent to

$$T = 600 \text{ psi} \times 1 \text{ in}^2 \times 3 \text{ in.} = 1800 \text{ lb-in.}$$

More commonly, the torque produced by a fluid motor is expressed by the equation:

$$\text{Torque (lb-in.)} = \frac{\text{pressure (psi)} \times \text{displacement (in}^3/\text{rev.)}}{2\pi}$$

▶ **Example** A gear motor has a displacement of 0.628 in³/rev. and operates under 600 psi. It would generate a torque of

$$T = \frac{600 \text{ psi} \times 0.628 \text{ in}^3/\text{rev.}}{6.28}$$
$$= \frac{600}{10} = 60 \text{ lb-in.}$$

According to this equation, it is evident that the torque increases whenever the pressure or the displacement increase. It is important to understand that as the displacement in a variable motor is increased, the motor speed is decreased in proportion to the gain in torque.

Fluid motors are gaining greater acceptance in wider applications because of their excellent control of acceleration, operating speeds, deceleration, smooth reversals, constant torque, and small bulk in relation to output horsepower.

The relationships between output horsepower, torque, and speed of a fluid motor can be expressed by the equation:

$$\text{Horsepower (hp)} = \frac{\text{torque (lb-in.)} \times \text{rpm} \times 2\pi}{33{,}000 \text{ ft-lb/min} \times 12 \text{ in./ft}}$$
$$= \frac{T \times \text{rpm}}{63{,}025}$$

T is expressed in pound-inches. If T is expressed in lb-ft, the equation will be

$$\text{hp} = \frac{T \times \text{rpm}}{5252}$$

▶ **Example** In the previous example, where the torque was determined to be 60 lb-in., how much horsepower would the gear-type motor produce if a 10-gpm pump were coupled to it?

$$\text{hp} = \frac{T \times \text{rpm}}{63,025}$$

First, calculate the rpm of the fluid motor:

$$\text{Speed (rpm)} = \frac{10 \text{ gpm} \times 231 \text{ in}^3/\text{gal}}{0.628 \text{ in}^3/\text{rev.}}$$

$$= \frac{2310}{0.628}$$

$$= 3678 \text{ rpm}$$

$$\text{hp} = \frac{T \times \text{rpm}}{63,025}$$

$$= \frac{60 \times 3678}{63,025}$$

$$= 3.50$$

Various combinations of pumps and motors are used for applications that require rotary force. Knowing the speed, torque, and output horsepower requirements, a designer can select the most economical and compatible combinations of pumps and motors to meet the needs of a specific application.

Gear motors generally have an overall efficiency of approximately 70 to 75%, as compared with 75 to 85% for vane motors and 85 to 95% for piston motors. Piston motors are capable of operating at higher maximum pressures and speeds.

As indicated earlier, rotary motors are generally rated in terms of displacement and torque. The displacement of a hydraulic motor refers to the theoretical amount of liquid necessary to force the motor through one complete cycle; it is expressed in cubic inches. Assuming there was no slippage, or leakage, the theoretical and actual displacement would be the same. The torque rating is theoretical and is stated in pound-inches per unit pressure. The power output of a motor in terms of speed and torque is limited by the power

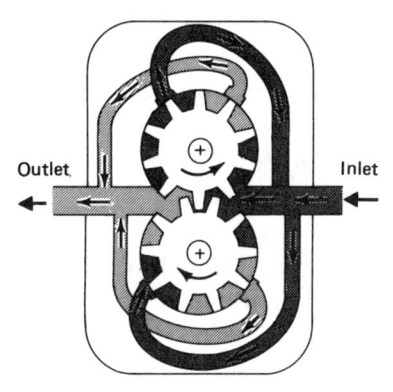

Figure 6.34 Balanced gear motor.

input in terms of rate of flow and maximum operating pressure. A motor's mechanical efficiency will influence the output torque for a given pressure.

Motors can be classified according to the type of internal element that is directly actuated by the flow. Rotary motors are classified as gear, vane, or piston.

Gear motors

A gear motor, like a gear pump, is by construction a fixed-displacement unit (Figure 6.34). The speed of rotation (rpm) depends on the volume of oil delivered to the motor. There are two types of gear motors that have wide application: external-gear and internal-gear (Figure 6.35).

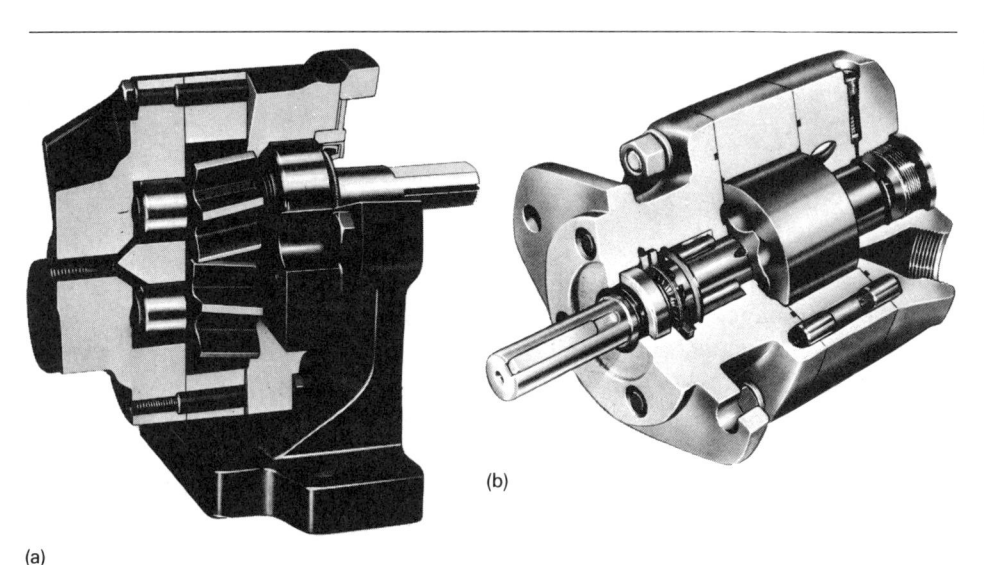

Figure 6.35 (a) Construction of a gear motor. (Courtesy Webster Electric Company) (b) Internal-gear or gerotor motor. (Courtesy Hydraulic Products)

Both gears are driven in a motor, but only one gear is connected to the output shaft. The operation is essentially the reverse of that of a pump. The fluid flows in either direction around the inside surface of the casing, forcing the gears to rotate. A gear motor must be hydraulically balanced if the close tolerances required in fluid motors are to be maintained throughout its life. Hydraulic balance can be achieved by having cored passages leading from the inlet and outlet ports to points diametrically opposite. Thus the hydraulic balancing on both sides of each gear eliminates the thrust against the bearings and on the opposite side of the gear, which would lead to uneven wear and eventual slippage.

External-spur gear motors are limited to peak operating pressures of approximately 1500 psi. They are available with a maximum capacity of 120–150 gpm, providing a top speed of 2400 rpm. Slippage losses are reasonably uniform for speeds in excess of 500 rpm, and fairly constant torque can be expected. Systems that require a hydraulic motor to start under load should incorporate a stall-torque factor of the motor during the design calculations. For gear motors, only 70 to 80% of the maximum torque rating can be expected to be available if the motor must start under load or operate at speeds below 500 rpm.

Internal-gear motors will generally operate at higher pressure, higher speed, and larger displacement than external-gear motors. Gear motors should be operated within the limits of the manufacturer's specifications for long and efficient performance.

Vane motors

A rotary-vane motor is designed so that the rotor and vanes are hydraulically balanced in much the same manner as a gear motor is, with the two inlet ports and two outlet ports diametrically opposed (Figure 6.36). Since no centrifugal force exists until the rotor begins to rotate, the vanes must have some means other than centrifugal

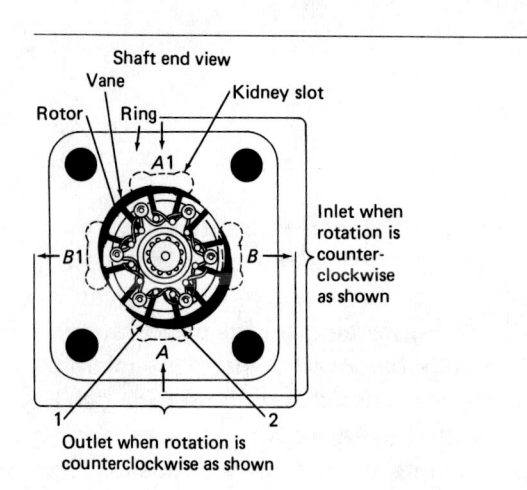

Figure 6.36 Rotary-vane motor. (Courtesy Vickers Division/Sperry Rand)

force to hold them against the cam ring. Mechanical springs are used in some motor designs to hold the vanes against the cam ring at low initial speeds. Other designs use pressure-loaded vanes.

A high performance by any hydraulic motor depends on the seal between its inlet and outlet sides. Internal leakage between the inlet and outlet, usually termed "slippage," reduces motor efficiency. The two critical seals are: (1) the points of contact between the vanes and the cam ring; and (2) the contacting surfaces between the pressure plates and the ring, rotor, and vanes. Most reversible hydraulic motors require external drain lines to conduct the leakage oil back to the reservoir (Figure 6.37). This drain line prevents the chamber from becoming pressurized, thus damaging the shaft seal. As was indicated earlier in the chapter, vane motors are fixed-displacement actuators. They are available in various sizes and capacities. The

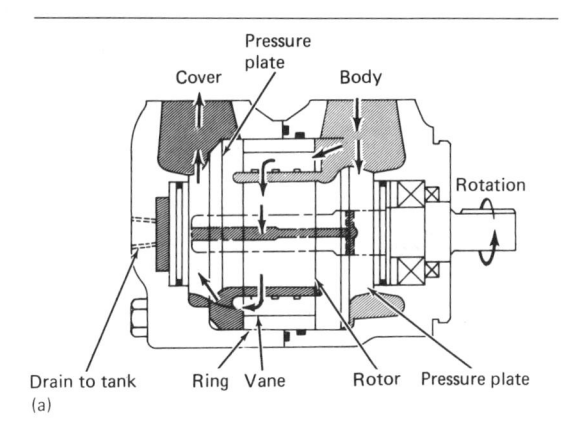

Figure 6.37 (a) Major parts of a vane motor. (Courtesy Vickers Division/Sperry Rand). (b) Rotary-vane motor. (Courtesy Webster Electric Company)

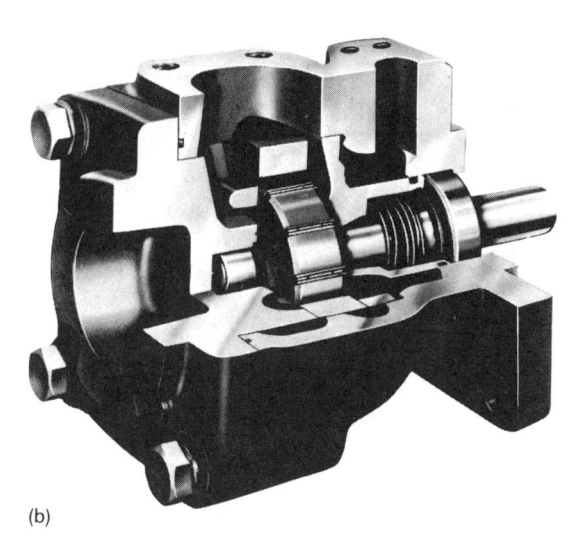

torque and horsepower capabilities depend on the delivery and fluid pressure supplied to them. Vane motors are available to operate at maximum pressures up to 2500 psi. Some designs will operate at maximum speeds of 4000 rpm and operate with maximum delivery of 250 gpm. Most motors run at values considerably less than these, particularly under continuous operating conditions. The stall torque of vane motors will generally be 80 to 85% of maximum running torque. Vane motors have an overall efficiency of 80 to 85%.

Piston motors

Rotary-piston motors can be designed to perform as either fixed- or variable-displacement units. The two types of rotary-piston fluid motors are axial and radial. Radial- and axial-piston motors are most often used in conjunction with radial- or axial-piston pumps as variable speed reducers or hydraulic transmissions.

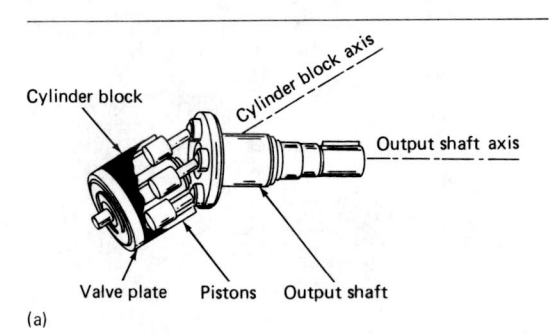

Figure 6.38 (a) Bent-axis axial-piston motor. (b) Inline, axial-piston hydraulic motor. [(b) Courtesy Webster Electric Company]

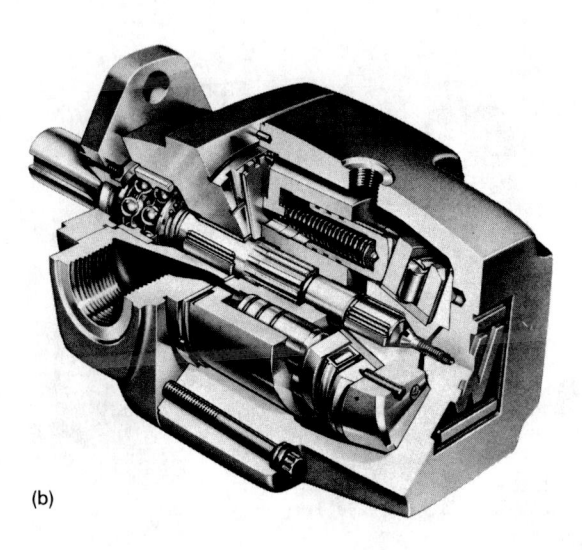

The *axial-piston motor* shown in Figure 6.38 operates on the principle that the fluid entering the inlet forces the piston outward, thus causing the cylinder block to rotate. Each sequential piston imparts a tangential force to the cylinder barrel while under pressure, causing it to continue to rotate. The rate at which the fluid is pumped to the motor for a given displacement determines its speed.

Piston motors are able to operate at speeds from 50 rpm to as high as 5000 rpm with stable torque output. Axial motors are known for their low inertia; they accelerate quickly and have excellent reversal capability. Axial-piston motors can be constant- or variable-displacement units. The fixed-displacement axial motor has a stationary cam plate; the variable-displacement unit must have some means of varying the cam angle to vary the piston stroke length.

The *radial-piston motor* has a cylinder barrel with an attached output shaft to transmit the force imparted to the pistons (Figure 6.39). The cylinder barrel has a number of radial bores with precision-fitted pistons. In a variable-displacement unit, the thrust ring is adjustable, while in a fixed-displacement unit the thrust ring is stationary.

As with all piston motors, when oil enters the cylinder bore, the pistons are forced against the thrust ring, imparting a tangential force to the cylinder barrel and shaft and causing it to rotate. Each piston is pushed inward by the thrust ring once it reaches the outlet port, thus pushing the fluid back toward the reservoir.

Radial-piston motors, like axial-piston motors, are equipped with external drains to avoid pressurizing the cylinder chamber. They usually have an odd number of pistons—five, seven, nine, or more, depending on the size of the unit. Radial-piston motors are high-performance units that run at 85 to 95% overall efficiency (Figure 6.40). Piston motors can be used for power output as high as 3000 hp or more and reach peak speeds of 12,000 to 14,000 rpm. Larger piston motors will operate with maximum delivery rates of 450 gpm.

Rotary piston motors are compact units, provide high torque and high acceleration, and have generally excellent life expectancy.

Hydraulic pumps, rotary motors, and valves can be combined in a number of ways to provide adjustable speed drives for almost unlimited motions and range of applications.

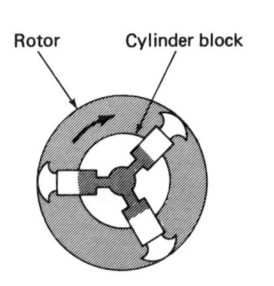

Figure 6.39 Radial-piston hydraulic motor.

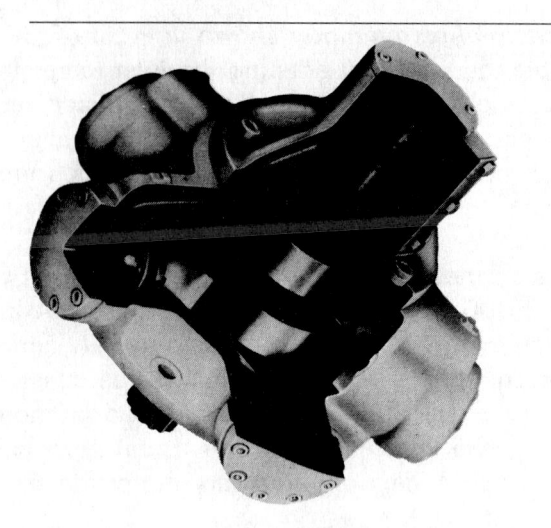

Figure 6.40 Radial-piston hydraulic motor. (Courtesy Double A Products)

Air motors

Air motors have a number of desirable characteristics, particularly for portable power tools. The advantages offered by air motors include low weight per horsepower output, compact size, low operating air pressures, indifference to overload and stall conditions, and unlimited reversals; they are cool running, relatively clean to operate, unaffected by dirty or explosive atmospheres, and shockproof.

Air motors start and stop instantly and provide infinitely variable control of torque and speed. The speed decreases as the load increases, until the stall condition is reached. When the air motor is nearly stalled, it will continue to produce high torque without damage to the motor. Air motors also have disadvantages. The initial cost is higher than for equivalent electric motors, and they have lower efficiency and higher maintenance costs.

Applications of air motors are found in powering conveyor belts, transfer systems, hoists, and mixers, driving hydraulic pumps on aircraft, and for many types of portable air tools and wrenches.

Vane motors

Rotary air motors are constructed with a variety of driving elements. The most popular element designs are rotary-vane and piston air motors. Other designs include the turbine and lobed-rotor. The turbine, perhaps the most sophisticated type of air motor, has extremely low starting torque for use in high-speed applications. A dentist might use such a motor for grinding or polishing teeth. Aircraft jet engines are also started by turbine air motors.

Rotary-vane air motors are similar in design to rotary-vane hydraulic motors (Figure 6.41). Such a motor contains a cylindrical rotor with sliding vanes positioned eccentrically in a cylindrical housing. Compressed air is admitted through the inlet port and fills the cavity between vanes, housing, and rotor. The entering compressed air

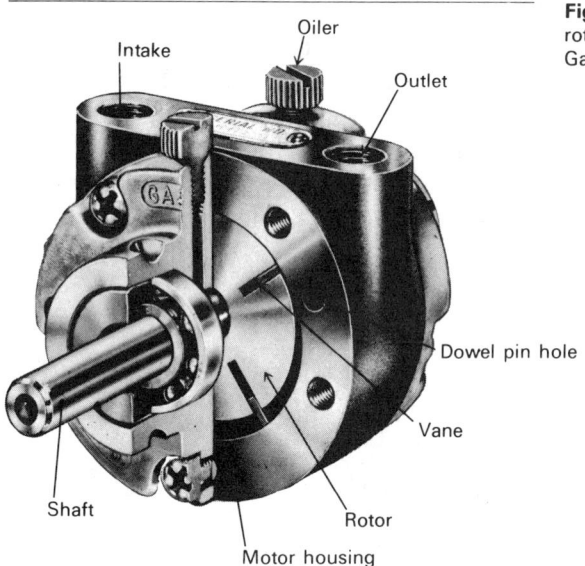

Figure 6.41 Cutaway section of a rotary-vane air motor. (Courtesy Gast Manufacturing)

acting on the vanes causes the rotor to turn. The torque T developed at the shaft is proportional to the product of the effective vane area A, pressure difference p, and torque radius d_t, measured from the center of the vane to the center of the rotor shaft: $T = Apd_t$. The horsepower of air motors can be relatively high since these units operate at high speeds. Horsepower is expressed by this equation:

$$hp = \frac{T \text{ (lb-ft)} \times \text{rpm}}{5252}$$

Positive air motors operate at speeds of 100 to 25,000 rpm and above, depending on the diameter of the rotor. Rotary-vane air motors are available in sizes from a $\frac{1}{8}$-hp unit up to a 25-hp unit. Pressure regulating and metering of air flow provide an infinitely variable torque and speed to air motors.

Air motors have fewer vanes than hydraulic motors, so that the pressure differential between vane enclosures is greater. The efficiency of rotary-vane air motors is generally low, in some cases as low as 25%. Internal frictional losses increase particularly with high speeds. The torque is also low compared with piston air motors. Rotary-vane air motors generally require lubrication in the air stream for optimum sliding efficiency on the housing.

Piston motors

Piston air motors are manufactured in axial- or radial-piston configurations. The operation of the *axial-piston* air motor is similar to that of the axial-piston hydraulic motor (Figure 6.42). Essentially, as the pistons reciprocate in sequence, they actuate the wobble plate, which in turn imparts a rotary motion to the output shaft. The compressed air energy is expended by moving a piston and this energy is converted to a mechanical force by the rotation of the output shaft.

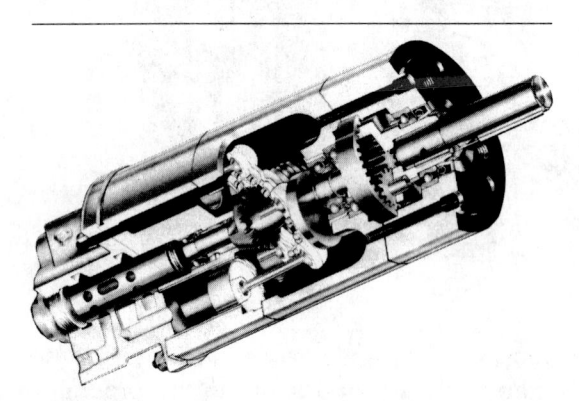

Figure 6.42 Rotary axial-piston air motor. (Courtesy Gardner-Denver Company)

Axial-piston air motors may have five or more pistons, generally of small size, using built-in lubrication, and are available in sizes from $\frac{1}{8}$ to $3\frac{1}{2}$ hp, with speeds between 220 and 4500 rpm. The horsepower developed by these air motors is dependent on the available compressed air pressure, the area of the pistons, the number of pistons, the stroke of the pistons, and the speed of the motor. Axial-piston air motors are compact, operate smoothly even at low speeds, provide high torque, and reach full speed in a fraction of a second.

The *radial-piston air motor* is a low-speed, high-torque output device with five or more reciprocating pistons (Figure 6.43). Compressed air is supplied to each piston by a timing valve that causes the piston

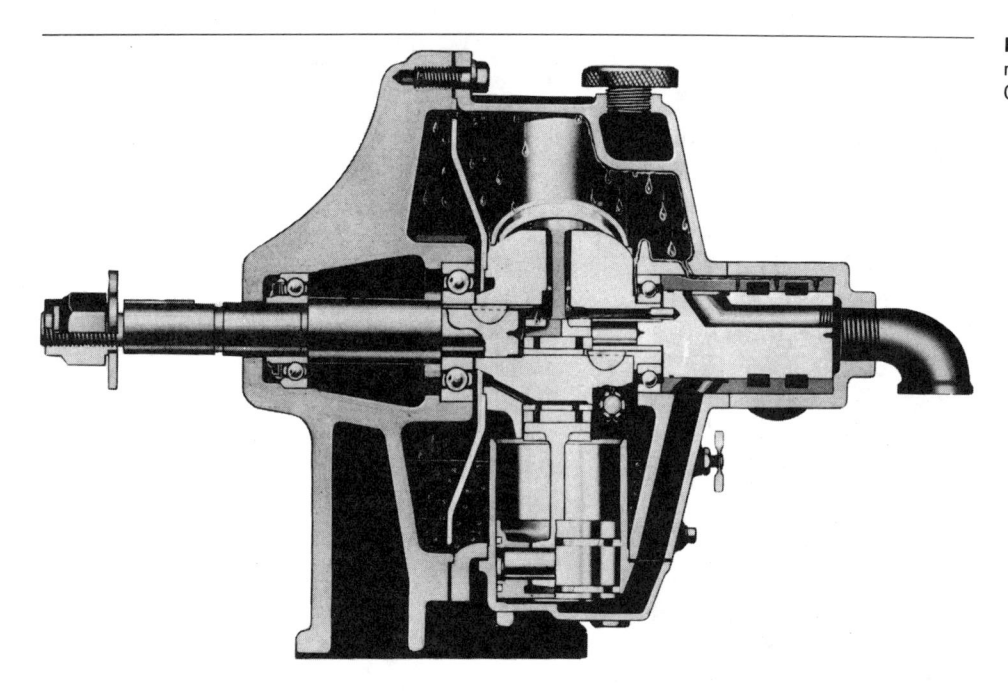

Figure 6.43 Radial-piston air motor. (Courtesy Gardner-Denver Company)

to move and turn the crankshaft through the connecting rod, which ultimately results in the mechanical rotation of the output shaft. The radial reciprocating-piston air motor is limited in speed because of the inertia of moving mechanical parts and operates at speeds of 3000 rpm as compared with 30,000 rpm for vane air motors. The power output of radial-piston air motors is also dependent on the inlet compressed air pressure, the number of pistons, the area of the pistons, the stroke, and the speed of the motor. More power can be obtained by selecting a motor with a higher speed or with a larger piston diameter, more pistons, or increased stroke. These motors are available in sizes from $\frac{1}{8}$ to 25 hp. Radial-piston air motors are particularly suited for low-speed operation where high starting torque is necessary.

The horsepower of an air motor is relative to the speed and to the inlet air pressure, as illustrated in Figure 6.44.

Speed

The speed of air motors is controlled by the volume and pressure of air admitted into the motor and can be regulated from minimum to maximum capacity without damaging the motor. These units are

208 Energy output devices

Horsepower versus rpm
Horsepower of an air motor is relative to rpm and to air line pressure.

Air Consumption versus rpm
Air consumption increases as speed and air pressure is increased.

Figure 6.44 Performance curves of air motors. (Courtesy Gast Manufacturing)

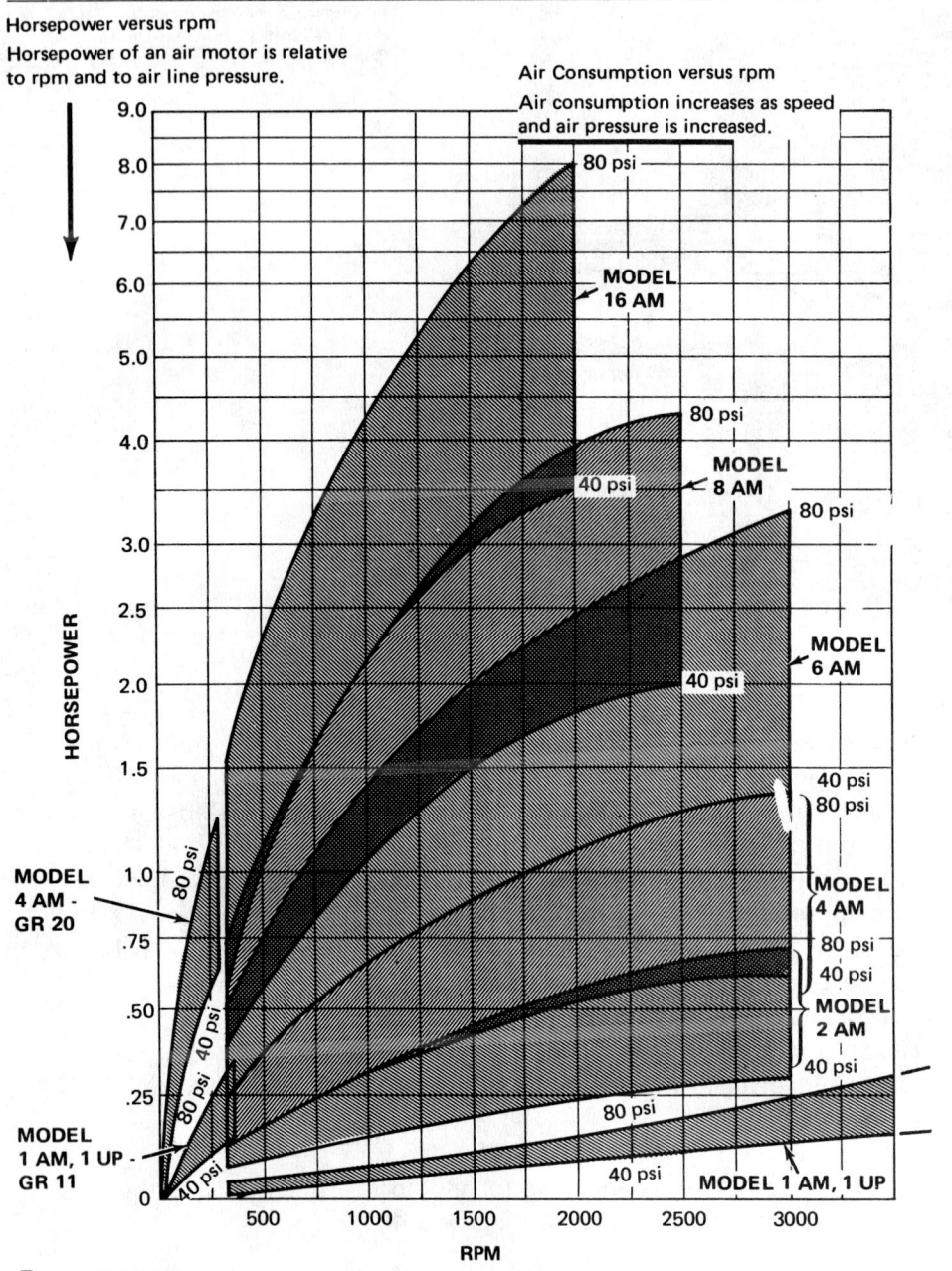

Torque versus rpm
An air motor slows down when load increases . . . at the same time its torque increases to a point where it matches the load. It will continue to provide increased torque all the way to stall condition. As the load is reduced, an air motor will increase speed and the torque will decrease to match the reduced load.

capable of stopping and starting with quick response without heating. For maintaining low noise level, silencers may be placed in control valves exhausting spent air from an air actuator (Figure 6.45).

Figure 6.45 Sintered-bronze air silencers. (Courtesy Scovill Fluid Power)

Limited-rotation motors

A limited-rotation motor is a device that rotates a shaft through a fixed arc, using a source of fluid under pressure to power the unit. The actuator's output torque varies directly with the fluid pressure applied to it. Limited-rotation motors can be used to agitate, feed, roll over, index, lock, transfer, or clamp (Figure 6.46).

Actuator designs

There are four basic design types of limited-rotation motors. One has one or more vanes attached to the drive shaft. The vanes are sealed in a cylindrical chamber and oscillate between stops integral with the casing. Rotation is caused by a differential pressure being applied across the vanes. These units are generally limited to 280° of rotation.

Another design type includes a helix machined in the drive shaft that meshes with a matching helix in a piston. Guide rods through the piston prevent it from rotating. Fluid pressure acting on either side of the piston causes it to oscillate the shaft. This type of unit may rotate beyond 180°.

The piston-chain actuator consists of two cylinders connected in

Schematic Applications

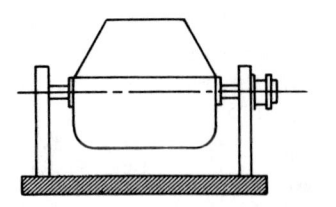

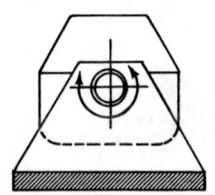

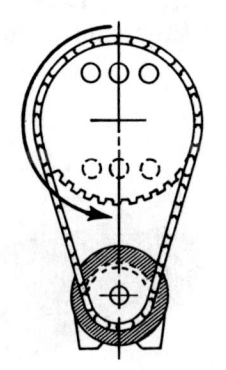

Dumping or tumbling

Fixture positioning or indexing

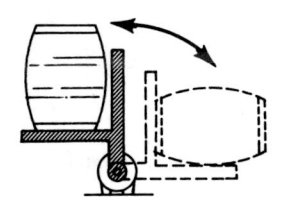

Roll over or upending

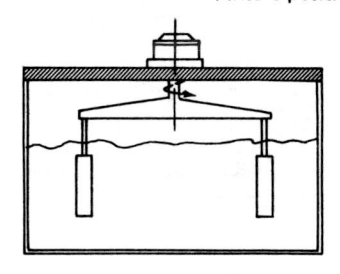

Mixing or agitation (oscillating action)

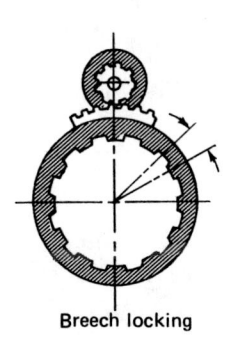

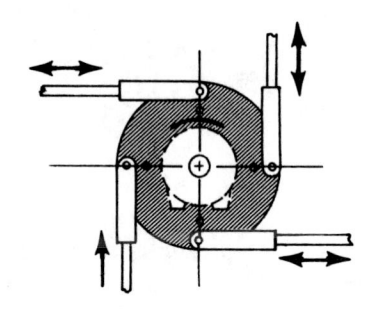

Breech locking

Trammel plate lock

Figure 6.46 Applications of limited-rotation motors.

parallel (Figure 6.47). A larger, primary cylinder provides the driving force while a smaller, secondary cylinder acts as a seal. Both pistons are fastened to a drive chain that rotates the shaft.

The fourth type of actuator uses gear racks to rotate a pinion attached to the drive shaft. The gear racks connect to the pistons, which extend and retract to rotate the drive shaft through a given arc.

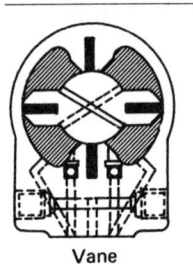

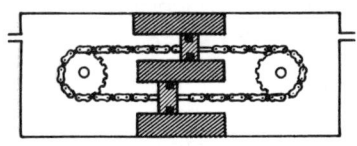

Figure 6.47 Types of limited-rotation motors.

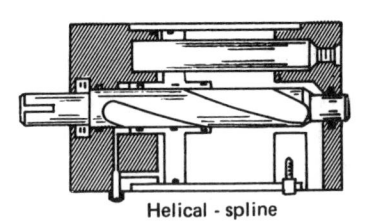

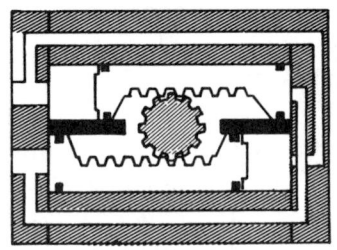

Vane

Piston - chain

Helical - spline

Rack and pinion

Construction

Limited-rotation motors generally use fixed or adjustable cushions to decelerate the load at the end of the stroke. Motion applications are either linear or rotary. Cylinders are used primarily to provide linear motions, while continuous rotation motors are used to provide rotary motions.

Applications requiring partial rotation generally call for a limited-rotation actuator. It is also possible to use other mechanical devices in combination with fluid power actuators to provide specific motions for given applications.

Summary

Energy output devices include cylinders and motors. Cylinders convert fluid energy into linear mechanical force and motion while hydraulic or pneumatic motors convert fluid energy into rotary mechanical force and motion.

Construction, as regards action (single acting or double acting), body styles, and cushioning, is an important factor in the selection of a cylinder. Selection factors include size, action, mounting, cushioning, rod size, cylinder packing, and space limitation to perform a specific job in a given environment. The selection of cylinders is generally determined by the job performance requirements.

Since cylinders have precision-machined and polished metal surfaces, it is important to use clean fluids and mechanical care in maintaining an efficiently operating cylinder. Hydraulic cylinders are made of steel because of the higher operating pressures while pneumatic cylinders are made of brass because of lower operating pressures and lower lubricity tolerance.

A fluid motor, either hydraulic or pneumatic, converts fluid energy into rotary mechanical motion. Such devices can be connected directly or through gears and clutches to a machine part to be actuated. A fluid motor works on the principle of consuming the fluid energy, put in motion by a compressor or a pump, to drive the element attached or geared to the shaft of the fluid motor. Just as in pump or compressor designs, fluid motors may be rotary or reciprocating. They can also be either fixed-displacement or variable-displacement devices. The pump or compressor delivery and the size of the fluid motor determine its speed (rpm). Fluid motors are gaining greater utilization in more applications because of their constant torque, control of acceleration and operating speeds, and small bulk in relation to output horsepower. Since fluid motors provide rotational motions their capacity is measured in pound-inches.

Review questions and problems

1 Define the term "actuator."
2 Discuss the operational difference between single-acting cylinders and double-acting cylinders.
3 Explain what is meant by a differential cylinder.
4 How is the effective area for the rod end determined when calculating the force or speed of a given actuator?
5 Why does the rod retract faster than it extends for a given delivery per minute?
6 What kind of cylinder can be used to produce equal forces and velocity in both directions?
7 What is meant by a cushioned-stroke hydraulic cylinder?
8 Name three body styles of cylinders.

9 List five factors that should be considered in selecting a cylinder.
10 What is the essential difference in function between a hydraulic pump and a hydraulic motor?
11 Given a fluid motor with a displacement of 2.31 in^3/rev. and a required rpm of 1000, determine the gpm of the pump.
12 Calculate the velocity of a piston with a cap-end area of 5 in^2 and a rod-end area of 4 in^2, using a 10-gpm delivery pump. Show your calculations for (a) the extension stroke, (b) the retraction stroke.
13 Write the equation for determining the torque of a hydraulic motor.
14 Calculate the torque of a hydraulic motor with a displacement of 2.31 in^3/rev. operating at 1000 psi.

Selected readings

- Basal, P. R., Jr. (ed.) *Mobile Hydraulics Manual M-2990-S.* Sperry Rand Corporation, Troy, Michigan, 1967.
- Bohn, Ralph C., and Angus J. MacDonald. *Power: Mechanics of Energy Control.* McKnight & McKnight, Bloomington, Indiana, 1970.
- Fitch, Ernest C., Jr. *Fluid Power and Control Systems.* McGraw-Hill, New York, 1966.
- Glenn, Ronald E., and James E. Blinn. *Mobile Hydraulic Testing.* American Technical Society, Chicago, 1970.
- McNickle, L. S., Jr. *Simplified Hydraulics.* McGraw-Hill, New York, 1966.
- Oster, Jon. *Basic Applied Fluid Power: Hydraulics.* McGraw-Hill, New York, 1969.
- Pease, Dudley A. *Basic Fluid Power.* Prentice-Hall, Englewood Cliffs, New Jersey, 1967.
- Stewart, Harry L., and John M. Storer. *Fluid Power.* Howard W Sams, Indianapolis, Indiana, 1968.
- Yeaple, Franklin D. *Hydraulic and Pneumatic Power and Control.* McGraw-Hill, New York, 1966.

Chapter 7
Fluid power circuits and systems

The material in this chapter is intended to illustrate typical circuits designed to accomplish specified control functions or to implement desired applications. Sample flow-control systems include the basic meter-in, meter-out, and bleed-off open-center circuits; flow-dividing and flow-integrating circuits; feed circuits; decelerating circuits; and synchronizing of motors.

Virtually all fluid power circuits require directional control in addition to the other two energy-control functions. Some of the sample circuits discussed in this chapter include special directional-control valve designs that allow for unusual or combined control functions. This chapter is intended to provide the student with an insight into circuit types and their implications for system performance.

Key terms

- **Closed-center circuit** An open-loop system in which the pump or accumulator source is blocked by the directional-control valve when it is in its neutral (centered) position.
- **Closed-loop system** A system that uses feedback to produce a self-adjusting or self-regulating system.
- **Counterbalance** Term applied to a control or circuit used to hold a load of constant reaction, such as a gravity load (weight), in place by maintaining a constant back pressure against the actuator or motor motion. A positive pressure must be applied to overcome a counterbalance control.
- **Feedback** The technique of using a transducer (form of instrument) to generate a signal proportional to, or representative of, the state of the output variable from the system, comparing this (feedback) signal to a

Portions of the following chapter appeared in Russell W. Henke, *Introduction to Fluid Power Circuits and Systems,* Addison-Wesley Publishing Company, Reading, Massachusetts, 1970; and in Russell W. Henke, *Introduction to Fluid Mechanics,* Addison-Wesley Publishing Company, Reading, Massachusetts, 1966. Prior to those publications, portions have also appeared in various articles in the magazines *Tooling and Production, Machine Design,* and *Hydraulics and Pneumatics.*

command signal, and adjusting the system to make the output conform as closely as possible to the desired level.
- **Open-center circuit** An open-loop system in which the pump is unloaded (bypassed back to the tank) by the directional-control valve when it is in its neutral (centered) position.
- **Open-loop system** A system in which feedback is *not* utilized to make the system self-adjusting or self-regulating.
- **Reducing** Term applied to a control or circuit designed to provide fluid from a primary circuit to a secondary circuit at a pressure level below that in the primary circuit.
- **Sequencing** Term applied to a control or circuit that uses pressure level as a signal to initiate switching of fluid from a primary to a secondary circuit branch.
- **Unloading** Term applied to a control or circuit designed to reduce pump pressure to a minimum when bypassing to the tank during stand-by periods in the work cycle.

A circuit represents the synthesis of a design requirement. Basically, these requirements fall into the three principal functions—directional control, pressure control, and flow control. Most circuits perform combinations of two or more of these functions. The sample circuits discussed are pressure-control circuits for limiting maximum system pressure, sequencing circuits, pressure reducing, load counterbalancing, unloading circuits, accumulator applications, fail-safe circuits, 2- and 3-pressure systems.

Open- and closed-loop circuits

An *open-loop circuit* or system is one in which feedback is not employed. The performance characteristics of the circuit are determined by the characteristics of the individual components used and their interaction in the circuit. Most so-called industrial circuits are of this classification (Figure 7.1). A *closed-loop system* is one in which feedback *is* employed.

Feedback is the technique of sampling the state of the output from the system, generating a signal proportional to this output, and comparing it with an input or command signal. If there is a difference between the command signal and the feedback signal, action is taken automatically to correct the output so that it matches the require-

Figure 7.1 Open-loop circuit.

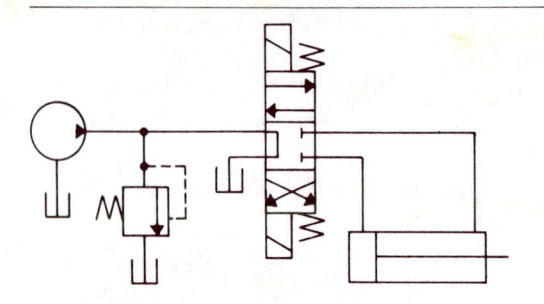

ments of the command. A servo system is a feedback system in which the output is a mechanical position (Figure 7.2).

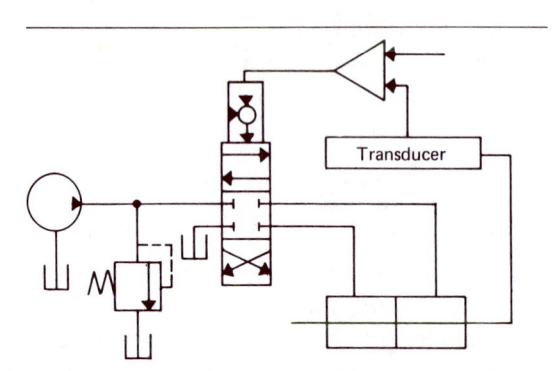

Figure 7.2 Closed-loop system.

Open-loop circuit classification

There are many ways to classify open-loop circuits: according to the *function* that they perform, according to the *method* by which they achieve control, by *system type,* or by *application.* All four circuit classifications are in common use today.

Function

Functional classifications are related to the basic areas of control:

Flow control The purpose is to control the energy transfer *rate* by controlling the flow rate in a circuit or a branch of a circuit.

Pressure control The purpose is either (1) to control energy transfer by controlling pressure, or (2) to use a pressure level as a signal to initiate a secondary action.

Directional control The purpose is control of the distribution of energy in a fluid power circuit.

Control

Control can be achieved in a fluid power circuit by one of three fundamental methods:

- **Valve control,** in which valving is applied to give the desired means of control.
- **Pump control,** wherein the pump itself, almost of necessity a variable-volume pump, provides control.
- **Actuator control,** in which the actuator, most frequently a rotary motor, is varied to provide control.

System

There are two basic types of open-loop circuits, open center and closed center. A typical open-center circuit was shown in Figure 7.1. In most *open-center circuits,* a directional-control valve unloads the pump, bypassing fluid to the tank when the valve is in the centered or neutral position, and a fixed-displacement pump is most commonly used. Energy transfer starts from a low level—essentially zero—when the valve is in neutral and builds up as the valve is shifted, so that the fluid stream is directed to the actuator and against the load resistance through it. Internal leakage is minimal when the valve is centered, unless the actuator is supporting a load in an elevated position. In general, an open-center circuit will be the minimum cost circuit, if it can meet performance requirements.

Closed-center circuits are characterized by the following features. All ports are blocked when the directional-control valve is in its centered, or neutral, position. When a fixed-displacement pump is used, an accumulator is also generally used, and an unloading valve is required. Energy transfer starts from a high level (the maximum pressure setting of the system) and is available to the actuator as rapidly as the valve can be shifted. Internal leakage is more likely to be a problem since the valve is holding against full system pressure at all times. A second version of the closed-center circuit is shown in Figure 7.3. This circuit uses a pressure-compensated, variable-volume pump in place of the fixed-displacement pump, accumulator, and unloading valve.

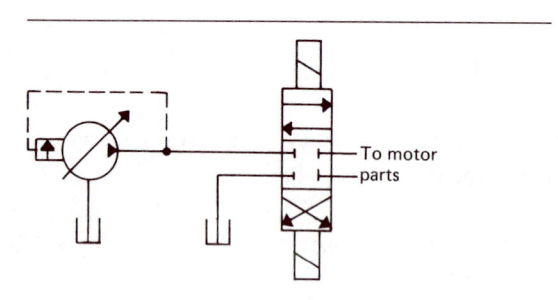

Figure 7.3 Variation of closed-center circuit.

Open-loop circuit design

An open-loop circuit was characterized at the beginning of this chapter as one in which there is no feedback. Thus the degree of control exercised by the circuit on the transfer of energy from the input to the output depends on the characteristics of the components used and their interaction in the circuit.

Criteria for selecting circuits

Before an intelligent approach can be made to the design of fluid power circuits, an analytical technique must be developed. A design check list for fluid power circuits might well include the following:

- definition of the **job** to be done,
- determination of the **time** per work cycle,
- determination of the **flow-rate** pattern over the work cycle,
- determination of the **pressure** pattern during the work cycle, and
- determination of the **horsepower demand** pattern.

Circuit load requirements

The design of a circuit begins with defining the job to be done. Every fluid power circuit involves the transfer of energy to an output device with the intent of accomplishing some useful work. This implies that a load is applied to the actuator. Before you can attempt to design a circuit the concept of *load* must be completely understood. A system of classification defines the three types of load as:

Resistive load (sometimes called a *positive load*) The load reaction on the actuator opposes the motion of the actuator (Figure 7.4). In a *constant resistive* load, the load magnitude does not vary. In a *variable resistive* load, the magnitude does vary.

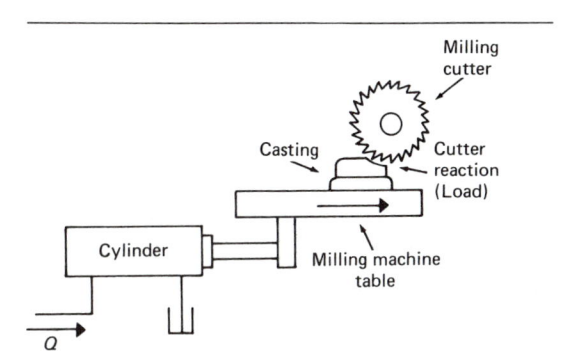

Figure 7.4 Resistive load.

Overrunning load (sometimes called a *negative load*) The load reaction on the actuator has the same direction as the motion of the actuator—the reaction motion of the actuator (Figure 7.5). In a *constant overrunning* load, magnitude does not vary. In a *variable overrunning* load, load magnitude does vary.

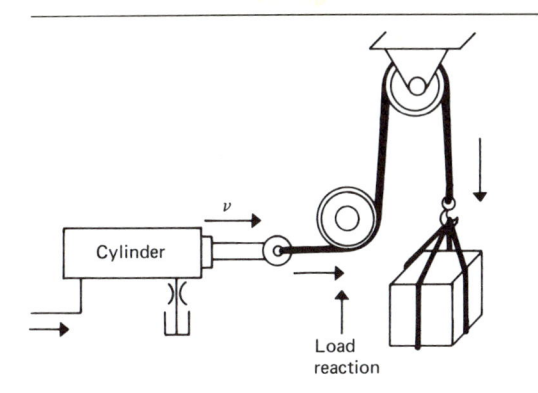

Figure 7.5 Overrunning load.

Inertia load The load reaction on the actuator is predominantly characterized by Newton's second law of motion, $F = M \cdot a$ or $T = J \cdot w$ (Figure 7.6).

These three types of loads, or combinations thereof, define all the

Figure 7.6 Inertia load.

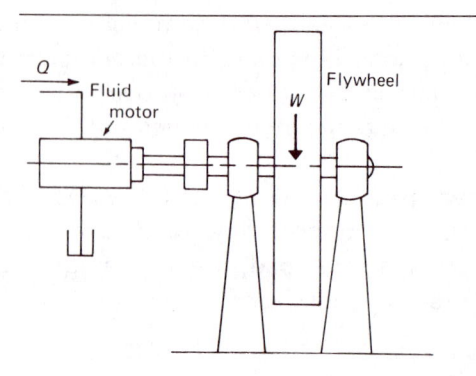

possible situations encountered in fluid power work. It is worth noting that a purely resistive load is encountered only under conditions of constant velocity. This is also true of an overrunning load. On the other hand, all fluid power systems undergo changes in velocity, and during these periods of change the load is of the inertia type. This point will be amplified later.

Cycle time requirements

Cycle time must be determined before other circuit parameters can be evaluated. Time has two basic implications for the design of a circuit: (1) it sets the *flow-rate* requirements relative to the cycle displacement pattern; and (2) it sets the *horsepower* requirements of the circuit or branch. Cycle times are determined on the basis of the following criteria:

Minimum overall cycle time The fundamental purpose is to accomplish the job as rapidly as is reasonably possible. This criterion is likely to result in the highest horsepower requirements.

Critical sector The one most important part of the cycle dictates the time span. This might be the time during which the useful work of the cycle is performed, the rest of the time being allocated to such functions as transfer, clamp, retract, etc.

Constant power The desire is to utilize a constant flow of energy during periods of varying load. For example, shorter time is allocated during low load levels (higher flow rate, lower pressures) such as occur during "rapid traverse" in a feed circuit, or longer time intervals are used during heavier load levels (low flow rate, higher pressure) such as feeding through a milling cutter, etc.

Constant speed–variable power This could be the case with a constant-displacement pump and a variable load, assuming there is no loss over a relief valve which would introduce a discontinuity into the cycle.

The determination of the actual time cycle will be a compromise between the desire to achieve the fastest, most efficient cycle possible and the cost of the equipment and controls necessary to realize this goal. In a practical circuit design, it is likely that more than one of these time-cycle criteria will be brought into play. A complete cycle profile would typically disclose the time-based relationships between force or torque, pressure, flow rates, speed, and horsepower during all phases of the work cycle and in all parts of the circuit.

In accordance with current practice in the technology, an open-center circuit is one in which the pump output is bypassed to the tank when the tandem center directional-control valve is in its neutral position. Figure 7.7 illustrates this for a simple circuit with only a single directional-control valve. Figure 7.8 shows this condition for

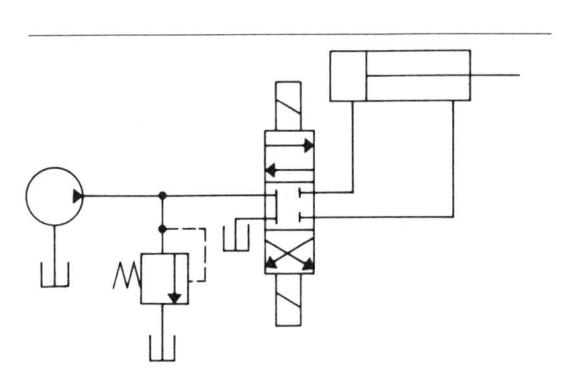

Figure 7.7 Basic open-center circuit.

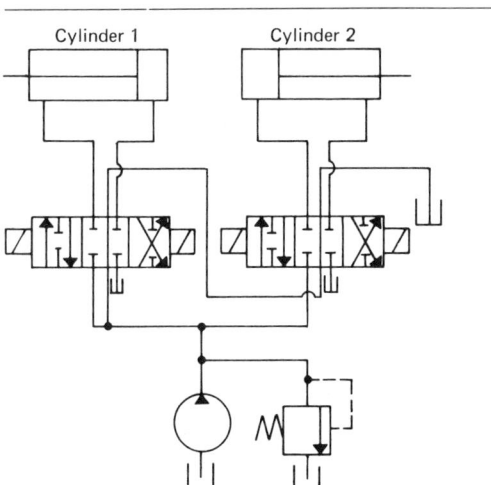

Figure 7.8 Open-center, stack-valve circuit.

a multiple, or stack, valve installation. Note that functionally the pump flow is returned directly to the tank with *both* valves in neutral. If one of the valves is shifted, the normal four-way valve function is initiated.

Pump pressure requirements

Pump pressure is another criterion that must be considered in designing an open-center circuit. In open-center circuits, the pump discharge pressure is a function of the load resistance encountered by the actuator. The load resistance is reflected across the actuator. The pressure demanded by the load is a function of actuator geometry and/or speed requirements. If the prime mover is capable of supplying the energy demand, it will do so. If not, it will either stall or, as is more likely in actual practice, the relief valve will open to allow fluid to bypass to the tank.

The relation of pump discharge to actuator speed is another factor that is critical to the design of an open-center circuit. In open-center circuits the pump delivery is *not* determined by the actuator's instantaneous speed requirements. This can be explained as follows. The open-center technique most commonly employs a fixed-displacement pump, in which the discharge rate is a function of pump displacement and the prime mover speed. If load inertia precludes rapid acceleration to a steady-state speed dictated by the combination of pump output and actuator displacement ($v = Q_p/A_p$), the excess flow from the pump must be bypassed over the relief valve. This is shown graphically in Figure 7.9. At time t_0, the control valve is shifted, porting pressure oil to the actuator. There is a very slight time lag due to the compressibility of the oil in the circuit, throttling during shifting of the valve element, etc. Then pump flow, Q_p, goes to its full rated output (the dotted line in Figure 7.9). At time t_0 the actuator is at zero velocity. At time $t_0 + \Delta t_0$ (the time corresponding to full build-up of pump output), the actuator has not yet started to move. Therefore, Q_a is zero. Looking to the pressure plot that has been superimposed over the flow plot in Figure 7.9, observe how the pressure tends to rise at a very rapid rate. Depending on the internal slippage in the pump, valve leakage, actuator slippage, and, primarily, the response time of the relief valve, the pressure will peak at some level above the relief-valve setting. Once the relief valve

Figure 7.9 Curves for pump Q_p versus cylinder Q_a.

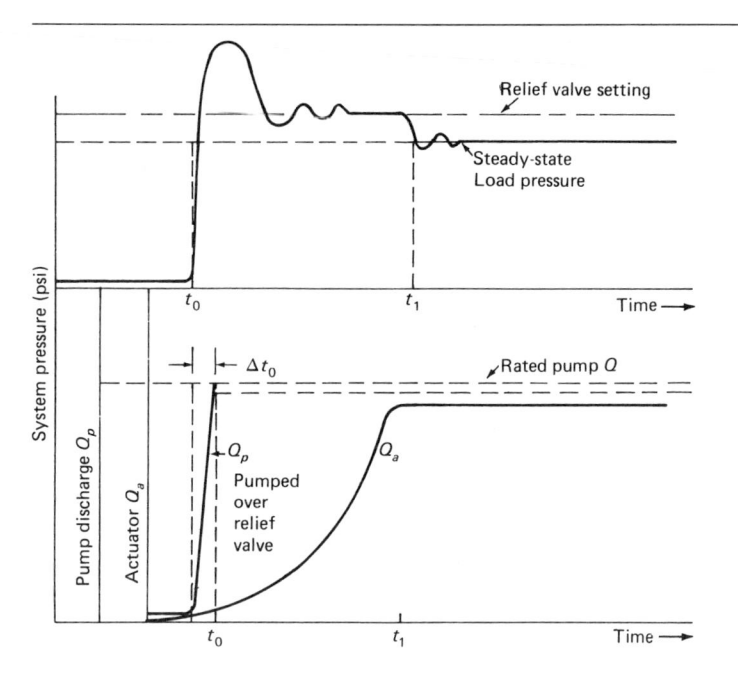

opens, the system pressure levels out at the relief-valve setting. Now look at the plot of actuator flow rate, Q_a. In a well-designed system, assuming a single pump and actuator, the pump output will just match the actuator input requirement at design speed; thus $Q_p = Q_a$, under steady-state conditions. However, at time Δt_0, Q_p is at the rated setting while Q_a is zero. Also the actuator and load must be accelerated from zero velocity to design velocity, which takes a finite interval of time, from t_0 to t_1. During this time Q_a is increasing along some curve until, at time t_1, $Q_a = Q_p$ and the actuator is up to speed. Note that the system pressure drops to the design steady-state level and the relief valve closes at this point. The shaded area between the two flow curves represents the oil bypassed over the relief valve during the acceleration period.

This somewhat complex sequence of events takes place in a fraction of a second so that it is difficult to detect under normal circuit operating conditions. In most open-center circuit applications, it would not even be a matter for consideration. Only in high-performance applications or if some inexplicable trouble occurred would this sort of analysis be needed. Another reason why pump flow rate and actuator flow may not match is that the pump must frequently be sized for multibranch circuit operation. Thus its capacity is designed for peak flow requirements and is too large for individual actuator needs.

Actuator sizing

Sizing of the actuator is another critical factor. In designing open-center circuits, an attempt is made to size actuators to meet speed requirements as a function of pump delivery. A cylinder should be selected such that $A(s/t) = Q_p$, where A is the piston area, s is the stroke, t is the time, and Q_p is the pump flow rate. In some instances this may give a cylinder larger than that required for force output alone. A motor is selected such that its input flow rate equals the pump output:

$$Q_p = V_p(N_p) = V_m(N_m) = Q_m$$

Circuit designers find various ways of unloading the pump. In open-center circuits, the control valve unloads the pump when it is in the neutral position. This is an advantage in that no auxiliary controls are required to unload the pump. Unloading the pump reduces unnecessary energy dissipation during passive time in the cycle; thus heat generation is minimized. Care must be taken that the selected valve has sufficient capacity to bypass full pump output without excessive pressure drop.

Output speed control

One other criterion that must be considered is output speed control. In an open-center circuit, actuator speed is controlled by throttling, or restricting, flow by means of a metering or flow-control device. The primary method is to use one of the many types of flow-control valves in combination with one of the basic flow-control methods. An alternative approach is to use the throttling characteristics of the directional-control valve. This is particularly true with manually operated valves. Any flow-control method involving throttling entails energy loss with attendant heat generation. A complete analysis is beyond the scope of this text.

In light of this discussion, recall the characteristics of open-center circuits previously named:

1 Pump discharge pressure is a function of load resistance.
2 Pump delivery is *not* determined by actuator speed requirements.
3 Actuators are sized to meet speed requirements as a function of pump delivery.

The last two statements make sense only with the connotation that the pump in a multiple-branch circuit is sized on the basis of *maximum system demand,* no matter when or where it occurs. During parts of the cycle other than peak demand, pump delivery will be greater than that required by one actuator.

Closed-center circuits

By definition, closed-center circuits are those in which the inlet and tank ports are blocked when the closed-center directional-control valve is in its neutral position, using either a fixed-displacement pump and an accumulator (Figure 7.10) or a pressure-compensated pump (Figure 7.11).

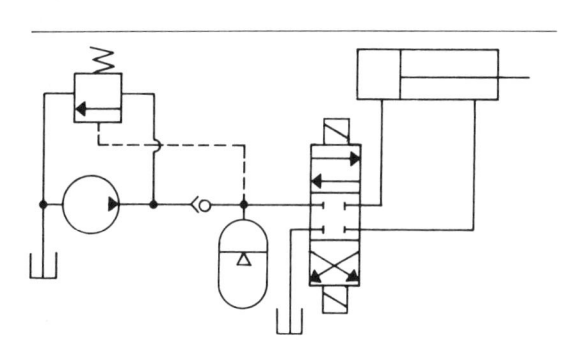

Figure 7.10 Basic closed-center circuit.

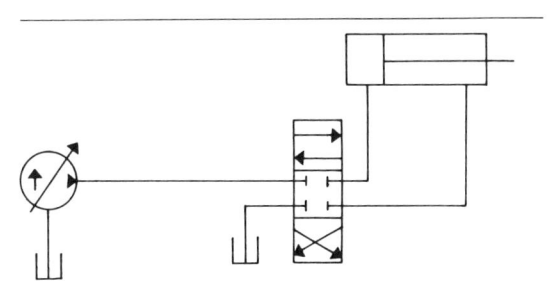

Figure 7.11 PV-PC closed-center circuit.

Fixed-displacement pumps are frequently used in closed-center circuits. With a fixed-displacement pump and an accumulator, pump pressure is *not* directly determined by actuator force requirements. As illustrated in Figure 7.10, the pump charges the accumulator to design pressure when the valve is centered. Design pressure is controlled in the circuit by the unloading valve setting. When this setting is reached, the valve opens and bypasses oil back to the tank at a low pressure drop across the valve. Note that pressure is sensed beyond a check valve placed between the pump and the accumulator. This check valve prevents unloading of the accumulator.

When the directional-control valve is shifted, porting oil to the actuator, full design pressure as stored in the accumulator is immediately available. As the cylinder moves, oil is ejected from the accumulator by the compressed gas behind the oil charge. After a time interval, the pressure will drop due to the expansion of the gas charge in the accumulator. At some pressure level that is a function of the unloading valve design, it will close, causing the pump discharge to re-enter the system rather than bypass to the tank. Depending on many factors, the pump will: (1) combine output with that from the accumulator at the lower pressure; or (2) recharge the accumulator to a higher pressure.

In some accumulator circuits, the design is such that the accumulator supplies the entire quantity of oil used during the active part of the cycle. However, it cannot do so at constant pressure, since the pressure of the gas charge drops as it expands during ejection of the oil. The load cycle must be such that it can still function at the lowest pressure delivered by the accumulator. This method is used where the active part of the cycle is rather short and is followed by a relatively long passive part, during which the pump recharges the accumulator. In such instances, the pump is sized so as to charge the accumulator during the passive time interval (Figure 7.12).

Figure 7.12 Pressure and flow relationship for accumulator circuits in which the accumulator supplies all the energy during the work cycle and the pump charges the accumulator during passive periods in the cycle.

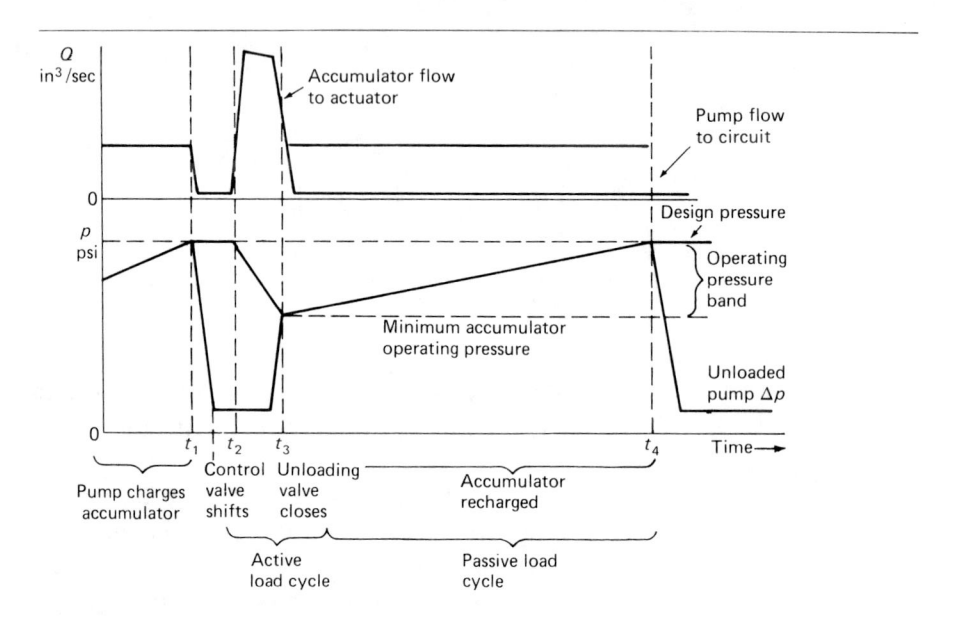

▶ **Example** In a circuit similar to that shown in Figure 7.10, an accumulator supplies oil to the circuit, discharging 924 in³ of oil in 10 sec. What would be the required pump discharge rate if the interval between work periods was 50 sec?

$$Q = \frac{924 \text{ in}^3}{50 \text{ sec}}$$

$$= 18.5 \frac{\text{in}^3}{\text{sec}} \times \frac{60 \text{ sec/min}}{231 \text{ in}^3/\text{gal}}$$

$$= 4.8 \text{ gpm}$$

In other circuits, the accumulator is used to supplement the pump during short, high-rate bursts. Thus, when a high flow rate is required for a short time interval, a smaller pump can be used together with an accumulator that it charges during the passive part of the cycle. When the valve is shifted, the accumulator output is added to the pump output and the two may be several times the pump output alone—but only for a short period of time. To achieve the same results by means of a pump alone might require a much more expensive pump.

▶ **Example** In the previous example, suppose the pump flow was added to the accumulator flow rather than bypassed to the tank. What would be the total flow to the system?

$$Q_t = Q_p + Q_a$$

$$Q_t = 4.8 \text{ gpm} + \frac{924 \text{ in}^3}{10 \text{ sec}} \times \frac{60}{231} = 4.8 + 24$$

$$= 28.8 \text{ gpm}$$

See Figure 7.13 for a graphic illustration of this situation.

Assuming 1000 psi operating pressure, what is the difference in horsepower required to drive the pump in the two examples?

$$\text{hp} = \frac{pQ}{1714} = \frac{1000 \times 4.8}{1714} = 2.8 \text{ hp}$$

$$\text{hp} = \frac{1000 \times 28.8}{1714} = 16.8 \text{ hp (an increase of 600\%)}$$

In this type of circuit, actuators are sized to meet force requirements as determined from a load cycle analysis. Frequently actuators, particularly cylinders, can be smaller than for an open-center circuit.

Figure 7.13 Pressure and flow relationships for accumulator circuits in which pump flow supplements accumulator flow during active portion of cycle.

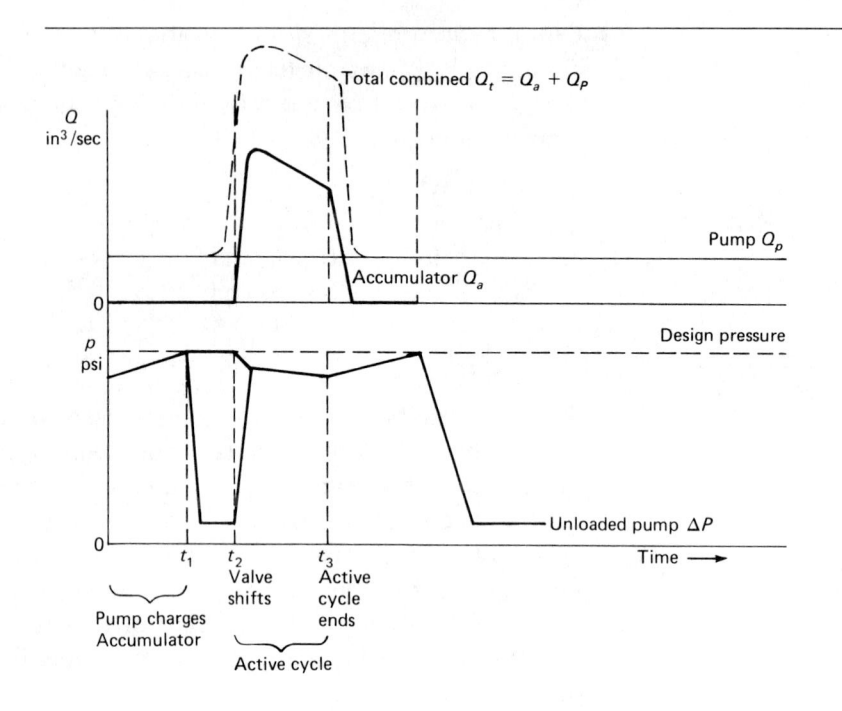

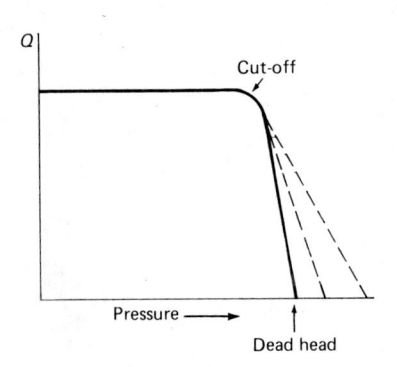

Figure 7.14 PV-PC characteristic curve.

Whereas in open-center circuits cylinders must be sized to give the required speed for the pump output available, in closed-center circuits, a given force is available. This causes the load to accelerate at a rate proportional to the mass. Thus the cylinder demands oil from the accumulator in proportion to its instantaneous velocity. The accumulator will deliver only on demand since it is not a positive-displacement device, like a pump.

Pressure-compensated pumps are also used in closed-center circuits. The simplest form of closed-center circuit using a pressure-compensated pump was shown in Figure 7.11. In a closed-center circuit with a pressure-compensated pump, maximum circuit pressure is governed by the compensator setting (Figure 7.14). The pump output flow is constant until a pressure is reached called the cut-off point. At this pressure, the force on the compensator begins to overcome the control spring holding the pump on stroke. As pressure increases, the pump starts to move off-stroke, thus reducing the displacement. The slope of the curve can be controlled somewhat by the spring rate of the compensator spring. Thus there can be a sharp cut-off or a more gradual one, as the application requires. When a certain pressure level (the *dead-head pressure*) is reached, the pump output is zero. The only power consumed by the pump at dead head is that required to overcome mechanical losses and make up internal leakage, which is a relatively low consumption. The pump will maintain full dead-head pressure on the system at this low power input.

Pump delivery depends on the ability of the actuator and load to respond, up to the maximum capability of the pump. Recall the discussion of load response with the fixed-displacement pump. It would require infinite acceleration for the load to absorb the entire pump output the instant it was switched to the actuator. With a compensated pump, full pressure force can be brought to bear on the actuator and load, but no flow rate will ensue until the load starts to accelerate. Thus it is a *demand system,* as was the case with an accumulator. In effect, a pressure-compensated pump acts like an accumulator of infinite volume.

A pressure-compensated pump functions as its own relief valve, shifting to dead-head conditions if an excessive load is applied. Never use a relief valve set near maximum design pressure in a circuit with a pressure-compensated pump. Instability and resultant oscillation can result under some conditions. If fail-safe protection is required, use a hydraulic fuse or set the relief valve substantially higher than dead-head pressure.

In a closed-center, pressure-compensated pump circuit, delivery is related to actuator speed requirements. This is illustrated in Figure 7.15. From time zero to time t_0, the control valve is centered and the pump is dead-heading—maintaining maximum pressure at zero delivery. At time t_0, the control valve is shifted, porting pressure oil

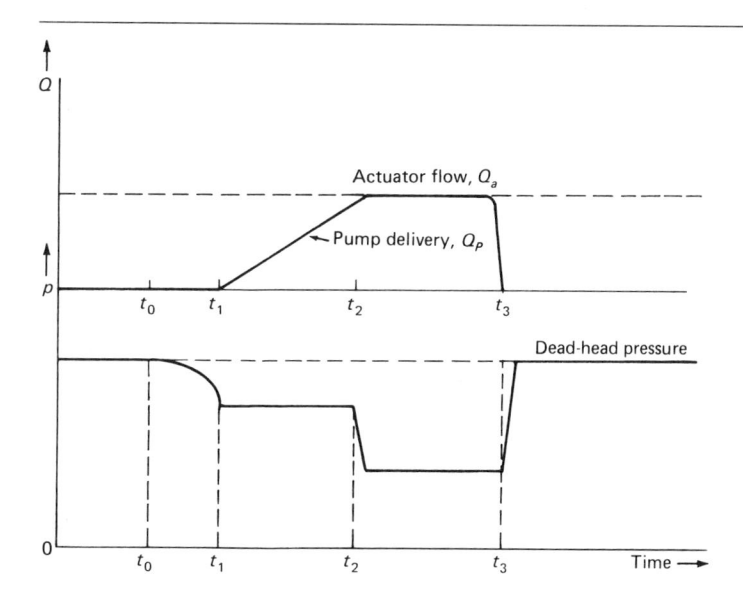

Figure 7.15 Q_m versus N_m characteristics.

to the actuator (see Figure 7.11). Full dead-head pressure is brought to bear on the actuator. The actuator cannot accelerate instantaneously, and so pump delivery remains at zero for a short interval of time, t_0 to t_1. During this interval the actuator begins to move. The pressure drops to some level below dead-head, which is required to accelerate the load. Simultaneously, the pump moves on-stroke. If the acceleration force required pressure greater than cut-off, the pump would compensate by operating at some reduced flow rate. This would be below that corresponding to cut-off but above the dead-head point. In this manner, the system is self-regulating.

From t_1 to t_2 the load is brought up to speed. At t_2 the pump has moved on full stroke, the load stops accelerating, and the operating pressure drops to some value corresponding to the steady-state resistive load. At t_3 the actuator hits a mechanical stop, or the end of its stroke, and the pressure immediately rises. The pump goes off-stroke and its delivery drops to zero. It then dead-heads until the control valve is reversed to retract the cylinder. In such a circuit there is no bypassing of pressurized oil over a relief valve. The pump supplies precisely what is demanded by the system.

Comparison of open-center and closed-center circuits

I. Application factors
  A. Open-center circuits
     1. Easier to design
     2. Greater variety of components available
     3. Most used on mobile equipment
     4. Less costly components, unless controls become complex
  B. Closed-center circuits
     1. Most useful in multibranch circuits
     2. Useful when there is a wide difference in flow-rate requirements from one branch to the next
     3. Useful when there are a large number of flow-rate differences between many branches
II. Functional factors
  A. Open-center circuits
     1. Control valve unloads pump to tank when in its neutral position.
     2. Pump discharge pressure is a function of load resistance encountered by the actuator.

3. Pump delivery is not determined by actuator's instantaneous velocity requirements.
4. Actuators are sized to meet average velocity requirements as a function of pump delivery.
5. Actuator speed must be controlled by throttling.

B. Closed-center circuits
1. Control valve, when in its neutral position, blocks all ports.
2. System pressure is a function of
    a) unloading valve setting, in an accumulator pump circuit
    b) dead-head setting of compensator control, in a pressure-compensated pump circuit
3. Delivery is on a demand basis, that is, a function of the ability of the mass of the system to accelerate and also overcome load resistance.
4. Actuators are sized to meet force requirements.
5. Actuator speed is controlled by throttling.

Circuits for effecting pressure control

Up to this point, we have discussed some of the basic factors in circuit design and performance. A knowledge of these factors is necessary to understand how circuits function; or why they don't, when they do not function. The balance of this chapter will be devoted to examples of typical circuits designed to perform specific functions within the energy transfer and control mission of fluid power technology.

Maximum pressure–limiting

Figure 7.16 illustrates the basic pressure-limiting circuit, using ANSI fluid power graphic symbols. This circuit involves use of a relief valve

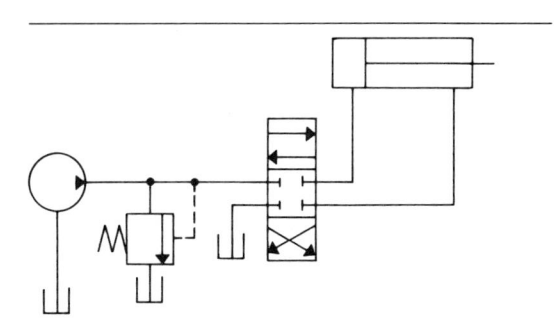

Figure 7.16 Basic pressure-limiting circuit.

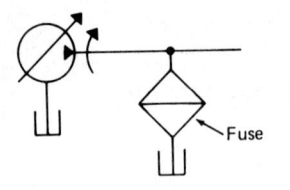

Figure 7.17 Hydraulic-fuse circuit.

in parallel with the main pressure line. When system pressure reaches the relief-valve setting, pump discharge is bypassed over the valve at full system pressure. During this period all of the energy input to the system is converted to heat.

The circuit of Figure 7.17 is similar to a relief-valve circuit except that a hydraulic fuse is used. This is a one-shot device, since the rupture disk blows out and must be replaced to reset the circuit. It is used mainly with pressure-compensated pumps for positive overload protection in case the compensator control on the pump fails to operate.

A pressure-compensated pump also serves to limit maximum system pressure (Figure 7.18). The maximum attainable pressure corresponds to the dead-head pressure setting of the pump control.

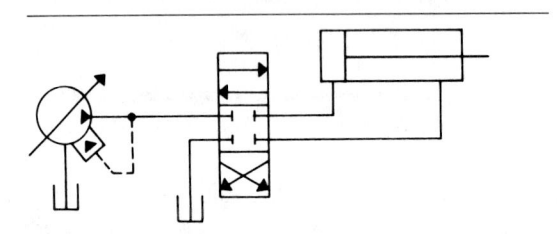

Figure 7.18 Pressure-compensated pump circuit.

Unloading

Some of the techniques used to unload pumps are illustrated in this section. Unloading circuits differ from maximum pressure–limiting circuits primarily in the pressure drop across the control device. In contrast to maximum-pressure control, unloading circuits bypass pump flow at minimum pressure drop across the control device. Thus heat generation during the bypass period is minimized.

Figure 7.19 shows a simplified unloading-valve circuit with a fixed-

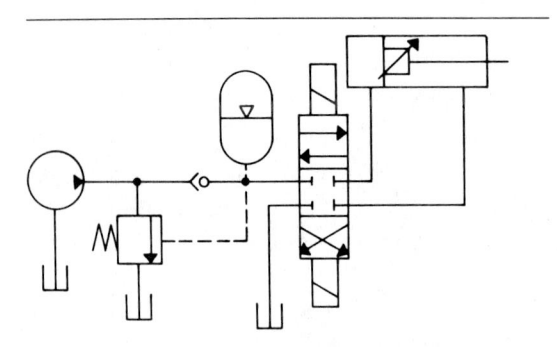

Figure 7.19 Unloading circuit, with fixed-displacement pump.

displacement pump. The unloading valve is externally pilot-operated from the main pump. In the illustrated application, the unloading valve would open when the cylinder bottomed out at the end of its stroke in either direction. Switching the directional-control valve to reverse the cylinder direction would cause a pressure drop in the circuit, which should allow the unloading valve to reset.

A typical pressure-switch control is shown in the circuit of Figure 7.20. Here the pressure switch actuates when the system pressure reaches a certain preset level and energizes the solenoid operator of a normally closed, two-way valve. The pump is unloaded when this two-way valve opens.

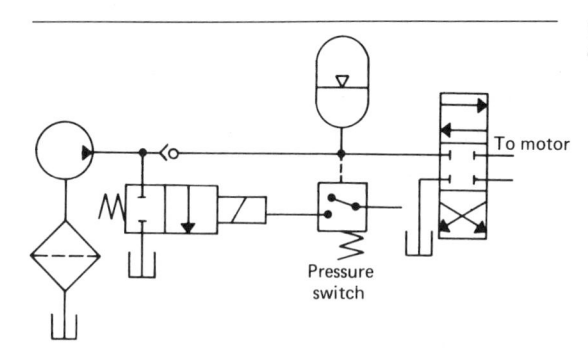

Figure 7.20 Pressure-switch circuit.

High-low

High-low circuits are really a special type of unloading circuit in which the intent is to provide two output speeds by changing the pump flow. Figure 7.21 illustrates the basic high-low circuit using

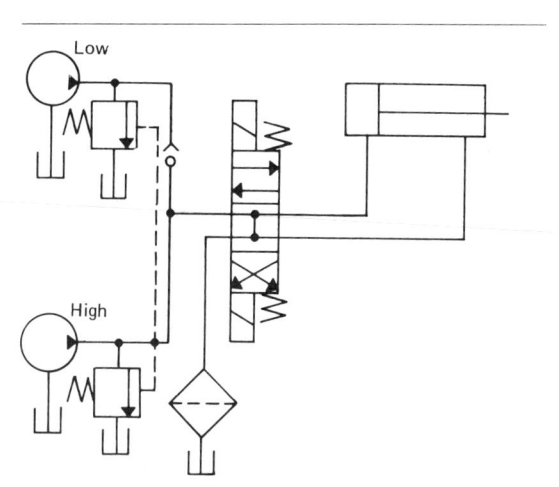

Figure 7.21 Basic high-low circuit.

two fixed-displacement pumps. The unloading valve unloads the low-pressure pump on attainment of some predetermined system pressure. Before this point is reached, both pumps feed the circuit. After unloading, only the high-pressure pump feeds the circuit. This type of application is frequently used for the rapid traverse–feed cycle on machine tools or the rapid approach–squeeze cycle on presses.

Controlling cylinder pressure

The typical circuits described in this section provide for pressure control under specific conditions. Figure 7.22 illustrates a technique for obtaining two different controlled pressure levels, one in the blank (cap) end and the other in the rod end of the cylinder. Relief valve RV_1 is placed in the line from the control valve to the blank end port and is set at one pressure level. Relief valve RV_2 is placed in the other line and set at a second level. This is a good illustration of the

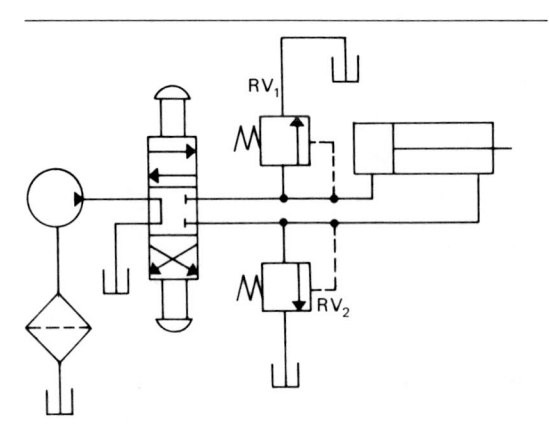

Figure 7.22 A circuit for obtaining two different controlled pressure levels.

point that placing a pressure-control valve ahead of the directional control causes control to be exercised over both actuator lines. Placing a pressure control downstream from the directional-control valve causes control to be effected only in the line in which it is placed. If, in Figure 7.22, valve RV_1 were placed ahead (upstream) of the directional-control valve, it would exercise priority control over the pressure level in both cylinder lines. Valve RV_2 would be effective only if its setting was less than that of RV_1, and then only in the rod-end line.

Figure 7.23 shows the application of a piloted relief valve to provide

235 Circuits for effecting pressure control

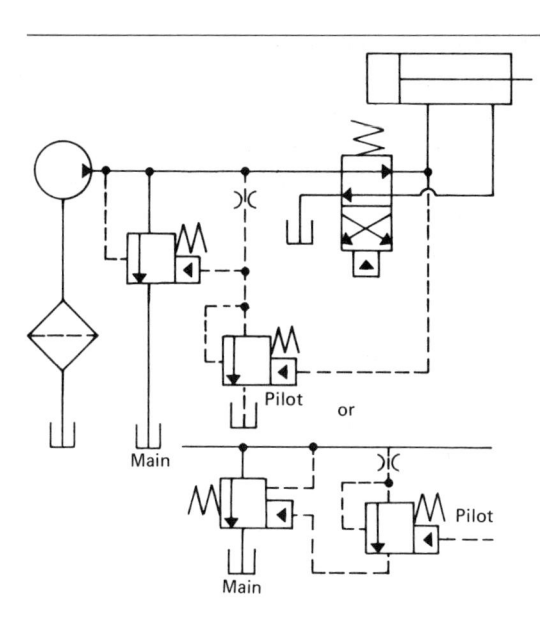

Figure 7.23 Piloted relief valve; cylinder external-pressure control.

high-pressure control for the extension stroke of the cylinder and low-pressure control for the retraction stroke. The pilot line is connected in parallel with the blank-end cylinder line. With the blank end pressurized, pilot pressure holds the pilot relief valve closed. It, in turn, holds the main relief valve closed against the high-pressure level. When the directional-control valve is shifted to cause retraction of the cylinder, the pilot line is dropped to return-line pressure. This enables the pilot relief valve to take over and control the main relief valve at whatever pressure the pilot valve has been set to operate.

Figure 7.24 shows a three-pressure circuit, in which the main relief

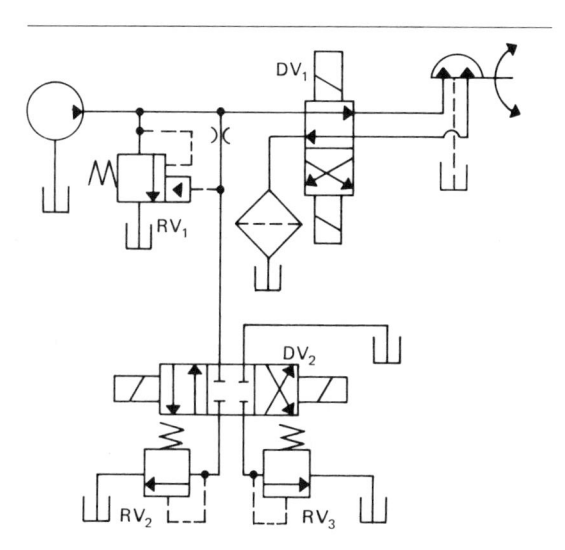

Figure 7.24 Three-pressure circuit.

valve, RV_1, can determine the limiting pressure, or either of the two pilot relief valves can do so. The auxiliary directional-control valve, DV_2, controls which of the pilot valves, RV_2 or RV_3, will be in the venting circuit of the main relief valve. When the directional-control valve is in its neutral position, neither RV_2 nor RV_3 is active and the main relief-valve setting controls the system pressure. The main directional-control valve, DV_1, serves to control the direction of motion of the piston rod.

The circuit of Figure 7.25 shows the use of a pressure-reducing valve to achieve reduced pressure control in a secondary branch of a multiple-branch circuit. The pressure level in the branch containing Cylinder 1 is controlled by the main system relief-valve setting. Pressure in the branch containing Cylinder 2 is controlled by the reducing-valve setting RD_1, provided the level is below that of the primary branch.

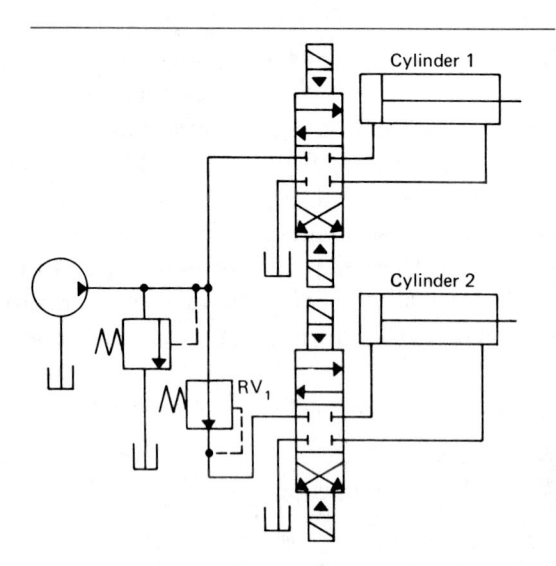

Figure 7.25 Pressure-reducing circuit.

Accumulator circuits

Accumulators are components capable of storing energy and returning it to the circuit on demand. They are used for two purposes in fluid power circuits: (1) to store energy and provide pressurized fluid to a circuit; and (2) to reduce pressure shocks in a system.

Figure 7.26 shows a circuit in which the accumulator is applied for

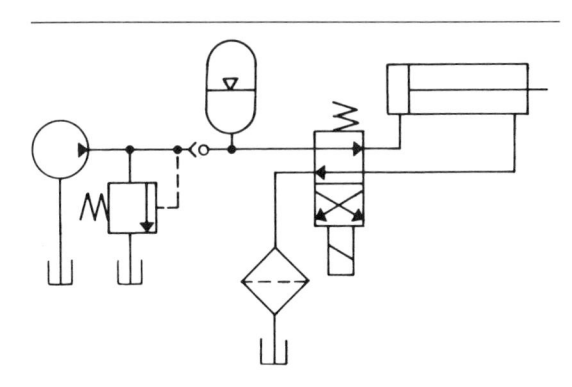

Figure 7.26 Accumulator used to maintain pressure on circuit.

the purpose of holding pressure on the circuit. The circuit looks identical to that in Figure 7.10, until one realizes that a relief valve is used in place of the unloading valve. This circuit is used for applications where it is necessary to keep the pressure level up, yet desirable to shut off the pump.

Figure 7.27 illustrates a modification of the pressure-holding circuit, in which the accumulator is applied in the cylinder line downstream from the directional-control valve. When a control device is so placed, it exercises control only in the one line. Thus, the accumulator shown in Figure 7.27 would hold pressure only on the blank end of the cylinder, while that of Figure 7.26 would hold pressure on both cylinder lines.

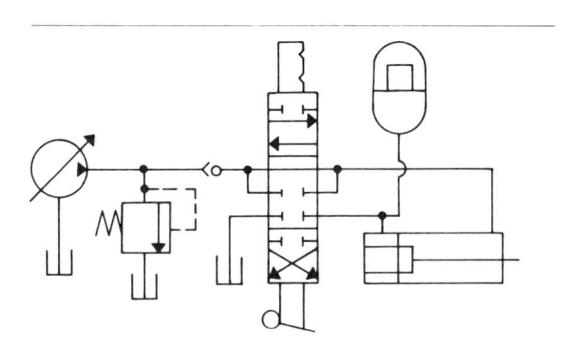

Figure 7.27 Accumulator pressure-holding circuit; one motor line.

Figure 7.28 shows the application of an accumulator in a "fail-safe-retract" mode. In case of pump failure, the accumulator would provide the necessary pressurized fluid to retract the cylinder. There are many applications in which this safety feature would be desirable.

Figure 7.29 shows an application of the accumulator to reduce pump pressure surges. Recall that all positive-displacement pumps exhibit a pulsating output due to the cyclic mechanical input to the pumping mechanism. In some applications, these pulsations must be damp-

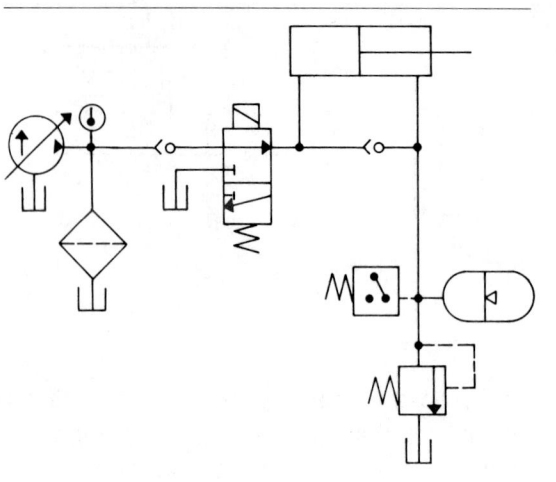

Figure 7.28 Fail-safe-retract circuit.

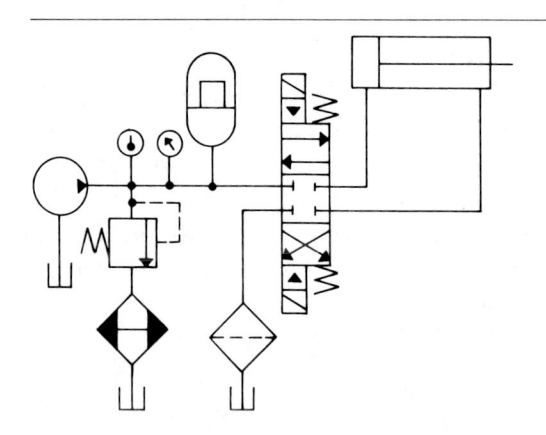

Figure 7.29 Accumulator used in pulsation-reduction application.

ened out. An accumulator is sometimes successful in performing this function.

Flow-control circuits

Flow control in a fluid power circuit or system presumes a requirement for speed (or velocity) control. Flow-control circuits are methods for effecting speed control.

Meter-in

The flow-control valve is placed in the line connected to the controlled output port, for example, the blank end of the cylinder in

Figure 7.30. A valve placed in this position would control the flow rate into the cylinder and the (output) velocity of the piston rod.

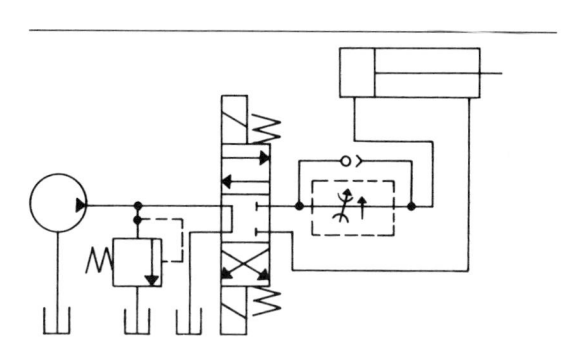

Figure 7.30 Meter-in, flow-control circuit.

This technique is best used with *resistive loads.* Control of a load by a fluid power circuit is predicated on control of the output member of the circuit. In the illustration, the output member is the piston–piston rod combination. The piston-rod combination can be thought of as a mechanical interface between the load reaction and the energy-charged fluid. Control of this interface consists of always keeping it locked-up between a load reaction and the pressurized fluid. Any time this locked-up condition is lost, control of the load is lost. An example would be load reversal—from resistive to overrunning.

Unless a bypass type of flow-control valve is used the pump is always working against relief-valve pressure in a meter-in circuit.

Meter-out

The flow-control valve is placed in the line connected to the return-flow port of the actuator. This is the rod end in the example of Figure 7.31. The output velocity of the piston rod is controlled by limiting the rate at which fluid can exit from the cylinder.

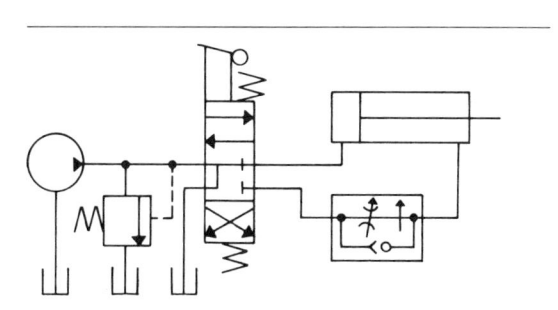

Figure 7.31 Meter-out, flow-control circuit.

This technique is used with *overrunning loads*. The pump is operating against a relatively constant pressure that is the sum of the load reaction and the flow-control back pressure. The circuit is rather inefficient. At low load levels or overrunning load cycles, the back pressure exceeds the pump pressure, owing to the area differential across the piston. This area differential produces an "intensifier" condition.

Bleed-off

In this form of flow control, part of the pump delivery is bypassed to the tank at system pressure. This provides for adjustment of speed around some average value. It does not allow speed control over the entire range. A further disadvantage is that it does not provide positive control of the load, as discussed under meter-in and meter-out circuits (Figure 7.32).

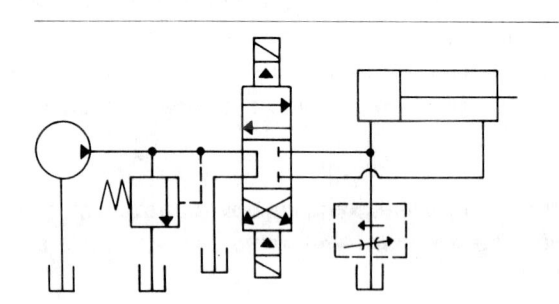

Figure 7.32 Bleed-off, flow-control circuit.

Regenerative

The regenerative circuit, sometimes called the differential circuit, is illustrated in Figure 7.33. A directional-control valve is used. This valve has one position in which both cylinder ports are connected in parallel to the supply port. Pressure is applied equally to the blank and rod ends of the piston. This would appear to lock up the piston hydraulically. An analysis of the pressure forces on both sides of the piston shows that this is not the case:

1 The sum of forces acting on the piston are:
   a. Blank-end pressure force: $F_1 = p \cdot A_1$
   b. Rod-end pressure force: $F_2 = -p \cdot A_2$ (The force carries a negative sign because it acts in a direction opposite to the blank-end force.)

Figure 7.33 Regenerative circuit.

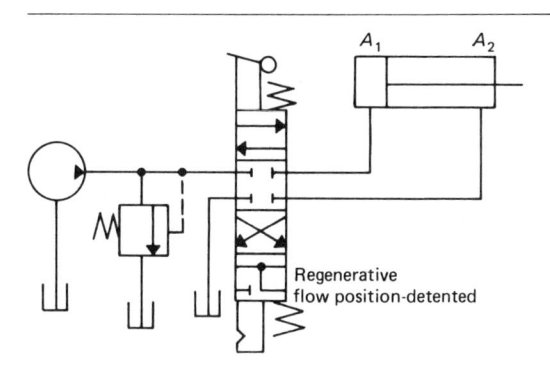

c. Load reaction: $F_L = \pm$ (Load sign depends on whether the load is resistive or overrunning.)

d. Then $F = F_1 + F_2 \pm F_L$

2 It can be shown that the effective area in a regenerative cylinder is the piston rod area, A_R. The net force developed by the cylinder to overcome a resistive load is

$$F_c = pA_R$$

3 As long as p is below the relief-valve setting the piston will be able to extend against the load reaction. That is, the cylinder will operate if

$$p = \frac{F_L}{A_R} < P_r$$

When the piston is moved in the cylinder, the fluid on the rod end is displaced. Normally, it would return to the reservoir through the four-way, directional-control valve. In the regenerative set-up, the fluid must bypass around to the blank end of the cylinder. The displaced fluid has the effect of adding to the pump delivery. If the area ratio (blank end to rod end) is 2:1, the piston velocity will be the same in both directions.

Intermittent-feed control

Meter-in and meter-out circuits can be converted to intermittent-control circuits by the addition of a cam-operated or limit switch–controlled, solenoid-operated, two-way valve, as shown in Figure 7.34. Return flow is bypassed through the normally open two-way valve during those parts of the cycle in which no flow control is

Figure 7.34 Intermittent-feed control circuit.

desired. The bypass valve is closed at the proper time during the work cycle. Return flow must then pass through the flow-control valve. When the requirement for flow control is past, the two-way valve is reopened. This takes the flow-control valve out of the circuit again.

Deceleration control

A specially designed, cam-operated, two-way valve can be used to decelerate a high-speed cylinder or heavy load. The deceleration may be thought of as a variable orifice that is gradually closed by the action of the cam. As the orifice is closed, a back pressure is built up that acts to slow down the cylinder. The circuit is illustrated in Figure 7.35.

Some deceleration valves are pressure-compensated to provide a

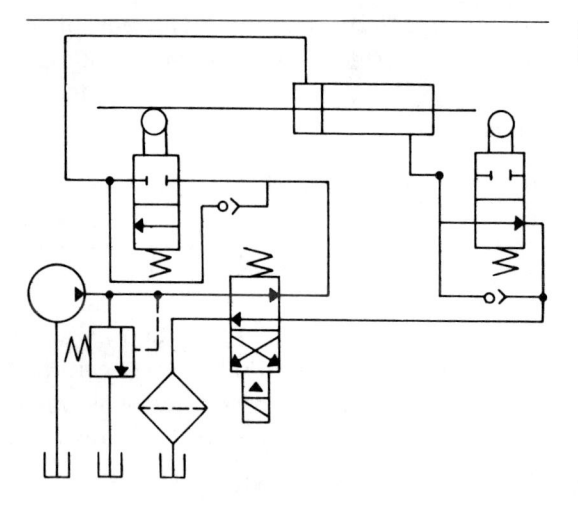

Figure 7.35 Deceleration-control circuit.

constant back pressure with change in flow rate. The integral check valve allows free-flow bypassing of the deceleration valve for reversal of the piston.

Flow divider

A flow divider may be thought of as a fluid bridge circuit. It will split a single input into two outputs. These outputs may be, although they need not be, equal.

In order to synchronize actuators in an open-loop circuit it is necessary to bring them up against mechanical stops to realign them. Otherwise, the only positive way of maintaining synchronization is to use a servo system.

Figure 7.36 shows the flow-divider circuit intended to synchronize rotary motors. It is subject to the same problem just described for linear motors (cylinders). Figure 7.36(a) shows a meter-in circuit; Figure 7.36 (b) illustrates a meter-out circuit for motor control.

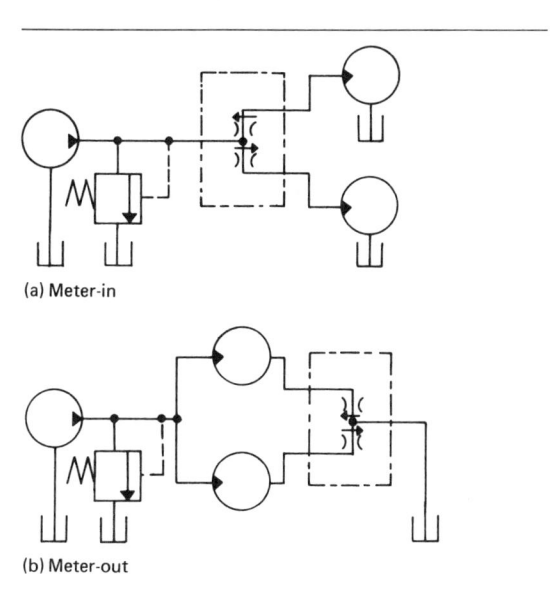

Figure 7.36 Flow-divider circuit. (a) Meter-in. (b) Meter-out.

(a) Meter-in

(b) Meter-out

Sequencing

The classic cylinder-sequencing circuit is illustrated in Figure 7.37. This covers the case in which two cylinders are to be operated se-

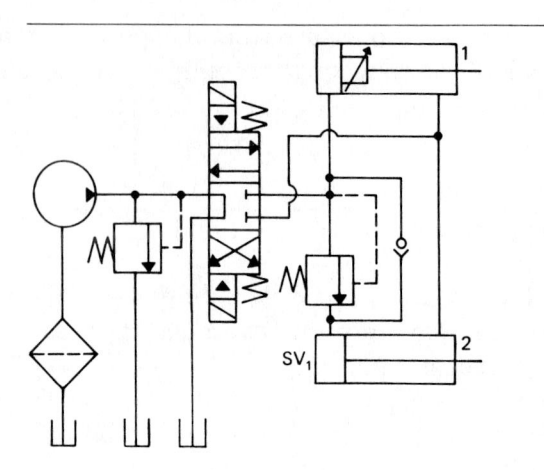

Figure 7.37 Cylinder-sequencing circuit, single motion.

quentially, that is, one after the other. Cylinder 1 is to complete its stroke; then Cylinder 2 is to move out on its stroke. The main directional-control valve has control over the direction of motion—in or out. A sequence valve, SV_1, is placed in parallel with the blank end of Cylinder 1, with its outlet port connected to the blank-end port of Cylinder 2. When Cylinder 1 has completed its stroke, pressure will tend to build up in the line. This increase in pressure, above that necessary to outstroke Cylinder 1, will cause the sequence valve to open. When the sequence valve opens, it ports fluid to Cylinder 2. This design is frequently used on machine sequences such as clamp work, where Cylinder 1 would operate the clamping mechanism and Cylinder 2 would operate the work mechanism.

The circuit shown in the preceding example would give positive sequencing on the outstroke part of the cycle, but the retract function would be random. That is, no control would be exercised over the retract strokes of either cylinder. If retract sequencing is required, as well as extend sequencing, the circuit of Figure 7.38 can be used to accomplish it. Note the addition of a second sequence valve, SV_2, in the rod-end line to Cylinder 1. This is the reverse order of the extend sequence.

Synchronizing actuators and motors

Figure 7.39 illustrates one of the most deceptive circuit diagrams in fluid power technology. It shows the amazing simplicity with which two cylinders can (apparently) be synchronized. The diagram suggests

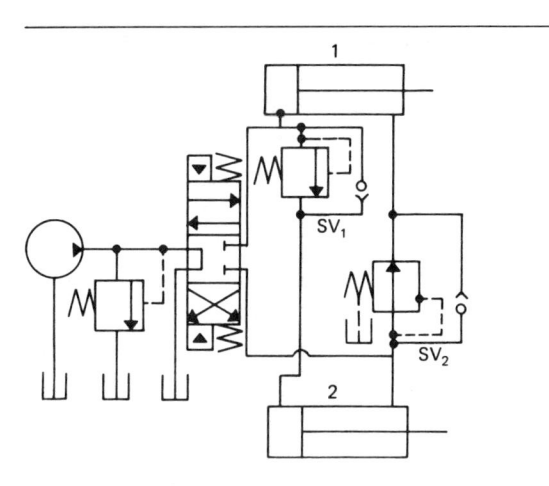

Figure 7.38 Cylinder-sequencing circuit, dual motion.

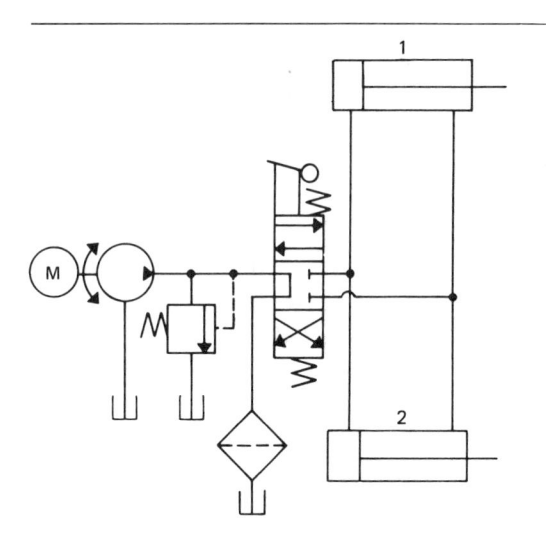

Figure 7.39 Cylinder-synchronizing circuit; parallel hook-up.

that two cylinders of equal displacement can be synchronized merely by piping them in parallel and using a pump of sufficient displacement. Nothing could be further from the truth!

Even if the two actuators of Figure 7.39 could be made identical—which they can't be—it would be necessary that the loads on the two piston rods also be identical in order for the two to extend in synchronization. If the loads are not the same, the cylinder with the lower load reaction will extend first, because it requires less pressure to move it. Only after the first cylinder completes its stroke can the second start to move, and then only after the pressure has built up to a new, higher level as required by the larger load reaction.

In addition to these obvious problems, a circuit designer must consider the more subtle, random variations in characteristics of the two cylinders. No two cylinders have the same packing friction; clearances are not precisely the same; internal leakage varies, etc. As a result, the two cylinders that were started out in synchronization would soon get out of phase. In many applications, lack of synchronization can be dangerous; in others, it is merely an inconvenience.

There is only one method that will assure absolute synchronization, the use of a servo system. All other methods will fall short in some degree. The tolerance on the process being controlled dictates allowable deviation from absolute synchronization.

The circuit of Figure 7.40 shows a useful open-loop circuit technique for synchronizing two or more cylinders. Pumped fluid is delivered to the blank-end port of Cylinder 1. Rod-end fluid from Cylinder 1 is delivered to the blank-end port of Cylinder 2; that is, the cylinders are piped in series. The rod-end port of Cylinder 2 is connected to the second cylinder port on the directional-control valve. In order for motion of the piston rods to be approximately synchronized by means of this method, the rod-end area of Cylinder 1 must equal the blank-end area of Cylinder 2. The pump must be capable of developing a pressure equal to the sum of the pressure differentials across *both* pistons. Note that the blank-end pressure of Cylinder 2 becomes the rod-end (back) pressure of Cylinder 1.

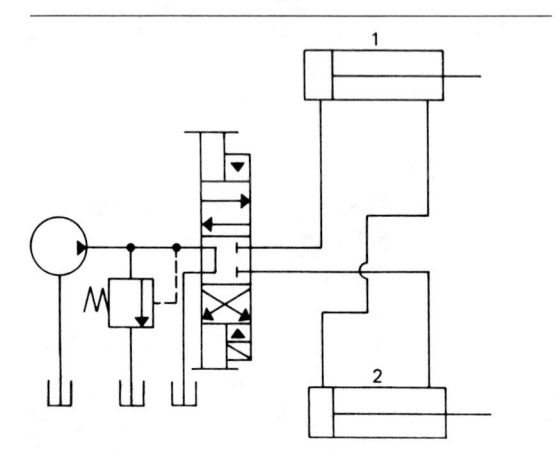

Figure 7.40 Cylinder-synchronizing circuit; series hook-up.

Even the method of Figure 7.40 will not guarantee piston-rod synchronization over a large number of cycles. This is due to the difference in leakage in the two cylinders. Where the method has been successfully applied, some provision has been made for resynchronizing the pistons periodically. This is done by providing (1) a mechanical stop for both of the pistons or rods, and (2) a method for replenishing fluid that has leaked out between the two pistons while they are both against the stops.

This series-actuator method will not work with rotary motors. Slippage (internal leakage) will vary from motor to motor. Assume, for example, that one piston motor's volumetric efficiency was 96%, while that of a second was 97%. It would take only a few minutes of operation for these two motors to get several revolutions out of synchronization.

Attempts are made frequently to use flow-control valves to synchronize motors, as shown in Figure 7.41. The success of such techniques depends on the tolerances on performance of the flow controls

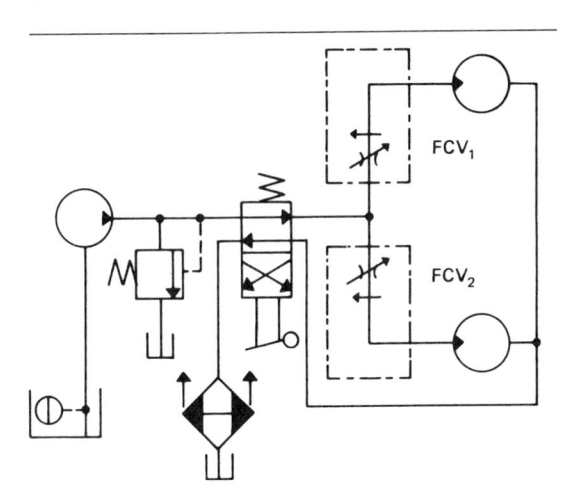

Figure 7.41 Flow-divider synchronizing circuit.

and on the allowable deviation from synchronization. It is a fact that cylinders or motors cannot be precisely synchronized over a large number of cycles of operation under varying load conditions using open-loop control. Fortunately, many processes enjoy wide enough performance latitude that flow controls will provide satisfactory synchronization. However, each case must be evaluated on its own merits.

The use of flow-divider valves is also sometimes proposed for motor synchronization. They are subject to the same limitations as flow-control valves.

Fail-safe circuits

Fail-safe circuits are designed to protect the operator and/or the machine against power failure, overload, carelessness, etc. Generally, they are designed to return the load to some "safe" condition should one of these unscheduled events occur.

A typical load-locking circuit is shown in Figure 7.42. Applications of cylinders frequently involve lifting a load against the acceleration due to gravity. Accompanying this action is the danger that the load might drop should anything happen to reduce the pressure on the piston. Power or pump failure, a broken line, inadvertent shifting of

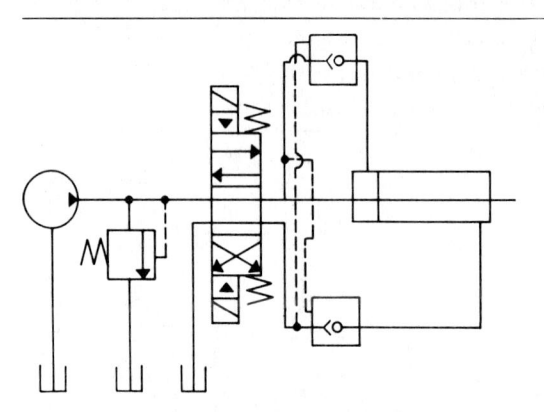

Figure 7.42 Load-locking circuit.

the wrong control valve, or valve failure are among the possible causes of load drop. The load-locking circuit makes use of pilot-operated check valves, placed as near to the actuator as possible. Some designs incorporate the load checks into the cylinder itself. There are directional-control valves on the market with load checks built in, but these have the disadvantage of not protecting against failure of any component beyond the directional-control valve. Their offsetting advantage is that such valves are easier to pipe into the system. When neither cylinder line is pressurized, the check valves remain seated, providing positive locking of the hydraulic circuit beyond. If either cylinder line is pressurized, a pilot connection transmits the pressure to the check in the opposite line. A pilot piston unseats the check element, allowing return flow to pass through the valve. The area ratio between the pilot piston and the check element determines the level of the pilot pressure needed to open the valve against a given back pressure.

The operator-protection circuit of Figure 7.43 shows a version of the classic two-hand, or palm button, type of safety circuit. It is

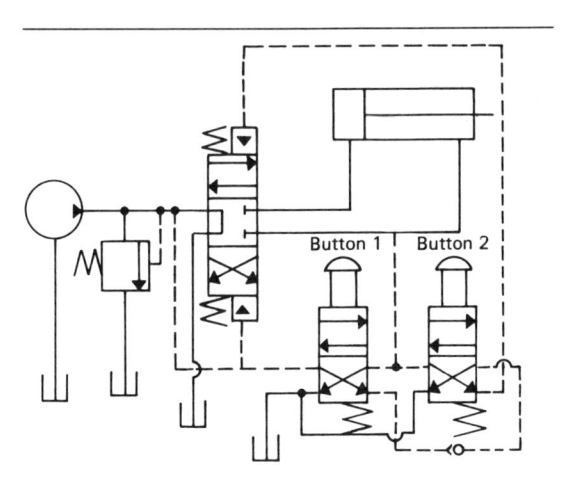

Figure 7.43 Two-hand safety circuit.

necessary for the operator to depress both manual valve actuators (buttons) on the two-position, four-way, directional-control valves simultaneously in order for the circuit to function. The operator cannot tie down one of the buttons because it is necessary to release both of them to retract the cylinder. Actuation of the two manual directional-control valves causes operation of the main hydraulic pilot-operated, directional-control valve.

Pneumatic power circuits

It is the intent in this section to discuss the differences between pneumatic and hydraulic systems. The thesis is that *functionally* the two are similar. Therefore, what has been said about the functional design of hydraulic circuits is valid for pneumatic circuits. The areas of difference lie in the handling of the fluid medium—liquid in one case, gas in the other.

Effect of fluid characteristics on motor performance

The performance of the fluid in a hydraulic or pneumatic system is related primarily to compressibility. Figure 7.44 illustrates this point. The load-cycle plot is identical for either the hydraulic or the pneumatic system. The pressure plot is determined from the load plot. Assuming that the hydraulic cylinder is driven by a fixed-displacement pump, the volumetric flow rate to the actuator is constant. This is represented by the horizontal flow curve in Figure 7.44. A constant

Figure 7.44 Comparison of hydraulic and pneumatic actuator performance.

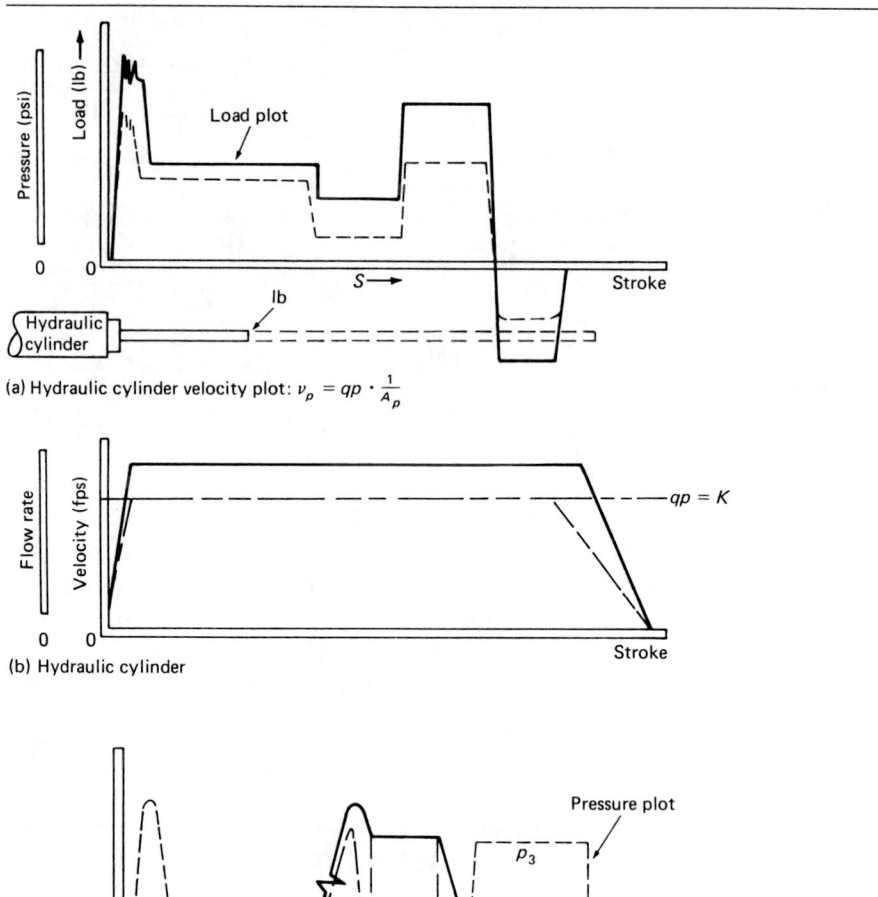

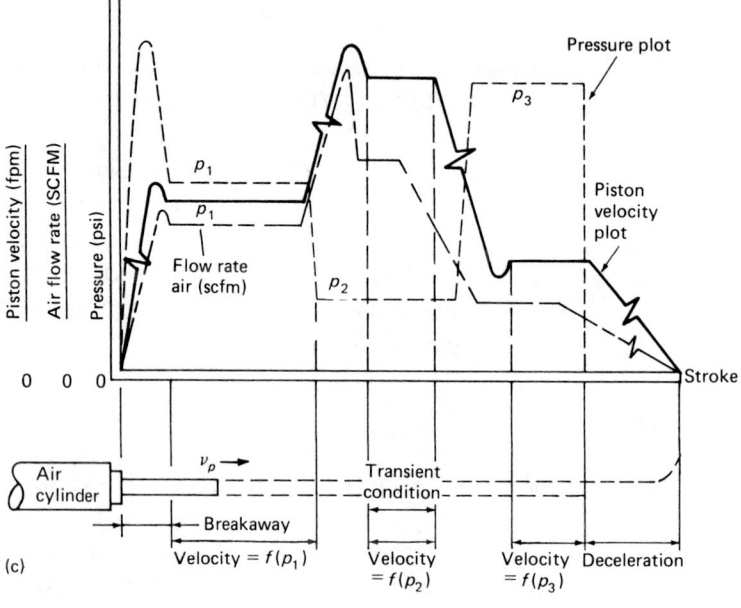

(a) Hydraulic cylinder velocity plot: $v_p = qp \cdot \frac{1}{A_p}$

(b) Hydraulic cylinder

(c)

input to the cylinder results in a constant output velocity (of the piston rod). Because of the relative incompressibility of the hydraulic fluid, fairly accurate velocity control is possible.

The pneumatic system pressure plot, shown in Figure 7.44, will have the same form as that for the hydraulic cylinder because the same load cycle is involved. At this point the similarity between the two systems ends.

It is very difficult to determine what the air flow rate to the cylinder really is. For instance, with an initial load pressure of pfp_1, there will be a corresponding initial flow rate of air. The density of the air is a function of the pressure, p_1, and the temperature, T_1. The source of the compressed air is the central compressor station and it is delivered through a pressure regulator, which is a throttling device. Since the compressor is assumed to be capable of delivering an unlimited quantity of air to the cylinder, the factors that limit flow rate to the cylinder are: load, resistance of the connecting pipes to flow, valve orifices, the regulator, etc. It is important to realize that *all* of these factors contribute pressure drops of one sort or another. And every change in pressure brings about a corresponding change in volume of the gas.

Consider the change in load pressure from p_1 to p_2, as indicated in Figure 7.44. This is dictated by the load reaction on the piston rod, not by the pressure of the air in the cylinder. At the instant the load pressure drops to p_2, a force imbalance in the cylinder results. This occurs because there is air in the cylinder at pressure p_1. The gas does the only thing it can—it expands until it reaches a new equilibrium pressure, p_2. This sudden expansion of the gas in the cylinder is evident from the lunging forward, or jerking, of the piston rod.

The load-pressure curve of Figure 7.44(c) indicates that there must be a pressure increase, from p_2 to p_3. The reverse of the process just described will occur. That is, the piston rod slows down or even stops momentarily, to allow incoming air to recompress the gas already in the cylinder to the new equilibrium pressure, p_4.

These momentary changes in piston-rod velocity will occur every time there is a change in load pressure, when a pneumatic system is used. Thus we cannot speak in terms of a flow rate to the actuator, in the same sense as we do for a hydraulic system. We must be concerned with instantaneous piston velocities.

The point of this discussion is that it is extremely difficult to obtain controlled output velocities with pneumatic systems. It is also difficult to maintain an accurate position with a pneumatic system because of the compressibility of the gas medium (Figure 7.45). As

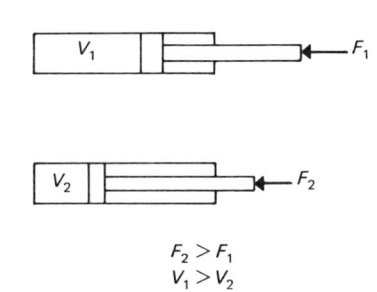

$F_2 > F_1$
$V_1 > V_2$

Figure 7.45 Pneumatic cylinder under varying load.

the load changes from F_1 to F_2, the gas in the cylinder is reduced in volume. This causes a change in the position of the piston rod.

From these comments, it is easy to deduce why hydraulic systems have taken the lead over pneumatic ones in applications that require accurate position or velocity control. Up to the present time, pneumatic systems have been used mainly for sequential circuits in which the end conditions were of prime importance, that is, the rod was either fully extended or fully retracted. Transfer and clamping press circuits, etc., are typical applications. That the picture for pneumatics is changing will become apparent from later discussions on logical circuit design.

Where the economics of the situation dictate the use of a pneumatic power system, yet control requirements are greater than those attainable with a straight pneumatic system, an air-oil system may be utilized. In air-oil systems, compressed air provides the source of potential energy and hydraulic oil provides the incompressible fluid characteristics requisite to achieving the desired degree of control.

Examples of pneumatic power circuits

The basic directional-control circuit for a single-acting cylinder (Figure 7.46) illustrates some of the differences between such a circuit and its hydraulic counterpart. Note the absence of an input device (pump). Most pneumatic circuits use the main plant compressors as their source of energy. The input to the circuit consists of "hooking into"

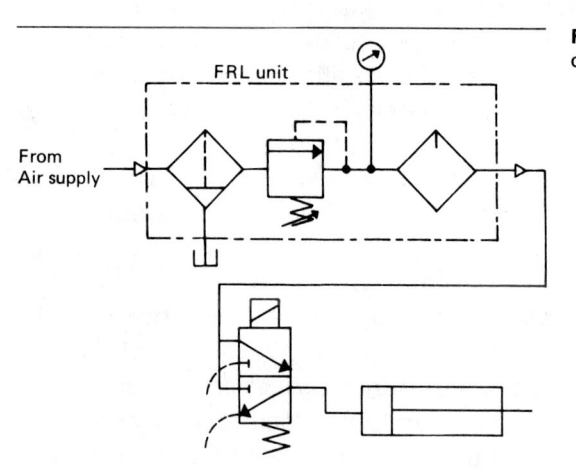

Figure 7.46 Basic pneumatic directional-control circuit.

the air manifold at a convenient location. This does not preclude the use of an individual air compressor located at the machine. Conversely, the use of individual power units for most hydraulic systems does not mean that a central system cannot be utilized.

A characteristic of pneumatic circuits is the use of a filter-regulator-lubricator (FRL) unit at the source, the point where the air manifold is tapped. The purpose of this unit is to provide clean air at a regulated pressure and to add enough lubricant to the air to minimize wear of component parts.

Note the absence of a return line to the tank in the circuit diagram of Figure 7.46. The short dashed line leading from the exhaust port of the control valve indicates exhausting of the air to atmosphere.

A single-acting cylinder control circuit with time-delay control of the directional-control valve is shown in Figure 7.47. The use of the time-delay control injects a lag, or time delay, between the instant at which the actuation signal is applied to the circuit and the instant at which the control valve responds. In most cases, the time-delay control is adjustable so that the lag can be varied over a given range.

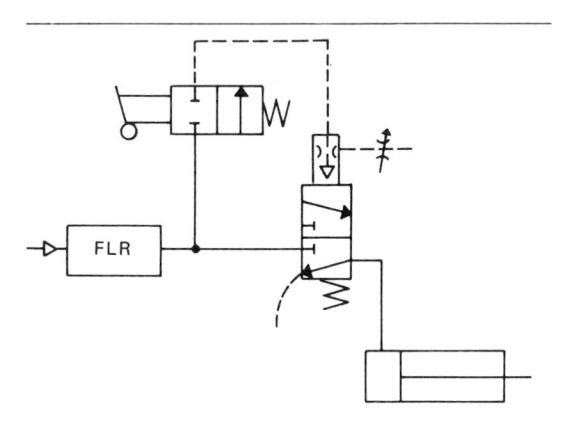

Figure 7.47 Time-delay pneumatic circuit.

Accurate flow control under conditions of varying load is difficult to obtain with a pneumatic circuit. Pneumatic flow controls are available and can be successfully applied when the load doesn't vary excessively. A typical flow-control circuit is illustrated in Figure 7.48.

The illustration of Figure 7.49 is of a multiple-branch pneumatic circuit that uses limit switches to effect control in a sequencing mode. The limit switches can be used to operate relays that, in turn, initiate secondary effects such as reversing directional-control valves.

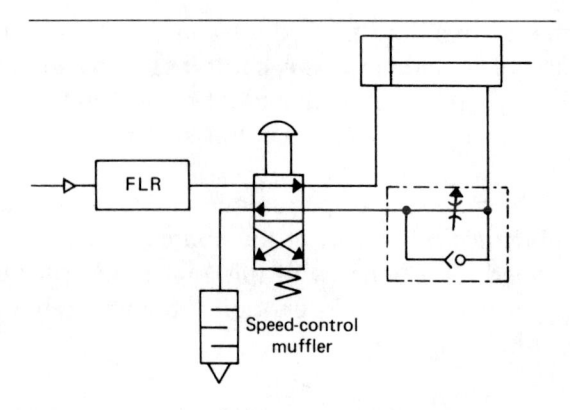

Figure 7.48 Flow-control pneumatic circuit.

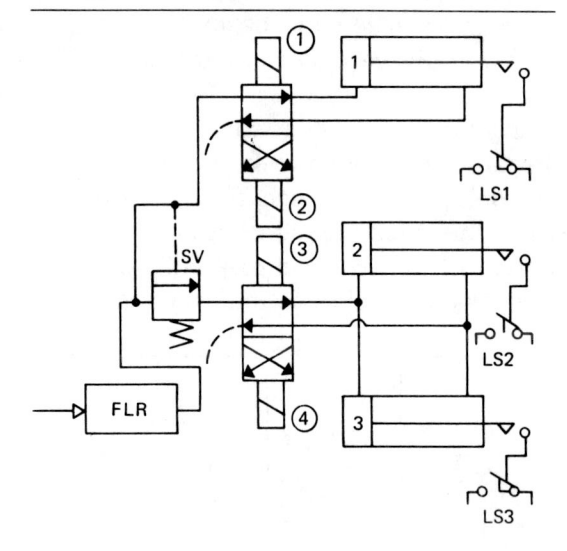

Figure 7.49 Limit-switch sequence circuit.

Figure 7.50 shows a circuit designed to achieve multiple speeds of the actuator. A shaped cam connected to the piston rod operates limit switches. These control the solenoid operators on the directional-control valve, DV_2. Each of the ports of DV_2 is connected to a variable-flow control valve. The latter can be set to control the flow of air out of the cylinder and thus the speed of motion of the piston rod in both directions.

Figure 7.51 illustrates a dual-pressure circuit. In this case, two pressure regulators are used to supply a single input to the circuit through a selector valve. Thus the circuit can be fed at one of two input pressure levels.

The pneumatic equivalent of the hydraulic closed-center circuit is shown in Figure 7.52. It must be re-emphasized that this type of

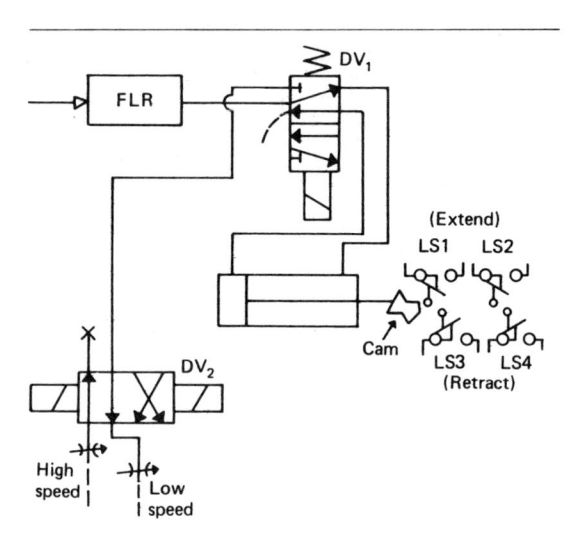

Figure 7.50 Multiple-speed pneumatic circuit.

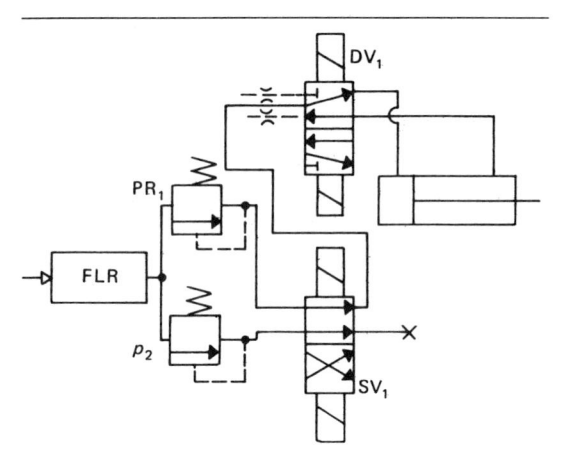

Figure 7.51 Dual-pressure circuit.

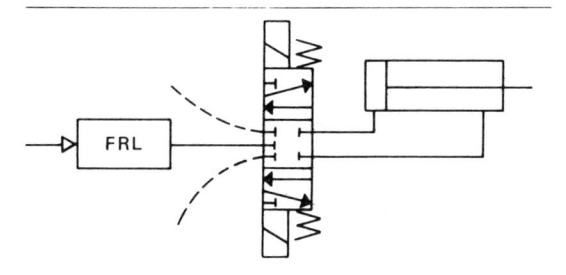

Figure 7.52 Closed-center pneumatic circuit.

pneumatic circuit will not have the same stiffness as its hydraulic counterpart.

A pneumatic safety circuit is shown in Figure 7.53. Both pilot valves, A and B, must be actuated to cause extension of the piston rod.

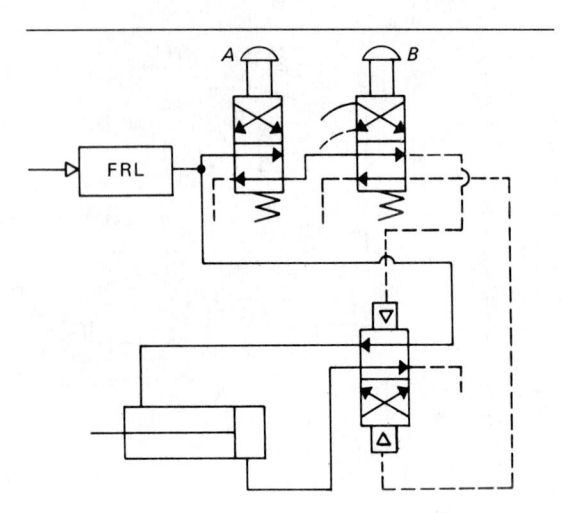

Figure 7.53 Pneumatic safety circuit.

Both pilot valves must be released to allow retraction of the piston rod.

Combination circuits using both air and oil have been devised to obtain the advantages of each medium. Figure 7.54 shows an air-oil feed circuit. Shop air supplies the energy required; oil provides the control capability. The pressure-reducing valve PR$_1$ is set at a higher pressure than PR$_2$. When the directional-control valve is shifted so as to vent the rod end of the cylinder, air pressure in the surge tank forces oil out of the tank, which causes extension of the cylinder. The cam-operated, two-way valve is shifted. Shifting the directional-control valve ports air to the rod end of the cylinder. This causes retraction of the piston rod, and the oil is forced back into the surge tank.

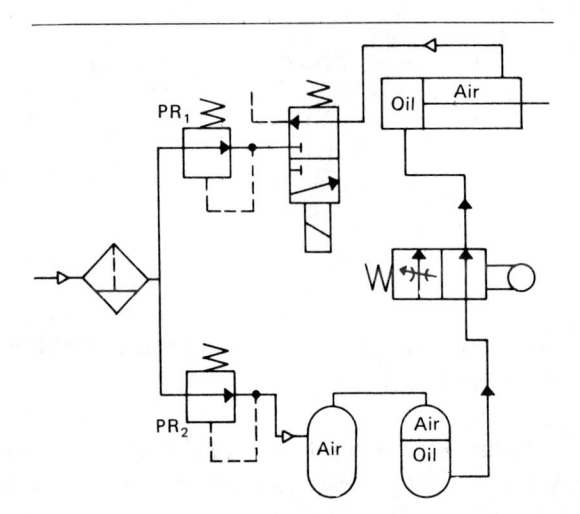

Figure 7.54 Air-oil feed circuit.

The circuit of Figure 7.55 shows a means of air powering a hydraulic cylinder and locking the cylinder at any point during its stroke. Two surge tanks are used to provide the air-oil interface and storage volume for the oil. The air is supplied through an open-center, directional-control valve. In the center position this valve bleeds both of

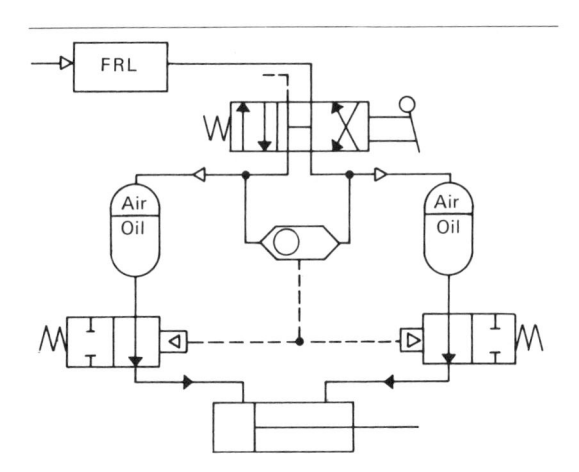

Figure 7.55 Air-powered hydraulic cylinder.

the surge tanks to atmosphere, removing pressure from the oil system. Shifting the directional-control valve pressurizes one of the surge tanks, the shuttle valve, and both air pilots on the hydraulic control valves. Oil is forced out of the pressurized surge tank into the cylinder. Return oil is forced into the unpressurized surge tank by the motion of the piston rod. Centering the four-way, directional-control valve bleeds off all air pressure. The air pilot–operated, hydraulic control valves shift, locking the cylinder in position.

The circuit of Figure 7.56 shows a method of obtaining faster cycling time in a hydraulic circuit by air piloting the main directional-control valve. Air pilots cause the valve to shift faster than it could with hydraulic pilots. The air pilot valve is solenoid operated.

Figure 7.57 shows a counterbalance system utilizing an air-oil circuit. Shop air is used to pressurize the surge tank. Oil is delivered to the cylinder through a check valve–orifice combination. The orifice provides controlled return flow of the oil when the control valve is shifted so as to vent air pressure from the surge tank. When the palm button is released, the load descends fully. The load cannot be stopped at any intermediate position.

These examples of circuits provide an insight into the diversity of

Figure 7.56 Air-piloted hydraulic circuit.

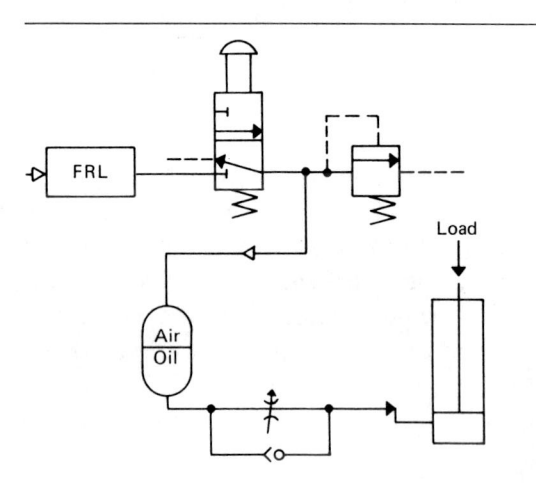

Figure 7.57 Air-oil counterbalance system.

control circuits that can be devised to solve application problems. The possibilities are limited only by the vision of the designer.

Summary

A fluid power circuit represents the synthesis of an application's design requirements. Essentially, circuit requirements are classified into three principal functions—namely, directional control, pressure control, and flow control. Most circuits are constructed with combinations of two or more of these functions.

Sample circuits were used in this chapter to illustrate variations of how to construct pressure-control or flow-control circuits. The functions of pressure-control circuits commonly applied and known in industry include limiting maximum system pressure, sequencing of circuit operations, reducing pressure, counterbalancing of loads, unloading circuits, accumulator applications, fail-safe circuits, and two- or three-pressure systems. It is helpful to remember that, within a fluid power system, pressure control represents force or torque control.

Other sample circuits include flow-control systems such as meter-in, meter-out, and bleed-off open-center circuits. Additionally, flow-dividing and flow-integrating circuits, feed circuits, decelerating circuits, and synchronization of fluid motors are illustrated.

In addition to the above two energy-control functions, virtually all fluid power circuits require directional control.

Understanding the theory of circuit design, the functional operation of components, and application design requirements should enable you to design basic circuits that can be constructed and tested within the laboratory. Infinite energy control is central to efficient and safe fluid power circuits. The starting point in designing any circuit is to have a thorough understanding of the job requirements. Once a circuit is initially formulated, examine ways to improve and simplify your solution. Eventually, you have to construct and put the circuit through a performance-testing phase.

Review questions and problems

1 What is the basic purpose of a fluid power circuit?
2 What do pressure drops in a fluid power circuit indicate?
3 Explain the difference between an open-loop and a closed-loop system.
4 What is the first consideration in the design of a fluid power circuit?
5 What are the three functional open-loop circuit classifications?
6 What classes of flow control are there by method?
7 It is possible to achieve pressure control by means of a pump. True or false?
8 It is possible to achieve directional control by means of a pump. True or false?
9 Distinguish between the types of open-loop systems.

260 Fluid power circuits and systems

10 What type of control circuit is a feed circuit? What are the types of this class of circuit?

11 Distinguish between a transfer circuit and a spindle-drive circuit as used on machine tools.

12 What is the purpose of decompression in a press circuit?

13 In an open-loop circuit, what controls the performance of the circuit?

14 What difference does it make if the flow-control valve is placed (a) in the pressure line ahead of the directional-control valve or (b) in one or the other of the cylinder lines beyond the directional-control valve?

15 To what do the terms "meter-in," "meter-out," and "bleed-off" apply? Do they refer to a particular type of control in each case (yes/no) or to the method of application of the *same* control (yes/no)?

16 What two basic effects does cycle time have on the design of a fluid power circuit?

17

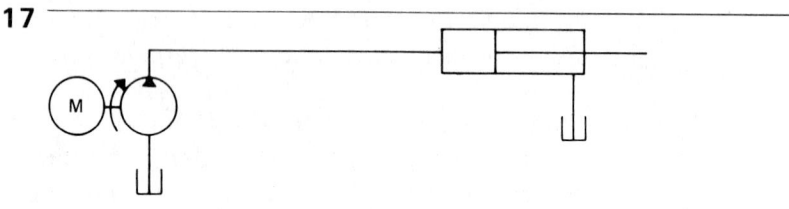

The circuit shown in the figure has the following dimensions: cylinder, 4-in. bore × 18.4-in. stroke; pump, eight pistons, each 0.25 in^2 × 1-in. stroke; volumetric efficiency, 95%.

Problem: If it takes 10 sec to extend the cylinder full stroke, (1) what is the flow rate or discharge rate from the pump? (2) what is the speed of the prime mover?

18

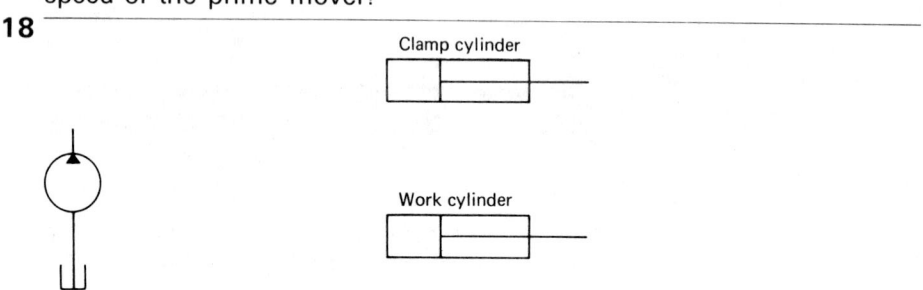

Properly complete the circuit diagram in the figure, using a pressure-control valve. The clamp cylinder moves out first to hold the work piece, then the work cylinder moves out to perform the work function.

19

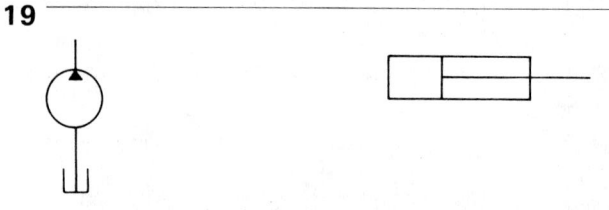

261 Review questions and problems

20. Complete the circuit diagram in the figure, for a closed-center system. Include all necessary components and controls.

21. What, if anything, is wrong with the circuit diagram in the figure?

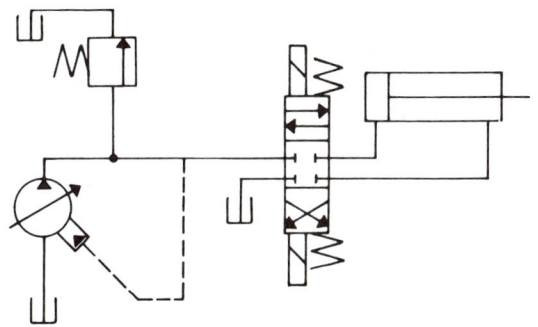

22. What is wrong with the circuit in the figure?

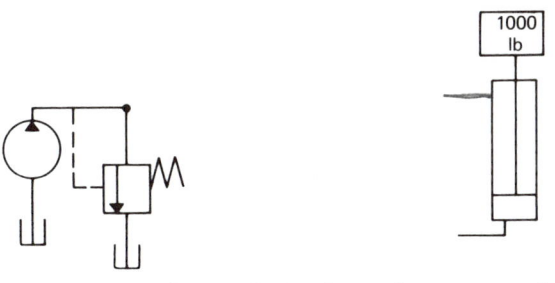

Using an ANSI symbol, show the proper directional-control valve for the circuit in the figure. What other type of control, if any, would help improve the operation?

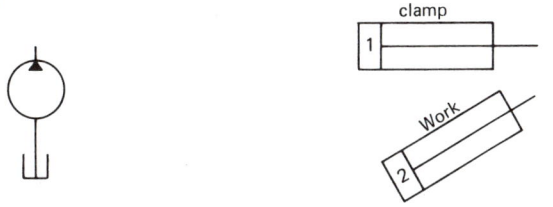

23. Complete the circuit in the figure for the following operation: Cylinder 1 extends first, stops against the work piece; Cylinder 2 moves out to

perform the operation. What is the name of the essential control in this circuit? What kind of control valve is it?

24

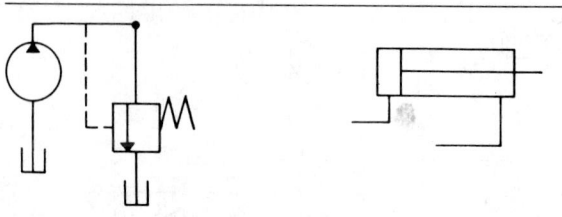

Complete the circuit diagram in the figure with the proper directional-control valve for limit switch control, unloading the pump in neutral, and automatic return to neutral when the electric signal is removed.

25

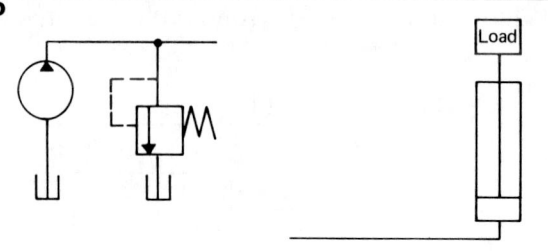

Complete the circuit in the figure by adding the symbol for the directional-control valve that will allow return of the cylinder under load.

26

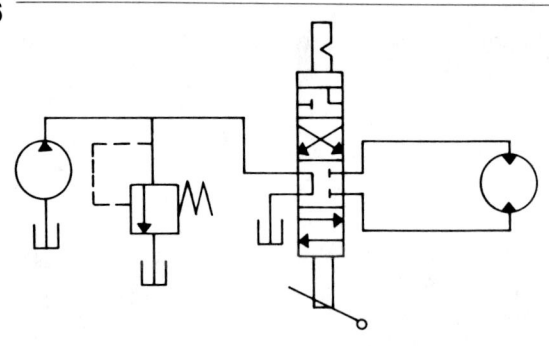

Describe fully the valve in the circuit in the figure. What, if anything, is wrong with the use of this valve in such a circuit?

27

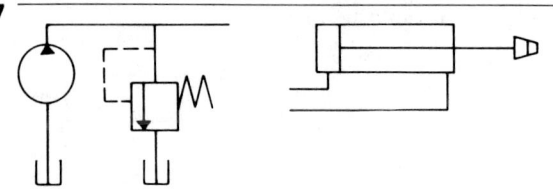

Complete the circuit in the figure, using a mechanically operated valve generally used to slow down the cylinder at the end of its

stroke. Also show the directional-control valve you would use if remote pilot operation were desired.
28 In each of the circuits illustrated, what type of pump has been used? What is the additional control valve shown in each circuit diagram used for?

Selected readings

- Basal, P. R., Jr. (ed.) *Mobile Hydraulics Manual* M-2990-S. Sperry Rand Corporation, Troy, Michigan, 1967.
- Henke, Russell W. *Closing the Loop.* Huebner, Cleveland, 1966.
- ———. *Introduction to Fluid Mechanics.* Addison-Wesley, Reading, Massachusetts, 1966.
- ———. *Introduction to Fluid Power Circuits and Systems.* Addison-Wesley, Reading, Massachusetts, 1970.
- Johnson, Olaf A. *Fluid Power—Pneumatics.* American Technical Society, Chicago, 1975.
- Pease, Dudley A. *Basic Fluid Power.* Prentice-Hall, Englewood Cliffs, New Jersey, 1967.
- Pippenger, John J., and Tyler G. Hicks. *Industrial Hydraulics,* 2nd ed. McGraw-Hill, New York, 1970.
- Pippenger, John J., and Richard M. Koff. *Fluid Power Controls.* McGraw-Hill, New York, 1959.
- Stewart, Harry L. *Hydraulic and Pneumatic Power for Production,* 3rd ed. Industrial Press, New York, 1970.
- Stewart, Harry L., and John M. Storer. *Fluid Power.* Howard W. Sams, Indianapolis, Indiana, 1968.

Chapter 8
Instrumentation

Instrumentation is a broad term that covers the hardware for making measurements and performing control functions in a system. As this chapter points out, there are two broad functions that instrumentation may perform: (1) signal generation—to represent the state of a physical variable of the system, or (2) signal processing—to do something with the signal such as perform a control function. The chapter discusses, qualitatively, some of the techniques used in signal generation and signal processing. A more sophisticated, quantitative discussion of instrumentation is beyond the scope of this chapter. In addition, some of the basic terms related to instrumentation are defined and explained as simply as possible.

Key terms

- **Accuracy** The "estimated uncertainty," expressed as a percentage deviation from the characteristic value, of an instrument reading (output signal).
- **Analog** Referring to proportional phenomena that have a continuous characteristic curve; instruments in which the output signal (dependent variable) is proportional to the independent variable in a continuous fashion.
- **Characteristic curve** A plot that relates the state of the measured variable to the output signal at all points over the range of the instrument.
- **Deadband** The minimum excursion of an instrument that produces no sensible output signal.
- **Dependent variable** The phenomenon internal to the measuring system that changes as a function of the independent variable being measured, usually the output signal.
- **Full-scale accuracy** The accuracy (tolerance on an output signal) of any (instrument) reading over the entire working range of the instrument.
- **Hysteresis** The phenomenon of variations in output signal level that occurs at the same operating point on an instrument's characteristic

---

Portions of the following chapter have appeared in the magazine *Machine Design* 40, (1968); Russell W. Henke, *Introduction to Fluid Mechanics,* Addison-Wesley Publishing Company, Reading, Massachusetts, 1966; and the magazine *Product Engineering*.

curve, depending on whether the point was approached in an increasing direction or a decreasing direction, relative to the independent variable.
- **Independent variable** The physical phenomenon that changes in response to stimuli outside the measuring system.
- **Instrument** A device to sense the state of a variable and produce an output that uniquely describes the state.
- **LVDT (linear variable differential transformer)** An electromechanical transducer in which electric properties of AC voltage output and phase are proportional to the magnitude and direction of mechanical displacement.
- **Piezoelectric effect** The phenomenon of change of electric characteristics as a function of stress. Can be used to measure pressure level and force level or functions thereof.
- **Point-to-point accuracy** The accuracy at any specified operating point on the characteristic curve.
- **Pressure** The distributed reaction of a force acting on a confined fluid.
- **Pressure gage** A device for measuring pressure level; generally uses the force-balance principle of storing energy in an elastically deformed member, the deflection of which is proportional to the pressure level.
- **Response time** Time required for any part, or all, of a physical system to respond to a change of state.
- **Sensitivity** A measure of an instrument's ability to detect changes in the independent variable being measured. The more sensitive the instrument is, the narrower its deadband.
- **Strain gage** A device that makes use of the phenomenon that the electric resistance of a wire (conductor) changes in proportion to its strain (elastic deformation). Since strain is proportional to stress, this phenomenon can be used in the measurement of force level.
- **Time** The standard second is equal to the interval occupied by 9 192 631 770 cycles of the radiation associated with the $(F = 4, MF = 0) - (F = 3, MF = 0)$ transition of the caesium 133 atom when unperturbed by external fields.
- **Transducer** A device to transfer energy from one system to another.

An *instrument* is a device for measuring and indicating the condition, or state, of a physical variable. The condition or state of the physical variables associated with a fluid power system may have to be determined for one of two purposes:

1 data acquisition—the collection and/or recording of data on system performance, or
2 control—maintaining system output within predetermined limits.

Many measurable variables are present in even the simplest fluid power system. For example, consider a circuit that contains a pump, a motor, and a control valve, all interconnected by tubing. Fluid enters the inlet side of the pump, driven by a prime mover. The pump input is torque T_1 at rotational speed N_1. The fluid emerges from the pump at an elevated level of potential energy, discharge pressure p_1, and flow rate Q_1. Measurements that could be taken at the pump include input speed, input torque, delivery pressure, delivery flow rate, delivery fluid temperature, inlet fluid temperature, and inlet pressure.

After leaving the pump, the fluid is forced through the conductor to the control valve. Resistance to flow through the line produces a line loss or flow-pressure drop. Thus entry conditions at the control valve are p_2 and Q_1. Because of internal leakage in the valve, the outlet flow rate is $Q_2 = Q_1 - q_{sv}$, where q_{sv} = valve leakage flow. Because of viscous flow losses through the valve, the fluid loses energy, which is reflected by the pressure drop, $p_2 - p_3$. Measurements that could be taken at the valve include inlet pressure, inlet flow rate, outlet pressure, outlet flow rate, and temperature. Line loss between the valve and the motor is represented by pressure drop, $p_3 - p_4$.

Conditions at the motor-inlet port are inlet pressure, p_4, and inlet flow rate, Q_2. The exit conditions are pressure, p_5, and flow rate, Q_3. The output from the motor is torque T_o at speed N_o.

Thus, in this simple system, there are two torque, two speed, five temperature, six pressure, and three flow-rate measurements that can be made. The number of measurements actually made depends on the requirements of the application. Table 8.1 summarizes the physical variables most commonly encountered in fluid power systems.

Signal-generating instruments may be considered interface devices between the physical variable in its own environment and the outside world. For example, a gage is exposed to pressure. It *transduces* this physical effect to produce a sensible output signal discernible to the observer—for example, a pointer indication.

Instrument output

The function of an instrument is to sense the state of the variable and produce an output that uniquely describes that state. Instrument

Table 8.1 Fluid power variables

Physical variable	Definition of variable	Dimension of variable	Units of variable
Displacement	Change of position of a body, linear or rotational.	linear–length rotational–angle	ft, in., m deg, rad
Velocity	Change of position with respect to time.	linear–length/time rotational–angle/time	in./sec, ft/sec deg/sec, rad/sec
Acceleration	Change of velocity with respect to time.	length/time^2 angle/time^2	in./sec/sec, ft/sec/sec, deg/sec/sec, rad/sec/sec
Force	External reaction on a body which tends to produce a change of position.	force	lb, dyne
Time	A basic descriptor of an interval during which a system may be subjected to a change in state.	time	sec, min
Pressure	Distributed reaction of a force acting through a fluid.	force/area	lb/sq in., lb/sq ft, dyne/sq cm
Flow rate	Volume of fluid passing a reference point per unit time.	volume/time	cu in./sec cu cm/sec
Mass flow rate	Mass of fluid passing a reference point in unit time.	mass/time	slug/sec lb-sec/ft
Torque	Moment of a force acting at a distance from an axis of rotation.	force × length	lb-in., dyne-cm
Temperature	Measure of the internal heat energy of a substance.	Btu, calorie	deg F, R, C, K
Viscosity	Measure of the resistance of a fluid to shearing.	force × time/length^2	lb-sec/ft^2, poise = dyne-sec/cm^2

outputs can be mechanical, electric, thermal, or fluid.

Typical *mechanical* instrument outputs include the motion of a pointer along the dial scale, a digital read-out, the position of a vane, the rotation of a propellor or turbine wheel, etc.

Electric outputs include a DC voltage proportional to the state of the variable AC frequencies, phase shifts, etc. Most electric indicating devices make use of both electric and mechanical phenomena to provide an output.

Thermal device outputs are generally in the form of the length of a column of gage fluid, related to temperature, or an electric signal generated as a result of thermoelectric effects. Radiation phenomena can also be used, as with infrared devices. The output from the latter is usually an electric voltage. Photoelectric devices also use radiation effects (light intensity) to produce a proportional electric voltage signal.

Fluid-effect devices generally employ a mechanical output. Some designs use an electric output proportional to the motion of some mechanical device.

A *transducer,* by definition, is a device that transfers energy from one system to another. Current practice in the field of instrumentation suggests a more restricted viewpoint. That is, a transducer is a device that senses a system variable and generates a signal uniquely describing its condition or state. In many instances, the terms "transducer" and "instrument" are used interchangeably. Perhaps a more useful distinction would be:

- an **instrument** is a device used for data acquisition, and
- a **transducer** is a device used in control situations.

▶ **Example** A simple illustration can be used to show the function of a variable-sensing device—either an instrument or a transducer. Figure 8.1(a) shows the relationship. The premise is that the instrument is designed to operate in the dimension in which the physical variable to be measured exists. That is, you do not try to measure acceleration with a device designed to measure torque nor attempt to measure AC current with an instrument designed to read DC voltage.

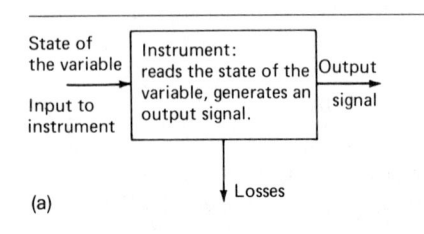

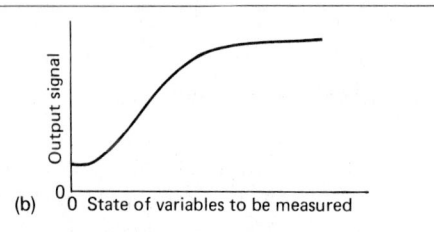

Figure 8.1 Relationship of instrumentation to physical variables.

The instrument is designed to accept the input variable, as indicated in Figure 8.1(a). It operates on the input signal to produce a unique output signal. Some losses are incurred in the processing of the signal, since no physical process is 100% efficient. *All data acquisition and control functions consume energy in order to perform the function itself.* This is a fact frequently overlooked.

Figure 8.1(b) illustrates how the instrument characteristic is most frequently displayed. It is a *characteristic curve,* relating the condition of the state of the measured variable to the output signal, at all points over the range of the instrument. The state of the measured variable, usually plotted along the abscissa, is sometimes called the *independent variable.* This is because the output, which is called the *dependent variable* and is plotted along the ordinate, depends on the variation in the measured variable.

The point was made previously that all instruments and transducers undergo energy losses during the performance of their measuring function; that is, they exhibit an efficiency somewhat less than 100%. One of the great problems in the field of instrumentation is to determine just what is the real efficiency of a particular instrument. In addition to energy loss, each instrument is unique unto itself as far as tolerances on its parts are concerned. Thus 10 apparently identical instruments are, in fact, not identical. One might say that each instrument has its own "personality"—it will react in an individual manner to a given input.

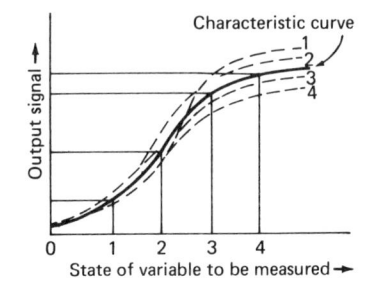

Figure 8.2 Family of characteristic curves measured by identical gages.

This is perhaps best illustrated by Figure 8.2. Assume you have four instruments, all supposedly identical. If identity did exist, the characteristic curve of Figures 8.1 and 8.2 would exactly describe the performance of all four instruments. You know from practical experience that this is not the case. Because of energy loss in the instruments, tolerance differences in their manufacture, varying degrees of wear, etc., each instrument will react slightly differently to the same input. At Point 1 in Figure 8.2, for example, Instruments 2 and 4 happen to fall right on the characteristic curve, while Instrument 1 falls below and Instrument 3 above the curve. This produces a range of readings for the same input. Similarly, there exists a range of readings for Points 2, 3, and 4 on the curve. Which of the instruments is reading correctly? It depends on the particular interpretation of "correctness."

If we assume that all readings must fall on the so-called characteristic curve to be acceptable, *none* of the instruments is acceptable over the full range of the independent variable V_1. In addition, only Instruments 2 and 4 are acceptable at Point 1; only Instrument 3 at

Point 2; none at Point 3; and none at Point 4. It is apparent that we must adjust our thinking as to acceptability, particularly if these are the best devices available for the type of measurement to be made.

Accuracy

This brings us to the concept of *accuracy*. Suppose that the deviation of the reading of Instrument 1 at Point 4 was 2% above the characteristic curve, and that of Instrument 4 was $1\frac{1}{2}$% below the curve. We might select limits of accuracy of + or −2% at Point 4 as acceptable performance. Thus all four instruments would meet our requirements.

But suppose our process dictated that no more than ±1% deviation from the characteristic curve could be tolerated. It would appear that only Instruments 2 and 3 could meet this requirement at Point 4. Thus Instruments 1 and 4 would have to be disqualified for use in this application.

We have established that a variation or tolerance on the ability of an instrument to indicate accurately the state of a physical variable must be allowed, if we are to have instrumentation at all. In the preceding example, we decided that ±1% at Point 4 would be acceptable. Does this mean that the instrument is accurate to ±1% over the entire range of the variable? Assume that the value of the output signal at Point 4 is represented by the number 10. Because of the shape of the curve, $Var_{s3} = 8$, $Var_{s2} = 5$, and $Var_{s1} = 2$. Then at Point 4, 1% of $Var_{s4} = 0.1$; at Point 3, it is 0.08; at Point 2, it is 0.05; and at Point 1, it is 0.02. The implication is that the instrument must be capable of generating signals of much smaller amplitude at the low end of the scale than at the high end, if it is to maintain a constant ±1% accuracy over the full scale.

Assume that an instrument is designed to produce an output signal accurate within 0.1 over its full scale. While this constitutes an accuracy of 1% at Point 4, on the high end of the scale, it means an accuracy of only 10% at the low end. It is readily seen that "accuracy" is an illusive thing and must be defined for each application of an instrument.

Each instrument has some form of lost motion inherent to its design and construction. In the case of mechanical devices, this has come to be known as "backlash." For electric and thermal devices, these are termed "hysteresis" losses. In the general language of instrumentation and control, the minimum excursion of the instrument that will produce *no* sensible output is called the "deadband." This characteristic is extremely important in control applications, perhaps less so in instrumentation situations. It is illustrated in Figure 8.3.

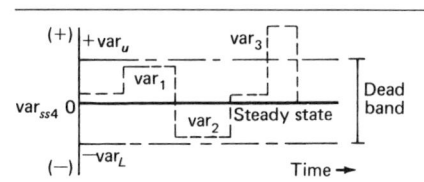

Figure 8.3 Deadband output.

Without considering the magnitude of the signals, assume that Var_{ss4} = steady state condition of the variable Var_1 at Point 4 of the characteristic curve of Figure 8.2. All that this means is that Var_{ss4} would remain constant over a period of time. As a practical matter, very few inputs are constant. Thus we might encounter an input signal variation such as that described by the dotted line in Figure 8.3. During part of the time interval in which we watch the instrument, the independent variable lies above the steady state value and it lies below this value at other times. Assume that the instrument is so designed and built that it cannot "see" variations in Var if they do not exceed $=Var_1$ or $-Var_1$, as indicated in Figure 8.3. The instrument's "deadband" is $2Var_1$ in width. Thus it will not register the signals Var_1 or $-Var_2$, since both of these lie within the deadband, but it will indicate the magnitude of signal Var_3. The greater the sensitivity of the instrument, the narrower the deadband.

Measurement standards

"If no instrument provides an absolute indication of the state of a physical variable, how do we know what is the correct reading?" The best answer that can be offered is that instruments are compared to arbitrarily defined standards. These standards are defined, frequently on an international basis, by law and become the "measuring sticks" against which all instrumentation is calibrated. In the United States these standards are maintained by the National Bureau of Standards, which is a part of the U.S. government. Ultimately, there is only *one* standard element for each defined variable—for example, a standard meter, a standard kilogram, etc.

There is also a set of carefully controlled primary standards maintained in selected locations about the country. These are maintained in as close calibration with the bureau standards as possible. They can be considered the working standards, against which a set of secondary standards—greater in number than the primary—are subsequently calibrated. The instrumentation and gaging equipment of the United States is then calibrated to these standards, within certain defined and known limits of accuracy and to specific tolerances.

Instrument classification

Before proceeding to a discussion of the means used for taking measurements, one further point should be considered. The field of instrumentation can be divided into two areas: instruments used for *signal generation;* and instruments used for *signal processing.* Signal-generating instruments may be thought of as interface devices between the physical variable in its own environment and the outside world. Signal-processing instruments, on the other hand, accept signals from generating devices and do something with them. One of the best known of this type is the cathode-ray oscilloscope. Another is the electronic recorder. In the controls field, the operational amplifier is a typical example.

Signal-generating instruments

One method used to classify instrumentation is based on the type of physical phenomenon employed to generate a signal—mechanical, electric, thermal, nuclear, or chemical. Of these types, mechanical, electric, and thermal instruments are the most often used in fluid power systems.

Mechanical types can be further subdivided into those utilizing rigid members (such as linkages, gears, cranks, shafts, springs, etc.) and those using fluid interactions. Electric types might be divided into those using voltage or current directly and those using magnetic effects.

Measuring pressure

Pressure is the distributed reaction of a force acting through a fluid. The force reacts against constraint, either the surfaces of a container or adjacent fluid. As illustrated in Figure 8.4, an applied force F_1 will cause a reaction to be distributed across the area of the piston surface A_1. The simple equation $p_1 = F_1/A_1$ expresses this relationship.

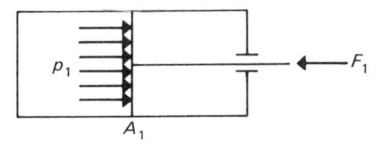

Figure 8.4 Reaction to force.

Pascal's law states that the pressure, p_1, will be transmitted undiminished in all directions throughout the confined fluid. Thus p_1 will act on surface A_1 and produce a force $F_2 = p_1 A_1$. The concept just described is the basis for pressure-measuring instrumentation.

Pressure is measured relative to one of two references, as illustrated in Figure 8.5. Atmospheric pressure is the most commonly used reference because we operate in the atmospheric environment here on earth. Pressure measured relative to atmospheric pressure is called

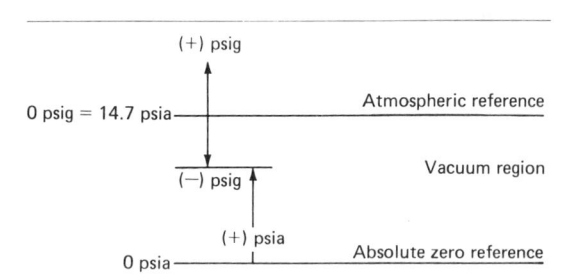

Figure 8.5 Reference points for measuring pressure.

gage pressure (expressed as psig). The second reference is absolute zero. Pressures measured relative to absolute zero are called *absolute pressures* and are expressed as psia. The difference between the two references is 14.7 psia, at sea level. That is, 0 psig = 14.7 psia. Absolute pressures are always positive. Gage pressures can be positive—this is what we mean when we say pressure exists—or they can be negative—which is what we mean when we say a vacuum exists.

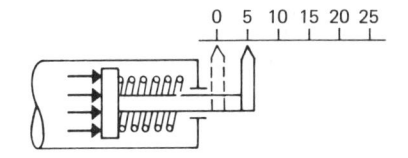

Figure 8.6 Pressure reacting against calibrated mechanical spring.

Consider the illustration of Figure 8.6, in which the piston A_1, of Figure 8.4, has had a spring placed behind it. Pressure acting on the piston surface causes the spring to be compressed until it generates a spring force equal in magnitude and opposite in direction to the pressure force acting on the piston. When this equilibrium position of the piston is reached, no further motion occurs unless the pressure changes. A pointer placed at the end of the piston rod and moving along a calibrated scale indicates the pressure level.

Mechanical

Mechanical pressure instruments—commonly called *gages*—are based on this principle of storing energy in an elastically deformed member. The elastic deformation of the springlike member produces motion that is proportional to the pressure level. This motion is harnessed to provide a read-out on a suitably calibrated scale. When the pressure is removed, the stored energy causes the instrument to return to its zero point.

Bourdon tube The best-known mechanical pressure gage is the Bourdon tube, which is illustrated in Figure 8.7. A metallic tube of a noncircular cross-sectional area is formed into an arc. When pressure is applied to the inside of the tube, it forces the tube toward a circular cross section. This action changes the arc radius of the tube, causing motion of its end. The end motion is transmitted through a linkage, rack, and pinion to a pointer, which indicates pressure level on a circular scale calibrated in the proper units.

Helical gage Another type of pressure instrument is the helical gage. It is based on a tube that is twisted into a helix, as shown in Figure 8.8. When internal pressure is applied, the tube tries to straighten out, causing the end to rotate. The degrees of rotation are proportional to the pressure level. A pointer attached to the end of the tube indicates the magnitude of the applied pressure.

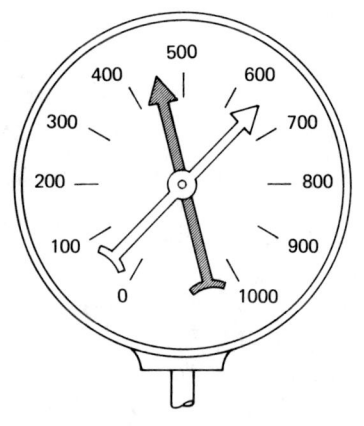

Differential pressure gage Differential pressure gages, generally employing two Bourdon tubes with separate inlets to each, are available. In one version, shown in Figure 8.9, two separate pointers indicate pressures at the two reference points. The lower reading must be subtracted from the higher to obtain the pressure difference between the two points. In another version, the motions from the two Bourdon tubes are fed to a differential mechanism, which subtracts the one from the other, and a single pointer indicates the pressure difference directly. Differential pressure indicators can also be made by opposing two piston units, such as the one shown in Figure 8.6.

In addition to the sensing elements shown, some gages employ tubes wound in a helical shape or wound spirally. Convoluted metal bellows (Figure 8.10) have been used in pressure instrumentation. The bellows acts as the spring and motion of its end is proportional to internal pressure. This motion can be transmitted to an indicating device similar to that described for the Bourdon tube.

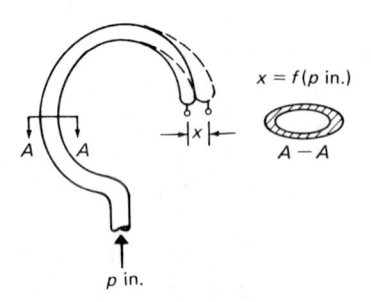

Figure 8.7 Bourdon tube pressure gage.

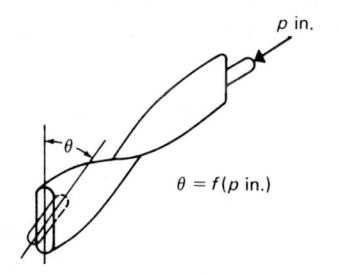

Figure 8.8 Helical pressure gage.

Figure 8.9 Differential pressure gage.

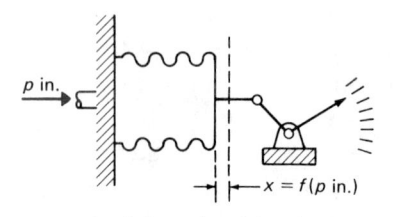

Figure 8.10 Convoluted metal bellows.

Diaphragm sensing device In the diaphragm sensing device illustrated in Figure 8.11, pressure is impressed on one side of a flexible diaphragm. The force generated operates against a spring, causing a proportional output motion. This motion is transmitted to an external indicator that registers pressure on a suitably calibrated scale.

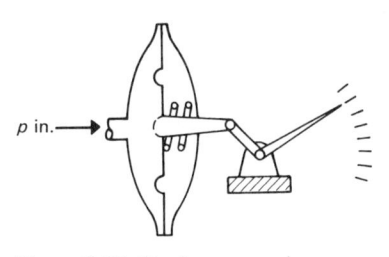

Figure 8.11 Diaphragm sensing device.

There are many variations on these basic types of pressure-measuring devices, but these are the more common mechanical elements used for sensing pressure.

Manometer Where pressure levels are too small to be sensed by a force-deflection type of gage, a manometer can be used. Figure 8.12 shows the basic manometer configuration. Pressure introduced in one leg of the manometer displaces fluid until the head difference, H_d, equals the pressure head. For very low pressure levels the gage fluid is water; mercury is used when pressures in the range of 1–15 psi are encountered.

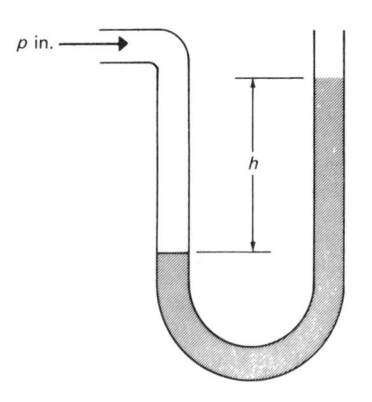

Figure 8.12 Basic manometer.

Read-out from mechanical pressure instruments Read-outs from the instruments illustrated in Figures 8.6 through 8.11 were shown as pointer indications along calibrated scales. The output motions could also be used to operate digital read-outs, indicating pressure directly (Figure 8.13).

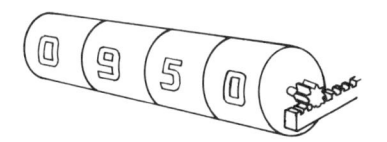

Figure 8.13 Digital pressure read-out.

Electric

Newer, and perhaps more useful, than the mechanical-output signals just described are electric-output pressure instruments. Electric signals offer greater scope for data acquisition and processing, and certainly for control. There are two types, electromechanical and solid-state electric.

Electromechanical pressure instruments The electromechanical types of pressure instruments make use of the force-deflection elements previously described. Instead of the motion driving a pointer or digital read-out it drives an electric device.

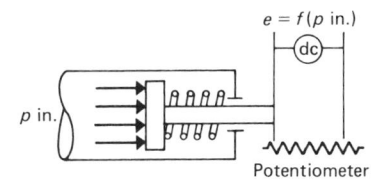

Figure 8.14 Potentiometric signal generator.

Figure 8.14 shows the simplest of these devices—the *potentiometric* signal generator. A potentiometer is connected to the output from the mechanical pressure-sensing element. The voltage output from the potentiometer is proportional to the pressure. This voltage signal

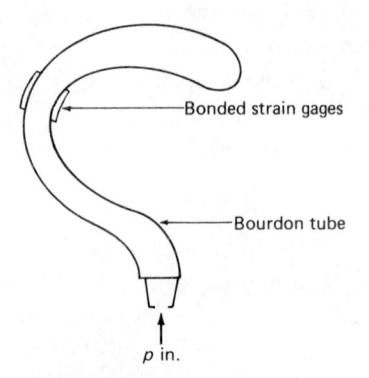

Figure 8.15 Strain gage bonded to a Bourdon tube.

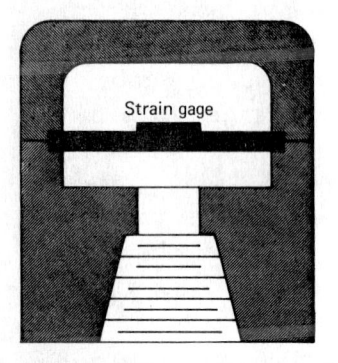

Figure 8.16 Strain gage attached to a metal diaphragm.

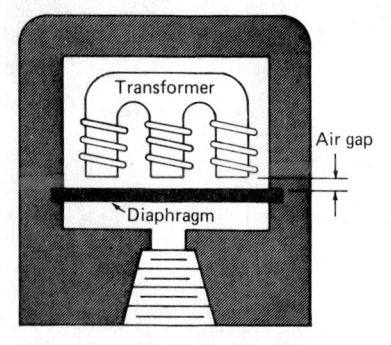

Figure 8.17 Variable-reluctance transducer.

can be used to drive an indicator, fed into a recorder, as a control signal, etc.

A similar version makes use of an electric device called an *LVDT* (*linear variable differential transformer*). The LVDT is more sensitive than the potentiometer. It also gives an AC signal, whereas the potentiometer's signal is DC. The magnitude of the signal is proportional to the pressure.

A third electromechanical type is the *strain gage* design. In some versions a Bourdon tube is used. Strain gages are bonded to the sides of the tube so that, when the tube is deformed by internal pressure, the gages produce an electric output proportional to the pressure level. Figure 8.15 illustrates a typical configuration. Another version uses a simple tube, closed at one end. In still another, shown in Figure 8.16, a metal diaphragm is used. The strain gage is attached to the center of the diaphragm and senses deflections due to pressure.

A newer type of electromechanical design is the variable-reluctance transducer. This is shown schematically in Figure 8.17. A diaphragm is positioned across the pole faces of a transformer core piece. The reluctance of the magnetic circuit is a function of the air gap between the diaphragm and the ends of the pole pieces. The diaphragm deflects as pressure is introduced behind it. Thus the air gap is proportional to the pressure. This type of transducer is gaining in popularity because of its relative simplicity, ruggedness, sensitivity, high response, and repeatability.

The most common of the purely electric devices is the *piezoelectric* pressure transducer. Certain types of crystals exhibit a characteristic change in their electric properties when subjected to stress. In a pressure transducer (Figure 8.18), the crystal is enclosed in a body to which pressure can be introduced. The pressure force acting on the crystal causes stresses, which in turn alter the electric characteristics of the crystal. The electric signal is proportional to the pressure level and can be used by an indicator or fed to a signal-conditioning instrument.

Measuring force

There is a strong parallel between instruments used for measuring pressure and those used to sense force level. A review of the pres-

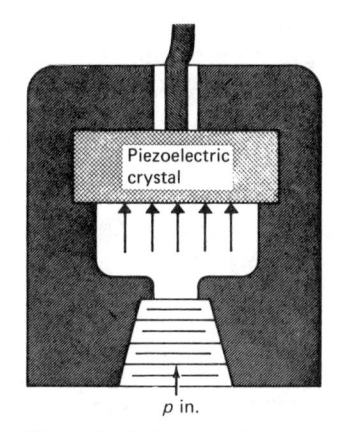

Figure 8.18 Piezoelectric pressure transducer.

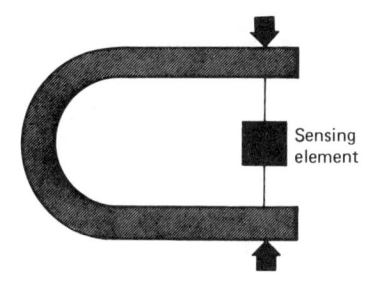

Figure 8.19 Force-measuring device.

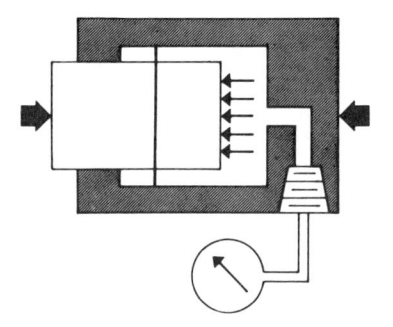

Figure 8.20 Calibrated cylinder.

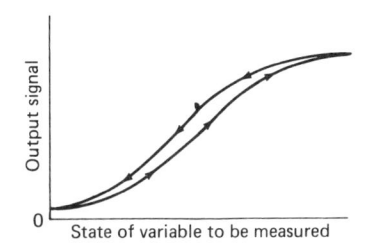

Figure 8.21 Hysteresis of output signals.

sure instruments already discussed will reveal that many of them are, in fact, force-sensing devices that are calibrated in terms of psi rather than pounds force. Any device designed to produce an output signal proportional to a linear motion can be adapted to sense force. Thus the techniques exposed in Figures 8.6, 8.14, 8.16, 8.17, and 8.18 can be considered as force-sensing methods.

A typical construction is illustrated in Figure 8.19. A C-member is subjected to a force reaction such that the legs are deflected. A suitable sensing device is placed between the legs to detect changes in the distance between them. Typically, an unbonded strain gage might be used. In some designs the deflection is large enough that a direct, mechanical read-out is used. For very large force levels the sensing element might be a solid steel bar with strain gages bonded to it. These strain gages would pick up deflections in the bar and convert them to proportional electric signals.

Hydraulic force-sensing devices have been used successfully. As illustrated in Figure 8.20, they consist of a calibrated cylinder to which the force is applied. A pressure is generated in the cylinder according to the relationship: $p = F/A$. This pressure is fed to an indicator calibrated to read in pounds force, rather than psi.

Of concern to designers and users of mechanical types of instrumentation, particularly that subjected to high force levels, are the inaccuracies that can result from varying deflection with different force levels, nonuniform friction characteristics, backlash, etc. All of these factors combine to give every instrument what is known as a *hysteresis* characteristic. This is unique to each instrument—even those of supposedly identical design. Hysteresis might best be illustrated by referring back to the characteristic curve of Figure 8.1, as reproduced in Figure 8.21. Note that there are now two curves instead of one. The lower curve represents the characteristic of an instrument when the readings of output signal are taken while the state of the input variable is changed in an *increasing* direction. The upper curve represents the characteristic curve obtained when the direction of the input variable is reversed; that is, values are *decreased* from a maximum to zero. Note that the instrument does not give the same readings for identical values of input, dependent on the direction from which the input was approached. The difference between these two curves is called hysteresis and it is the result of all the uncontrollable effects that can occur in instrumentation. It can be determined only by testing.

Mechanical instruments are not alone in exhibiting this hysteresis characteristic. Electric devices are also subject to the problem. Since many instruments that are thought of as electric are actually electro-mechanical, they are subject to mechanical problems, as well as those associated with electric phenomena such as residual magnetism, change in resistance with temperature, lumped capacitance, inductive effects, etc. This point is brought up here, not to discourage the reader, but to emphasize the need for extreme caution when approaching the problem of instrumenting a system.

Measuring torque

A torque is a force that acts at a specified distance from some axis of rotation. Thus you might expect that torque-measuring instruments would embody many of the same techniques covered under pressure- and force-measuring instrumentation. This is true, with the additional constraint that rotation of a shaft must generally be permitted in order for the torque to be transmitted.

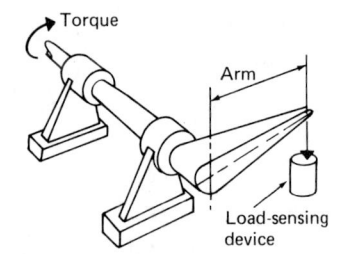

Figure 8.22 Simple torque-measuring device.

The simplest torque-measuring device is that illustrated in Figure 8.22. It consists of a shaft supported in bearings and a lever arm; thus the load cell could be calibrated to read out in torque units. This design does prevent rotation of the shaft.

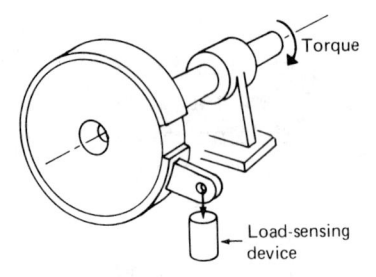

Figure 8.23 Prony brake.

The *prony brake* is the classic torque-measuring device. It is shown schematically in Figure 8.23. A brake is wrapped around a drum of known diameter; when the brake is applied, a friction force is generated between the brake and the drum surface. The magnitude of the friction force is sensed by a load cell placed at a known distance from the axis of rotation. The torque being absorbed by the prony brake is equal to the product of the load-cell reading and the moment arm. All of the energy is converted to heat. Thus, in high-energy systems, cooling of the brake is one of the biggest problems.

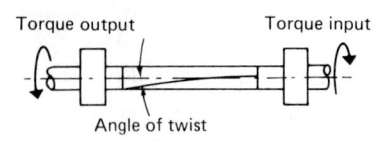

Figure 8.24 Principle of dynamic torque instrumentation.

A type of dynamic torque instrumentation has been evolved around the basic principle illustrated in Figure 8.24. If a shaft of known characteristics—material, diameter, length, etc.—is subjected to a torque loading, it will deflect through a predictable and controllable angle, θ. Stresses are set up in the shaft that are proportional to the angular strain and thus to the applied torque. Two approaches have been made: (1) instruments that measure the strain and put out a signal

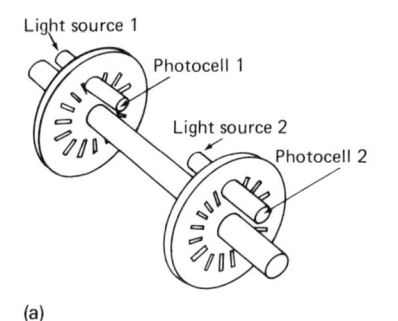

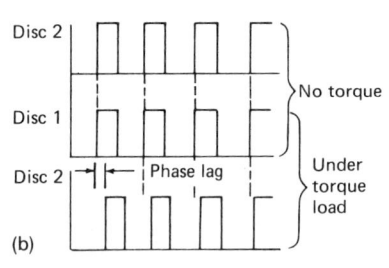

Figure 8.25 Optical torque-sensing signals. (a) The instrument. (b) The signal.

proportional to torque, i.e., strain-gage types; and (2) instruments that measure the actual angular deflection.

Strain-gage torque instruments require slip-rings to transmit electric signals in and out since the strain gages themselves are mounted on the rotating shaft. These slip-rings can be a source of "noise" that hampers the effective performance of the instrument. Efforts to eliminate slip-rings have led to designs that attempt to measure the angular deflection of the shaft. One of these, shown in Figure 8.25, uses optical sensing. Two disks with optically transparent slits placed radially around them are positioned at a known distance from each other. A light source and photoelectric cell are accurately lined up at each disk, so that the slits function as shutters between each light and cell. As the disks are rotated the output is a series of electric voltage pulses generated by the photocells. When there is no torque applied to the shaft, the series of pulses are in phase; that is, the pulses from Disk 2 occur at the same instant as corresponding pulses from Disk 1. When a load is applied the shaft is twisted. As a result, the slots in Disk 2 arrive at Photocell 2 *after* those in Disk 1—they are no longer in phase. The phase lag (the angle θ) is the amount of twist in the shaft. Since the twist is proportional to applied torque, phase lag is also a measure of the torque.

This kind of torque meter is an interesting example of a requirement to combine the signal-generation and signal-processing functions of instrumentation into one device. The output signal from the photocell is a voltage pulse. This would be useless without further processing. The two pulses are fed into suitable electronic equipment that compares them and determines the phase angle. On the basis of other known parameters the processor then provides a read-out in terms of torque.

Another version of the optical torque meter makes use of a reflective stripe placed parallel to the axis of the shaft. A pair of light sources and photocells are used to determine the phase angle by means of light beams reflected from the stripe as it passes beneath them.

The technique shown in Figure 8.25 could be converted to fluidic sensing by replacing the light source and photocells with fluid nozzles–receiver combinations. Pressure pulses would be generated, instead of voltage pulses. These pressure pulses would have to be processed by a fluidic system in a manner analogous to that employed by an electronic system, to determine the phase angle.

Some hydraulic instruments have been developed in which the torque reaction was converted to a hydraulic pressure. This was fed out to pressure-sensing instruments calibrated to indicate torque units. As with all instrumentation, there have been many variations developed on the basic theme. What we have covered here is a sampling of these techniques.

Other variable measurements of time and temperature are critical to the behavior of fluids. Since virtually all physical systems are time variant—that is, their state changes over a period of time—it has become increasingly important to accurately determine the *elapsed time* during which a measurement is taken.

In the early days of reading instruments a technician would literally watch a clock to determine the time interval. This gave way to the stopwatch, followed in due course by the electric timer. Each of these was an attempt to eliminate human error in reading instruments and to improve repeatability.

As requirements became increasingly stringent and the necessity for comparing time-variant data increased, timing functions moved into the domain of electronics. Electronic timers became available with a capability of counting in millionths of a second. Frequency-comparing techniques have extended this into the nanosecond range.

With the increased use of frequency as a timing or comparing function, it has become necessary to generate very accurate frequencies. In the lower ranges tuning forks have been employed. Electronic frequency generators have been designed. And atomic vibrations have been tapped as a fixed-frequency source under very exacting conditions. In many applications, the ability to measure time accurately has been a limiting factor on the results attainable.

Temperature is related to the atomic-energy level of the matter being observed. In a manner analogous to absolute pressure, as discussed earlier, absolute zero temperature is considered to correspond to no energy. We are not usually confronted with this scale of temperature measurement in industry and can content ourselves with a standard based on the properties of water.

For our purposes the freezing point of water, determined under specified conditions, is one reference. This corresponds to 0 °Celsius, 32 °Fahrenheit, 459.6 °Rankine, or 273 °Kelvin—all scales against which temperature is compared. The boiling point of water is taken

Figure 8.26 Thermometer for temperature sensing.

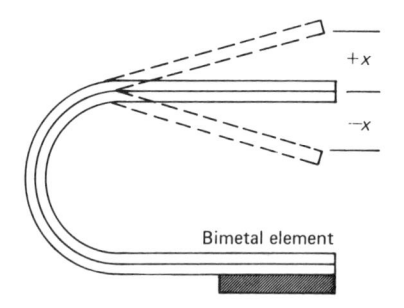

Figure 8.27 Bimetallic temperature-sensing device.

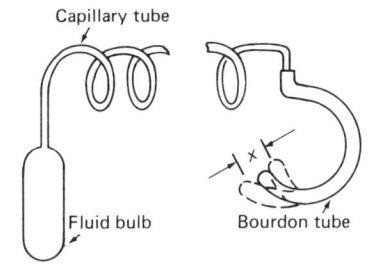

Figure 8.28 Gas-bulb thermometer.

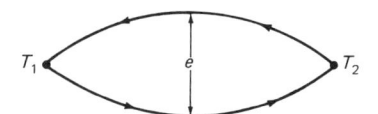

Figure 8.29 Thermocouple.

as another reference, corresponding to 100 °C, 212 °F, 373 °K, and 671.6 °R. The two references correspond to changes in phase of the substance known as water. Note that "temperature" is *not* the scale reading to which it is compared—degrees Celsius, Fahrenheit, Rankine, or Kelvin; temperature is the energy state of a substance and the scale reading is only the comparative indication of what that energy level is.

The most basic temperature-measuring instrument is the thermometer. In one form, a capillary tube is connected to a bulb at its base. The bulb contains a quantity of a gage fluid with a known coefficient of thermal expansion. This is shown in Figure 8.26. As the fluid temperature changes, the volume will vary correspondingly. The height of the column of fluid in the tube is a measure of the temperature of the fluid, and thus of its environment. The instrument of Figure 8.27 makes use of bimetallic material to produce a temperature-responsive output. Two metallic materials of different thermal expansion characteristics are bonded together. When a change in temperature is encountered, the composite material is deflected due to the uneven thermal expansion of the two materials. Since this deflection is predictable and controllable, it can be used to drive a read-out element such as a pointer, a digital read-out, etc., that is calibrated to indicate temperature. The bimetal element might also be used to operate an electric device similar to those described for pressure instruments.

The gas-bulb thermometer is another type of instrument in common use. It operates on the principle that the pressure of a confined fluid is directly proportional to its temperature. As shown in Figure 8.28, a sealed container is connected to a pressure-sensitive element such as a Bourdon tube or bellows by a capillary tube. As the temperature of the container (or bulb) varies, the pressure of the confined fluid varies. A suitably calibrated indicator is attached to the pressure element to provide a temperature read-out. The element could also be used to drive an electric device, as indicated previously.

One of the largest classes of temperature instruments is based on the thermoelectric properties of dissimilar metals in a circuit. These are the *thermocouples*. The Seebeck effect involves a circuit consisting of two dissimilar metals joined at either end, as shown in Figure 8.29. If one end is held at a temperature T_1 and the other end is at a temperature T_2, so that $T_2 > T_1$, an emf will be induced and a current will flow. By holding T_1 at a known constant temperature and measuring the electric effects existing in a known circuit, the temperature T_2 can be determined. In practice, it is usual to immerse the "reference junction'' in an ice bath to maintain T_1 at a constant zero degrees. In other applications, constant-temperature ovens are sometimes used to achieve a constant reference temperature.

The output from a thermocouple can be fed to a suitable meter to indicate temperature directly or it can be fed to a signal processor to perform a data-acquisition or control function.

Measuring displacement

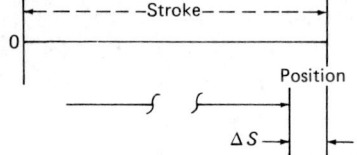

Figure 8.30 Linear scale for measuring displacement.

With the definition of displacement as the change of position of a body, the simplest instrument to measure this change would be the yardstick. It is capable of measuring linear distance between two points. However, something is missing from the concept—the body must be displaced relative to something else. There must be a reference, as shown in Figure 8.30. If a reference is fixed, the linear scale does become the simplest displacement-measuring instrument.

Similarly, in rotary motion the circular scale indicates angular displacement. A whole area of metrology has built up around refinements of these two simple devices—quality control, production gaging, machine tool set-up, etc. Scales, verniers, dial gages, and similar devices are used for measuring this variable.

In another sense the field of displacement measurement connotes signal processing for data acquisition and control. Here visual reference instrumentation is not adequate and an array of electromechanical, optical-mechanical, and other devices have been evolved.

While the electric types of transducers have become better known in recent years, you should not lose sight of the fact that position can also be sensed mechanically, pneumatically, optically, by radioactive means, etc. The reason for the emphasis on electric devices is their flexibility, small size, low power levels, and the general emphasis on electric technology in the past few decades.

Electric signal output

It may be useful, at this point, to further explain the difference between the kinds of electric signals that transducers generate. The first

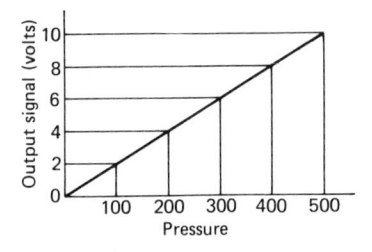

Figure 8.31 Pressure-voltage relationship.

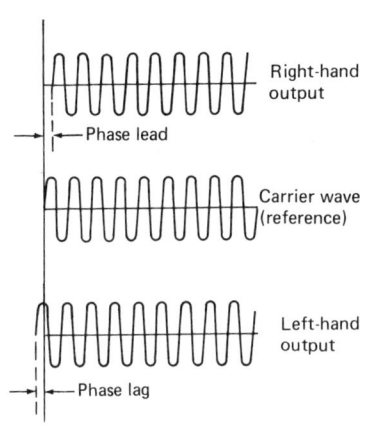

Figure 8.32 Carrier wave-frequency signal.

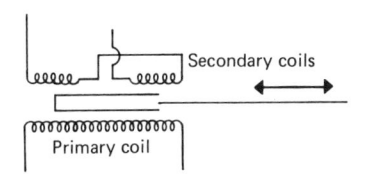

Figure 8.33 Linear variable differential transformer.

kind is a DC voltage that is proportional to the state of the physical variable being measured. The plot of Figure 8.31 could very easily relate the DC output signal as a function of the input variable. Ideally, this would be a straight line (that is, linear), as shown in Figure 8.31. This is a characteristic pressure-voltage line with a slope of 2 volts DC per 100 psi pressure. We can do many things with the DC output signal, from simple indication of pressure level—by feeding it into a voltmeter calibrated to read in pressure units—to initiating a control function.

The second kind of electric output obtainable is an AC signal, which is associated with the transformer instruments such as LVDT's and synchros. The basic input to such instrumentation is a carrier frequency—an AC input of known amplitude and frequency, as shown in Figure 8.32. Consider an LVDT: in Figure 8.33, with the armature centered, the coupling between the primary and both secondaries would be the same. Thus the output signals generated in each would be in phase with the carrier input. The output amplitude would depend on the ratio of turns between the primary and secondary and losses in the system. When the armature is shifted from the center position, the coupling between the primary and secondary on the side of the shift is improved; the coupling between the primary and the secondary on the side opposite the shift is lessened. Thus the signal amplitude of the close-coupled secondary winding is increased, while that of the opposite winding is decreased. Shifting the armature also produces a phase shift between the output signal and the carrier. The amplitude of the output signal is proportional to the displacement of the armature from center position; the phase shift enables us to tell in which direction it has been moved.

This discussion also involves two classes of devices known as analog and digital. An *analog* device is one whose output is proportional to the state of the measured variable. Thus the potentiometric, DC device described in Figure 8.14 is an analog instrument. So is the LVDT covered in the discussion of Figure 8.33, and indeed most of the instruments described so far in this chapter. The difficulty with analog instrumentation is primarily that of maintaining linearity over the desired range of the device. Repeatability can also be a problem, due to hysteresis losses and the like, as can drift of the instrument due to environmental changes such as temperature. In more complex instruments, stability can also become a problem.

These potential difficulties with analog devices have led to increased interest in digital techniques. The output from a *digital instrument* consists of discrete pulses, like those described in discussing Figure 8.25. The amplitude of the pulse is irrelevant as long as it is great enough to be distinguished. Measurement or control is effected by counting the number of pulses and comparing it with some control frequency or number.

A second class of digital instruments makes use of "coded" representations. Each discrete condition of the variable is represented by a unique combination of on and off signals, called the digital code. As before, amplitude does not play a part in the coding function.

Electric instruments for measuring displacement

We can now return more meaningfully to the subject of displacement instrumentation.

Figure 8.34 Electric position transducer.

Potentiometric

The simplest electric position transducers are the potentiometric types, as illustrated in Figure 8.34. The basic principle on which these devices operate is that resistance changes with changing length of conductor. Figure 8.34 shows this in its most rudimentary form—a single wire with a slider moving along it. There will be a voltage drop in proportion to the location of the slider along the wire. Commercial linear potentiometers generally use wound wire, thin film, or a printed circuit (Figure 8.34). By the use of these techniques, a much greater length of resistor can be put in the same linear length. Thus the changes in voltage drop per unit length will be greater. *Rotary potentiometers* are used to sense the rotational position of a shaft, with the general reservation that the number of revolutions is limited.

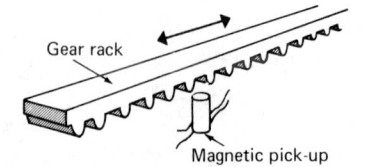

Figure 8.35 Digital device employing gear rack and magnetic pick-up.

Variable transformers

A second class of position transducers is the variable transformers. Essentially, these consist of coils of wire with an armature inside, as shown in Figure 8.33. The distance the armature is pushed into the coils controls the coupling between the primary and secondary of the transformer. Thus a stronger or weaker voltage signal is pro-

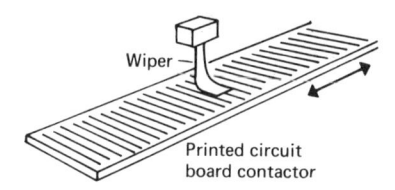

Figure 8.36 Printed circuit board.

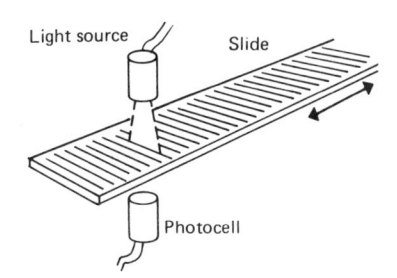

Figure 8.37 Optical version of transducer.

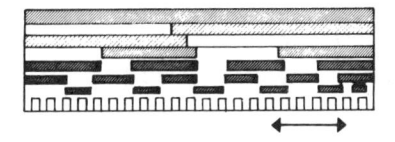

Figure 8.38 Digitally coded version of optical transducer.

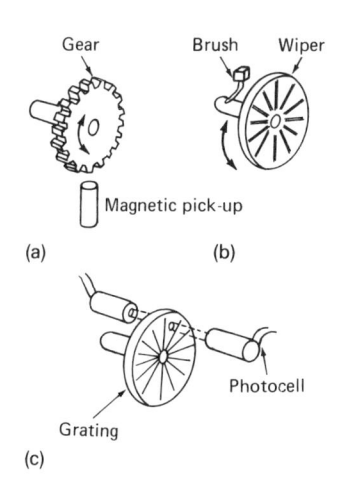

Figure 8.39 Rotary versions of optical transducers.

duced as a function of the armature being pushed in more or less. In the linear variable differential transformer (LVDT) the armature is moved back and forth in a straight-line motion within the coils. The rotary version is termed the RVDT and can be used to sense rotary motion of less than one revolution. The device can also be designed to indicate which way the motion is going.

Digital transducers

Figure 8.35 shows a digital device, which consists of a gear rack and a magnetic pick-up. As the teeth on the rack move past the magnetic pick-up a pulse is generated for each tooth. This type of device is obviously rather crude, since the best resolution that can be realized is the pitch of the rack. A more sensitive device is that shown in Figure 8.36, consisting of a printed circuit board with closely spaced "contacts" and a wiper. A pulse is sent out each time the wiper crosses one of the contacts.

Optical transducers

Figure 8.37 shows an optical version of the transducer. A grid is etched into the surface of a glass strip. A light source is positioned on one side of the strip and a photoelectric cell on the other. As the grid moves between the lamp and the cell, the light beam is interrupted by the grid lines. For each such interruption a pulse is generated. A digitally coded version of this transducer is shown in Figure 8.38. Each finite position of the circuit board is described by a unique digital code. The resolution of the device depends on how close together the minimum divisions are.

Figure 8.39 illustrates the rotary versions of optical transducers. In Figure 8.39(a) there is a gear with a magnetic pick-up, to sense the rotational displacement of a shaft; Figure 8.39(b) shows a printed circuit and wiper type; while Figure 8.39(c) is an optical version, using a grid, lamp, and photocell. The output of all of these devices is essentially the same—a series of electric pulses. Figure 8.40 shows how a pulse-signal type of position control might work. For simplicity's sake, assume that a scale of 100 pulses to 1 in. is selected and the reference, or zero point, is coincident with the fully retracted position of a cylinder. If the process called for moving the

Figure 8.40 Pulse-signal position control.

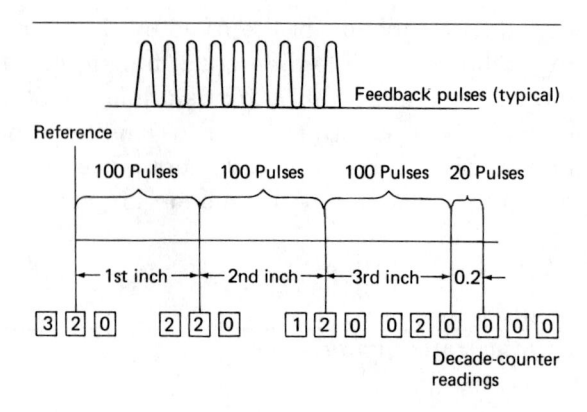

cylinder-piston rod through a stroke of 3.2 in., this would be equivalent to 320 pulses.

The input-to-command signal would be obtained by setting a decade counter, sometimes called a predetermined counter, for 320. As the piston rod extends, the transducer starts sending back pulses. For each one the counter receives, it decreases its setting by one unit until the counter is finally zeroed again. When this occurs, a switch might be actuated, causing a control valve to shift and stop the cylinder. The transducer must have sufficiently fine increments to be able to generate 100 pulses/in. of travel. If, for example, the transducer had only 50 increments/in., the cylinder would actually stroke 6.5 in. before 320 pulses were sent out. Note that the best resolution the system could achieve would be 0.01 in. since it cannot sense anything smaller than a one-pulse difference. A further complication is the response time of the control valve. Even after the counter is zeroed, a finite time is required to send out the closing signal, shift the valve, and bring the cylinder to a halt. These dynamics must be considered in designing a system to meet the application requirements.

Synchros

An electric device called a *synchro* has been employed extensively in rotary position-control applications. These transducers are more complex than the types considered so far. From a black-box viewpoint, they can be thought of as a rotating device that puts out an electric signal, which is a function of the rotation of the shaft. Shown schematically in Figure 8.41(a), a synchro is similar to a three-phase motor. A similar device, shown in Figure 8.41(b), is a *resolver*.

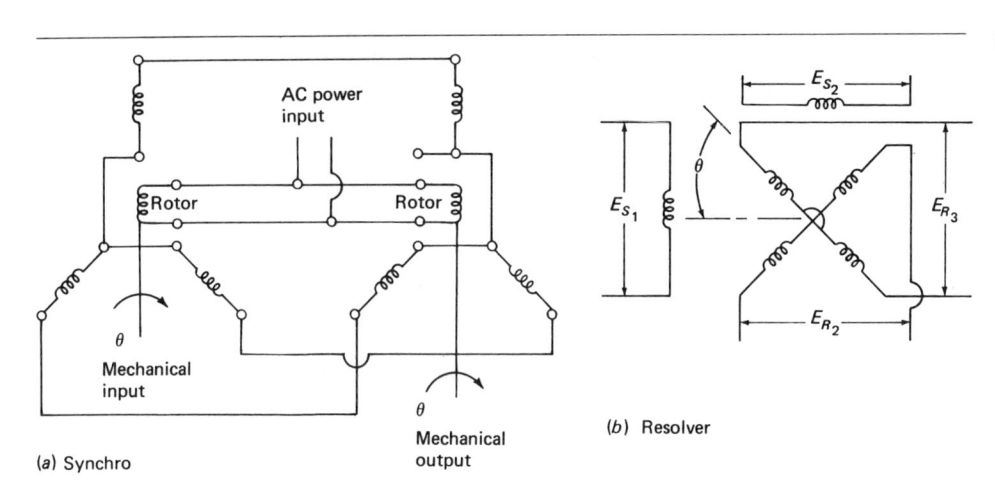

Figure 8.41 Synchro resolver.

(a) Synchro

(b) Resolver

Mechanical instruments for measuring displacement

The devices talked about so far have been electric transducers. It is possible to sense position by other means. A mechanical position feedback, most applicable where small displacements are involved, is shown in Figure 8.42. The control valve is shifted by an input signal. This ports oil to the actuator, causing the piston rod to stroke outward. The mechanical feedback linkage causes the control valve to be zeroed, or nulled, when the rod has moved through the desired distance. Zeroing the valve shuts off flow to the cylinder, and the system stops until another input signal causes the valve to be shifted again.

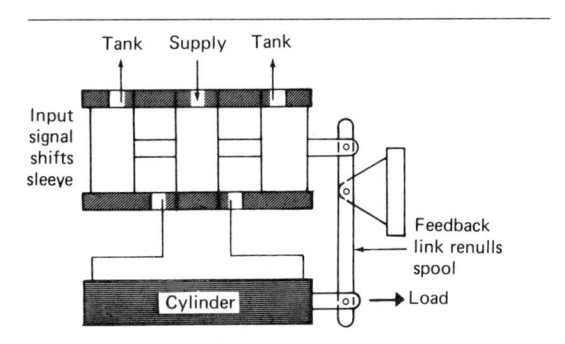

Figure 8.42 Mechanical position feedback.

Pneumatic transducing is another possibility. Figure 8.43 illustrates how a pneumatic nozzle and relay might be used to sense position and control a process. Air is supplied to the nozzle, and the pressure

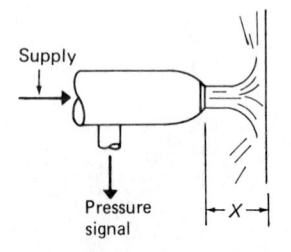

Figure 8.43 Pneumatic transducer.

built up depends on the distance between the nozzle and the surface being sensed. The sensing pressure can be fed through a pneumatic relay to operate a control.

With the advent of fluidics and related all-fluid digital logic, it is now possible to use the pulse techniques previously described for electric methods.

Measuring velocity

Velocity is the rate of change of position, $v = \Delta s/\Delta t$. It is a time-based phenomenon.

When displacement was discussed in the last section, no reference was made to the *time* it took to accomplish the change of position. Displacement is a quantity—a total change in position. It would appear possible to use much of the displacement instrumentation just described as velocity instrumentation, with the introduction of time. This is, essentially, what is done in practice.

The rack and magnetic pickup of Figure 8.35 are used as velocity transducers by time-integrating the pulses, that is, counting the number of pulses per unit time. If the distance between each gear tooth and the number of gear teeth passing the pickup per unit time are known, the linear velocity can be found. The linear pulse generators shown in Figures 8.36 and 8.37 are used in a similar manner.

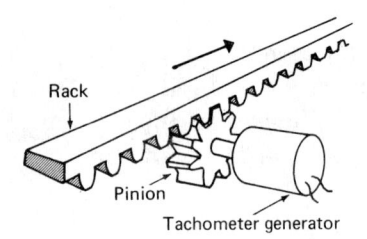

Figure 8.44 Tachometer generator.

The most widely used instrument for measuring velocity is the *tachometer generator* (Figure 8.44). This is an electric device that delivers a DC voltage proportional to its rotational speed. Thus the rack and pinion is required to convert linear motion to rotational motion for the tachometer generator. The tachometer generator is, obviously, one of the favorite angular-velocity transducers, due to its voltage-speed characteristic, simplicity, reliability, etc.

A *hydraulic pump* can be used as a rotational velocity transducer, since the output from the pump is proportional to rotational speed. There is at least one commercial servo system utilizing this principle on the market. The rotary position transducers shown in Figure 8.39(a), (b), and (c) can also be utilized as rotary velocity instruments by counting generated pulses for a fixed time interval.

Measuring acceleration

Acceleration is the rate of change of velocity—the second derivative of displacement with respect to time:

$$a = \frac{\Delta v}{\Delta t}$$

This class of transducers functions, basically, according to D'Alembert's principle

$$F = M\frac{\Delta v}{\Delta t}$$

The force is then used to develop an electric signal that is proportional to the acceleration.

Since the force generated is a function of a mass, acceleration transducers generally use a mass in combination with a sensing device that generates an electric signal as a function of the force exerted on it. One of the most common uses a strain-gage element (Figure 8.45). In the bonded strain-gage type, the mass-acceleration force causes the deflection of an elastic member. This deflection causes a change in resistance of the strain-gage element and thus a varying electric output signal that is proportional to acceleration. In the bonded strain-gage type the output signal is essentially the same.

A second type of accelerometer uses a piezoelectric crystal to develop the electric signal proportional to the force exerted. A piezoelectric material is one whose electric properties vary with changes in stress. Thus, when the crystal is loaded, as by an acceleration force, the resulting electric signal is proportional to the rate of change of force.

A variable-inductance transducer could be used, in which a calibrated spring would oppose the acceleration force. Deflection of the spring would be proportional to acceleration and motion of an armature into a coil would give the desired electric output signal (Figure 8.46).

The new field of fluidics has provided a type of rotational acceleration transducer based on the vortex amplifier. The output of this device is a pneumatic pressure proportional to the angular acceleration.

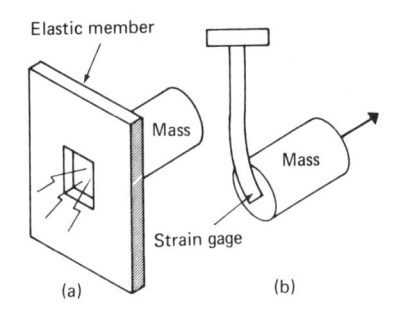

Figure 8.45 Acceleration transducer using a strain gage.

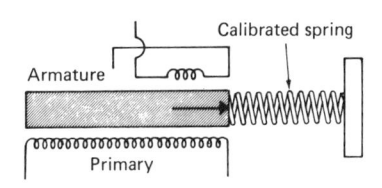

Figure 8.46 A variable-inductance transducer.

Measuring viscosity

Measurement of fluid viscosity on anything but a laboratory basis is difficult. There are two approaches to the problem. One results in a comparative output, measured in terms of time required to pass a sample of fluid through a standard orifice. This is the Saybolt Second Universal method; it compares fluids relatively, but does not give a quantitative answer as to what the viscosity really is. Another similar method employs a calibrated ball dropped into the fluid. The time it takes to fall a specified distance through the fluid is considered to be a measure of the relative viscosity.

The second method involves the use of a "drag-cup" viscosimeter. Essentially, this method consists of driving a thin metal can of specific dimensions, immersed in the fluid. The torque required to drive the can is a function of the surface area, surface velocity, can diameter, and viscosity of the fluid. Since all but the viscosity and torque can be predetermined, and the latter can be measured directly, the viscosity becomes the one unknown. Torque or power input to the drive member of the instrument can be calibrated in terms of fluid viscosity. If an electric motor is used to drive the device, the current demand of the motor can be used as an electric signal analogous to the viscosity.

Measuring flow of fluids

Quantities of fluids can be determined on one of two bases:

1 Volume of fluid (in^3, ft^3, gallons, liters, cc, etc.)
2 Mass of fluid
   a. pounds mass (lb/in^3, lb/ft^3)
   b. slugs mass (slug/sec = lb sec/ft kg)

Volume

One form of flow instrumentation is based on the sharp-edged orifice. It can be shown that the relationship

$$Q = C_d A_o \sqrt{2\,gh}$$

expresses the flow rate through such an orifice as a function of orifice geometry and head (pressure) difference across the orifice. If we consider all factors for a given orifice to be constant, this relationship reduces to

$$Q = K\sqrt{\Delta p}$$

Orifice instruments are designed to measure the pressure drop across the orifice and indicate flow rate as a function thereof. As shown in Figure 8.47, an orifice meter consists of a section of reduced cross section with properly located pressure taps upstream and downstream. The pressure difference, $p_1 - p_2$, can be applied to all the types of pressure-measuring instrumentation discussed in the early part of this chapter. Their read-outs would be calibrated to read in flow-rate units, instead of pounds per square inch.

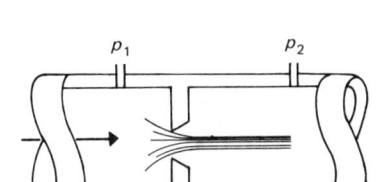

Figure 8.47 Fixed-orifice meter.

A variation of this technique is the variable-area orifice meter, sometimes known as a rotometer, shown in Figure 8.48. A weight of known mass is placed in a tapered tube. In any position along the length of the tube, a fixed pressure drop is required to hold the weight up against gravity. The relationship for this type of instrument might be expressed in the form $Q = K'A_o$, where A_o is the area of the annular orifice defined by the walls of the tapered tube and the surface of the weight. When the flow rate increases, the annular area must also increase. The only way this can occur is for the weight to rise to a position where the D_i of the tube is larger. Thus the vertical position of the weight is analogous to the flow rate. In direct-reading devices, a scale is provided adjacent to the tapered tube. The scale is calibrated in flow-rate units. In Figure 8.48, an LVDT provides an electric signal proportional to the flow rate, which can be used for remote indication or control purposes.

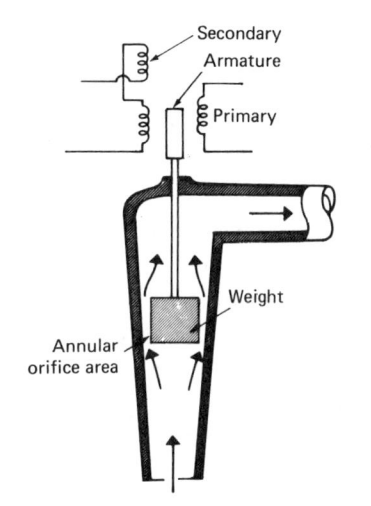

Figure 8.48 Variable-area orifice meter.

The venturi meter (Figure 8.49) is another type based on the generation of a pressure differential as a function of flow rate. In this respect, it is similar to the orifice meter of Figure 8.47. One pressure tap is located in the full-diameter section of the pipe, while the second is located at the throat. The pressure differential, $p_1 - p_2$, is a function of the flow rate Q. Note that these meters are based on the continuity of flow principle; that is, the flow rate is equal to the product of the cross-sectional area of the conductor and an average flow velocity, $Q = Av$.

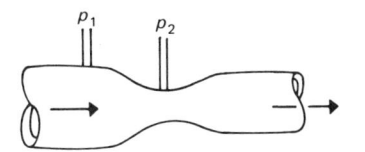

Figure 8.49 Venturi meter.

A second group of flow-rate meters function on the jet-impingement reaction principle. These are the propellor and turbine meters, as

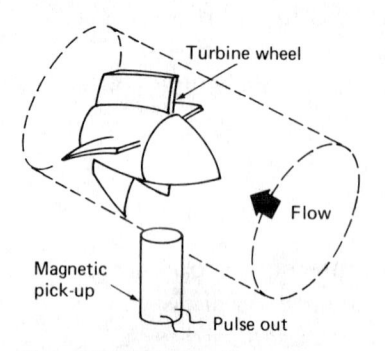

Figure 8.50 Turbine flow meter

shown in Figure 8.50. Fluid, flowing in through the meter, impinges on the turbine blades, causing a force reaction proportional to the mass flow rate. This, in turn, produces a torque reaction on the turbine wheel and it rotates at a speed proportional to the flow rate. The most common type of meter uses a magnetic pick-up to sense the blades as they pass by. This generates a series of pulses, the frequency of which is also proportional to the flow rate. These pulses can be processed in any of the ways previously discussed for digital instrumentation.

Another version uses a gear train to transmit turbine wheel rotation to a totalizing counter. This counter functions as an integrator and provides a digital read-out of the total volume of fluid that has passed through the meter. A tachometer generator could be used in place of the counter, in which case the output signal would be an analog DC voltage.

Positive displacement meters form an entirely different class of flow-measuring equipment. A positive-displacement meter is a mechanical device that transfers a fixed volume of fluid from its inlet port to its outlet port at every cycle. In this respect, it is not unlike the positive-displacement pumps and motors used in fluid power systems. In truth, a P.D. meter is really a hydraulic motor with exceedingly high volumetric efficiency. Since it need only drive a counter or similar low-force device, the pressure drop across a P.D. meter is quite low.

The nutating-disk meter is the most commonly used type. It consists of a flat disk mounted on a central ball retainer, in such a way that it can nutate about a central axis and drive an output shaft. Each rotation of the shaft is equivalent to an amount of liquid equal to the displacement of the metering element. A shaft-driven odometer (digital read-out), similar to that shown in Figure 8.13, acts as an integrator and indicates the total volume of fluid that has passed through the meter, regardless of time.

Reciprocating-piston meters, such as the one illustrated in Figure 8.51, are the most accurate type of P.D. meter over full flow range. Pistons, reciprocating within cylinders, cause a crank or swash-plate mechanism to be rotated. As was the case with the disk meter, the rotation operates an odometer to provide a visual read-out of total fluid volume.

In some applications the meter spindle is used to drive a rotary pulse

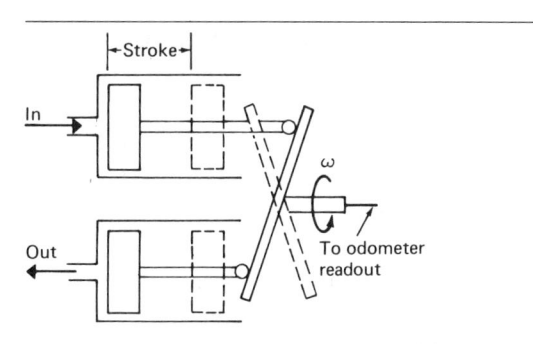

Figure 8.51 Reciprocating-piston meter.

generator like those illustrated in Figure 8.39. Under these conditions, the P.D. meter can be used as a flow-rate instrument since the pulse generator does not integrate. The rotational speed of the meter spindle is generally too slow to successfully drive a tachometer generator, although there have been units designed using speed-up gearing to provide tachometer generator speed ranges.

The advantage of a P.D. meter over one of the other types is that it is less sensitive to changes in fluid properties resulting from environmental conditions. The kinetic-energy types of flow meters are more apt to be affected by viscosity or density changes due to temperature variation, etc., than are the potential-energy (positive-displacement) types of meters. The latter can also be more readily switched from one fluid to another, as long as their construction materials are compatible with the fluid. The pure rate meters, on the other hand, are only good for the specific fluid for which they were originally calibrated, and then only under the conditions of calibration.

The fluid meters discussed so far are basically volumetric types. They can be used as mass-flow meters only if the approximation that fluid density remains constant is reasonably true. The premise is that $Q_m = \rho \, Q_v$, where ρ is the mass density.

Mass

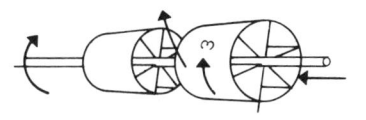

Figure 8.52 Mass-density meter.

There have been true mass-density meters devised, which are insensitive to changes in density with temperature, etc. Typically, they incorporate a mechanism for imparting a tangential component of velocity to the fluid flowing through the meter, as shown in Figure 8.52. The fluid then impinges on a reaction member. The torque reaction is proportional to the true mass per unit time impinging on the second member.

Hot-wire anemometry is a technique used to measure flow rate on the basis of heat dissipation (Figure 8.53). A probe of very fine wire, such as platinum, is positioned in the fluid stream. The wire is heated electrically and heat is carried away by the stream of fluid at a rate that is a function of the mass flow rate at that point in the stream and the thermal characteristics of the fluid. The rate is determined electronically and a suitable read-out provided to indicate flow velocity. This technique received considerable impetus from the new technology of fluidics, wherein it was necessary to measure flow velocities with minimum disturbance to the fluid stream itself.

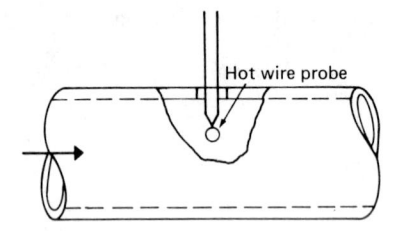

Figure 8.53 Hot-wire anemometry device.

Some work has also been done on the use of ultrasonic vibrations propagated in a fluid stream to measure flow velocity. As shown in Figure 8.54, an ultrasonic transducer sends out a beam of known frequency. It impinges on, and is reflected back to a receiver by, a target placed downstream in the fluid conductor. The flow velocity of the fluid stream contributes a Doppler effect time differential to the time it takes for the beam to hit the target and the time it takes to reflect back to the receiver. This time differential is proportional to flow velocity and thus to flow rate. The signal can be processed electronically to give either an indication of flow rate or as a control signal.

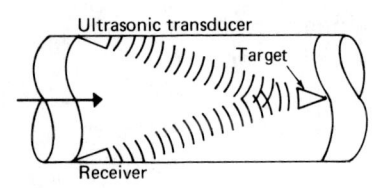

Figure 8.54 Ultrasonic vibration device for measuring flow velocity.

The flow-rate meters conforming to the rule

$$Q = K \sqrt{\Delta p}$$

that is, the square root law, have the problem that the relationship is nonlinear. We have already discussed the desirability of linearity between the variable and the output signal. This rule describes turbulent flow, which is what persists in most practical flow meters. If it were possible to maintain laminar flow in the meter, the flow rate versus pressure drop characteristic would be linear, that is, $Q_1 = k\Delta p$. Laminar-flow meters have been developed. The technique consists essentially in breaking up the flow path into a multiplicity of capillary flow tubes, as shown in Figure 8.55. The tubes are sized so that true laminar flow conditions hold over the entire range of the meter. Thus the pressure drop generated across the tube bundle is directly proportional to flow rate. This pressure drop becomes the output signal from the meter and is processed as previously discussed (Figure 8.56).

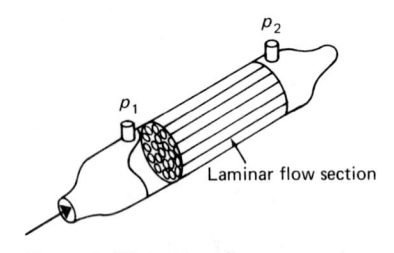

Figure 8.55 Laminar-flow meter.

Figure 8.56 Hydraulic circuit tester.

Summary

This chapter has presented a discussion of some of the more common techniques for measuring the state of fluid power system variables. Precision instruments are used for data acquisition and for control or maintaining system output within predetermined limits. Designers must be able to answer key questions such as: How much pressure, force, etc.? How high is the pressure, viscosity, etc.? How fast is the velocity, flow rate, etc.? Once these factors are determined, instruments must be available to test these physical variables within an operational fluid power system. The limits of accuracy and range of these variables have to be established for maximum efficiency and safety.

Instruments record the output, providing an indication of the state in terms of standard units of measure. These standard units of measure are not absolute values, but are arbitrarily accepted measures defined by international law.

It should be noted, however, that the *state* of the system variable is absolute, but the measurement unit, in terms of which it is expressed, is not. The measurement is not the state of the variable. This is a difficult distinction to grasp and appreciate.

Review questions and problems

1 What is a variable?
2 What is an instrument?
3 What is a transducer?
4 The measurement indicated by an instrument is not the physical variable being measured. True or false? Discuss your answer.
5 What two kinds of functions (with respect to their signals) can be performed by instruments?
6 What is the characteristic curve of an instrument? Why is it important?
7 What is a calibration curve for an instrument? How, if at all, does it differ from a characteristic curve?
8 What is meant by the accuracy of an instrument? How is it expressed?
9 What is meant by the resolution of an instrument?
10 What is meant by the repeatability of an instrument?
11 All instruments of a given type will indicate the same state of a variable in exactly the same way. True or false? Explain.
12 What is meant by hysteresis in an instrument? What causes it?
13 Explain the idea of deadband in an instrument.
14 What is the difference between deadband and hysteresis?
15 What is the relationship between the deadband and the sensitivity of an instrument?
16 Explain the difference between the accuracy of an instrument at an operating point and over the full scale.
17 What functional characteristic is common to all mechanical and electromechanical pressure-measuring instruments?
18 Explain the basic working principle of spring, diaphragm, piston, and bellows pressure-measuring devices.
19 Assume you want to design a simple pressure-measuring device of the hydraulic piston–spring force balance type shown in Figure 8.6. If you use a piston area of 1.5 in^2, want to limit the motion of the indicator pointer to 2 in. full scale, and need to measure pressure up to 1000 psi, what spring rate should be used?
20 How is the principle of question 18 used in strain-gage instruments?
21 Explain how the principle of question 18 is applied to vacuum gages as well as positive pressure-measuring devices. Do these instruments indicate gage or absolute pressure?
22 How does a variable-reluctance type of instrument vary from those discussed in question 18?
23 What is the principal difference between force and torque?
24 What is the basic principle used in most dynamic torque meters?

25 What is the difference between analog and digital instruments? Do they, or do they not, measure the same thing?
26 Is the principle of question 24 useful only in analog, digital, or both types of torque meters? Explain.
27 Discuss why time is so important in many measurement and control situations.
28 Discuss the similarity between the concepts of absolute zero temperature and absolute zero pressure.
29 What is the Seebeck effect? Where is it used?
30 Where can output from a thermocouple be used?
31 Displacement is a function of the basic dimension length. What is it?
32 Why is a yardstick (or meterstick) not a displacement instrument of itself?
33 Discuss how a potentiometric displacement instrument works.
34 Given a linear potentiometer that has a resistance of 300 ohms per inch of length, plot a typical characteristic curve for a length of 12 in. and a supply voltage of 24 vdc. What variable is actually used as the analog of linear displacement in this example?

Selected readings

- Glenn, Ronald E., and James E. Blinn. *Mobile Hydraulic Testing.* American Technical Society, Chicago, 1970.
- Henke, Russell W. *Introduction to Fluid Mechanics.* Addison-Wesley, Reading, Massachusetts, 1966.
- ———. *Introduction to Fluid Power Circuits and Systems.* Addison-Wesley, Reading, Massachusetts, 1970.
- Johnson, Olaf A. *Fluid Power—Pneumatics.* American Technical Society, Chicago, 1975.
- Mott, Robert L. *Applied Fluid Mechanics.* Charles E. Merrill, Columbus, Ohio, 1972.
- Pease, Dudley A. *Basic Fluid Power.* Prentice-Hall, Englewood Cliffs, New Jersey, 1967.
- Pippenger, John J., and Richard M. Koff. *Fluid-Power Controls.* McGraw-Hill, New York, 1959.
- Stewart, Harry L. *Hydraulic and Pneumatic Power for Production,* 3rd ed. Industrial Press, New York, 1970.
- Streeter, V. L. *Fluid Mechanics,* 5th ed. McGraw-Hill, New York, 1971.
- Yeaple, Franklin D. *Hydraulic and Pneumatic Power and Control.* McGraw-Hill, New York, 1966.

Chapter 9
Fluidics

This chapter provides an introduction to fluidic theory and devices. It presents the basic theory and operation of fluid logic devices with illustrations of how these control devices function. A list of the advantages of fluidic technology suggests reasons why these devices are capable of operating in hostile environments. Both digital and analog devices are described briefly, together with symbology for various devices.

Examples of fluidic circuit applications are provided to serve as an example for students to begin to explore other possible applications.

Key terms

- **Active** The general class of devices that require a power supply separate from the controls.
- **Amplifier** A component that enables one or more signals to control a source of power and thus is capable of delivering at its output an enlarged reproduction of the essential characteristics of the signal.
- **Analog** Of or pertaining to the general class of devices or circuits whose output varies as a continuous function of their input.
- **AND** Any device having two or more inputs and a single output, in which the output is on only when all inputs are on and off when any one or all inputs are off.
- **Bias Port** The port at which a biasing signal is applied.
- **Binary arithmetic** A number system of base 2. In this number system, only two possibilities are allowed. This makes it naturally applicable to digital design.
- **Bistable** Of or pertaining to the general class of devices that maintain either of two possible operating states in the presence or absence of the setting input.
- **Bit** A single binary character.
- **Boolean algebra** The mathematics of logic (named after George Boole), the algebra of propositions. Only two possibilities are allowed—true or false.
- **Capacitor** An enclosed volume fitted with one or more ports; often used to provide a time-delay effect.

- **Coanda effect** A fluid dynamic phenomenon wherein a jet of air can be attached to a wall in a direction different from the direction of origin of the jet.
- **Component** A device in a system or circuit that performs a given function.
- **Control jet** The stream of fluid within a device that acts to displace or switch the power jet.
- **Control port** The port at which a control signal is applied.
- **Counter** A device that will maintain a continuous record of the number of pulses received at its input. The output of the counter indicates the sum of the number of input pulses.
- **Device (assembly)** An element, component, circuit, or system normally fabricated as a packaged unit or integral assembly.
- **Digital** Of or pertaining to the general class of devices or circuits whose output varies in discrete steps (i.e., pulses or "on-off" characteristics).
- **Element** The general class of basic structures used to make up components and circuits, e.g., nozzle, volume passage, restrictor, etc.
- **Exclusive OR** A device in which an output is given only if either one of two input signals is present. If both input signals are present, there is no output.
- **Fan-in** The number of operating controls in a single device that individually and in combination will produce the same output.
- **Fan-out** The number of like devices to which operating controls are supplied by the output of the device.
- **Fluidics** The technology wherein sensing, control, information processing, and/or actuation functions are performed solely through the use of fluid dynamic phenomena.
- **Gate** A generic term referring to monostable devices such as the OR/NOR, AND/NAND; derived from the ability of these devices to allow one signal to control, or "gate," another.
- **Hysteresis** The difference, or dead zone, between the "on" and "off" input switching points of a device.
- **Impact-modulator amplifier** A fluidic device in which the impact plane position of two opposed streams is controlled to alter the output.
- **Interface** A point or component where a transition is made between media, power levels, modes of operation, etc.
- **Logic devices** The general category of components that perform logic functions, for example, AND, NAND, OR, and NOR. They can permit or inhibit signal transmission with certain combinations of control signals.
- **Logic state** Signal levels in logic devices are characterized by two stable states, the logic 1 (one) state and the logic 0 (zero) state. The designation of the two states is chosen arbitrarily. Commonly, the logic 1 state represents an "on" signal and the 0 state represents an "off" signal.

- **Memory or Flip-Flop** A device that has two stable states, either "on" or "off," and remains in one of these states, even after the signal has been removed, until a command to change its state is received. It is known as a bistable device.
- **Monostable** Referring to a device capable of only one stable state.
- **NAND** A device having two or more inputs and a single output, in which the output is off only if all the inputs are on. (This can be accomplished by combining AND with NOT. The term "NAND" is a contraction for NOT-AND.)
- **NOR** A device having two or more inputs with a single output, in which the output is on only in the absence of all inputs.
- **NOT** A device with one input and one output, in which the output is on only when the input is off. (The application of NOT to any other logic function inverts the output.)
- **OR** Any device having two or more inputs and a single output, in which the output is on whenever one, or more, of the inputs is on.
- **Output port** The port at which the output signal appears.
- **Passage** A connecting line between two points within a component.
- **Passive** The general class of devices that operate on signal power alone.
- **Resistor** A device designed to produce a pressure drop (linear or nonlinear) as a function of flow.
- **Sensor** A device capable of detecting some particular event or presence and indicating this detection with a fluidic signal.
- **Signal** A fluid flow and/or pressure that indicates the state of a device or a sensed event.
- **Stream-deflection amplifier** A fluidic device that utilizes one or more control streams to deflect a power stream, altering the output.
- **Transducer** A device that transforms a signal from one medium to another (i.e., fluidic to mechanical, electric to fluidic).
- **Trigger** A pulse used to initiate transition in a Flip-Flop or other circuit.
- **Truth table** A table showing the different output conditions for all possible input conditions.
- **Turbulence amplifier** A fluidic device in which the power jet is at a pressure such that it is in the transition region of laminar stability and can be made turbulent by a secondary jet or by sound.
- **Vent** A passage to a reference pressure (usually atmospheric).
- **Vortex amplifier** A fluidic device in which the angular rate of a vortex is controlled to alter the output.
- **Wall-attachment amplifier** A fluidic device in which the control of the attachment of a stream to a wall (Coanda effect) alters the output.

The promising and intriguing technology of fluidics has, in a relatively short time, produced devices used as digital switches, amplifiers, sensors, timers, detectors, and other binary logic devices. The National Fluid Power Association defines fluidics as *engineering science pertaining to the use of fluid dynamic phenomena to sense, process information, and/or to actuate.* The word "fluidic" was derived from the combination of the words "fluid" and "logic."

Historical development of fluidics

Fluidics technology—the technique of using the flow of fluids (gas or liquid) to accomplish sensing, power amplification, computation, or actuation without the aid of electronics or moving mechanical parts—had its beginning as an applied research laboratory project at the U.S. Army Harry Diamond Laboratory (known as the Diamond Ordnance Fuse Laboratories) in 1958.

Initially, research was directed to applications under government-sponsored work on military weaponry controls. Engineers were attempting to overcome the deficiencies of the existing energy systems—electronics, hydraulics, and pneumatics. Because fluidic devices have nonmoving mechanical parts, this technology offered great promise of providing control, computation, or timing that would not be vulnerable to electronic jamming, severe shock vibration, high acceleration, extreme temperatures, or radiation from a nuclear explosion. Engineers soon realized that fluids offer a sophisticated control means that can operate very reliably in an explosive atmosphere without special protection, in extremes of temperature without cooling or heating, under extreme shock loads and vibration without shock absorbers or vibration dampers, and under intensive nuclear, magnetic, or electromagnetic radiation without screening. It was not surprising that operation of control in these environments was among the early applications for fluidics. The lack of moving parts increases the reliability of fluidic devices.

The initial research for military purposes inspired other industries to examine the potential of this emerging technology. It was demonstrated early in the 1960's that fluidics could prove competitive in normal industrial environments, especially where electromechanical or pneumatic systems were frequently used, and so fluidics made considerable inroads into the field of machine controls. As machines become more sophisticated, the control system is more critical to the machine's productivity and reliability. Fluidic controls often use the same power source as the device they are controlling, namely, compressed air, thus eliminating pneumatic-electric interfaces. Today, fluidic devices and systems are used in such diverse applications as switching railroad engines, controlling the flow of gas through pipelines, preventing the overheating of electric motors, regulating plastic injection molding, smelting ores, counting items in packaging processes, controlling the water level and wash cycle of a household washing machine, controlling the temperature on an automotive air conditioner, and gyroscopically stabilizing missiles.

The elimination of moving mechanical parts in fluidic devices provides simplicity, high reliability, and practically unlimited life. In addition, fluidic devices can operate in hostile environments because these ceramic devices are unaffected by high temperatures, shock, vibration, and nuclear and electromagnetic radiation.

Advantages of fluidics

The features that contributed to an early interest in the development and practical applications of fluidic devices include:

- They have a practically unlimited shelf and operating life combined with high reliability, because fluidic devices have no moving parts. (Some fluidic systems do, however, incorporate some moving parts.)
- They have a high tolerance to nuclear and electromagnetic radiation.
- They have an outstanding tolerance to environmental shock, vibration, and temperature. Devices have been operated under test conditions of 50 g's at 5000 cycles per second. Ceramic devices can operate in a temperature as high as 400 °C. Some of these devices can operate at 700 to 1400 °C, which is well within commerical and military requirements.
- High reliability and low maintenance costs due to the simplicity in construction of these devices.
- Operating characteristics ideal for hazardous environments.
- Ability to operate at frequencies up to 5000 Hz.
- Low production costs for large quantities, particularly if injection-molded plastics or metal die-casting processes are used.
- Ability to be interfaced with other devices, including electronic, hydraulic, pneumatic, and mechanical.
- Circuits can be placed in layers and fused into a monolithic block. All interconnections, with the exception of power supply inputs and outputs, are within the block, thus eliminating tubing and external connections.

Disadvantages of fluidics

While fluidic devices have many advantages, they also have certain disadvantages that limit the scope of their applications:

- Fluidic devices have a slower response speed than electronic devices. Fluidic amplifiers respond in milliseconds, 10^{-3}, while electronic controls respond in nanoseconds, 10^{-9} (billionths).
- Fluidic devices have not been successfully miniaturized in comparison with solid-state electronic elements; therefore, these devices are not employed in applications that require a large number of elements such as sophisticated computers.
- Fluidic devices, because of their fine and intricate orifice channels, are more susceptible to contamination than other control systems. Filters on the order of 5-micron size are necessary for high reliability.
- Fluidic device outputs are not normally powerful enough to perform appreciable work.

The two major disadvantages of fluidics are the slower response speeds and the possibility of malfunctioning with air contamination. Fluidic switching speeds, typically 1 to 2 milliseconds with practical circuit frequency response up to 500 Hz (cycles per second), are much faster than relays and are successfully applied to the control of machinery in many industrial situations. Many manufacturers are turning their emphasis from theoretical research to design and improved product lines for field applications, as they recognize the virtues and limitations of fluidic devices and systems.

The potential applications of fluidic devices have extended from military and space purposes to industrial, railroad, hospital, business, and household uses. In such diverse applications, fluidic devices act as on-off switches, proximity sensors, signal amplifiers, counters, timers, and oscillators.

Basic fluidic principles

The tendency for a jet to attach itself to an adjacent surface was first noted in 1800 by a scientist of optics, Thomas Young. More than a century later, in 1910, Henry Coanda, a Rumanian-born aerodynamics engineer and inventor, observed that the flames of a jet engine he had built attached themselves to the flammable fuselage and set the plane on fire. At the time he was more concerned about the destruction of his plane than with the effect that caused it. Several decades later, in 1932, he recalled the incident, made several studies on the wall-attachment effect, and published his findings.

Coanda effect

Essentially, what Henry Coanda observed is that, as a free jet emerges from a jet nozzle, the stream tends to follow a nearly curved or inclined surface and will come in contact with and flow along, or even attach itself to, this surface if the curvature or angle of inclination is not too sharp. The Coanda effect is of major importance to fluidic technology. Although these effects are observable in nature, the wall-attachment phenomenon was not usefully employed until 1959 by the Diamond Ordnance Fuse Laboratories engineers.

It can be observed that particles of snow or rain have a tendency to come over the eaves of a roof and attach themselves to the wall of a building (Figure 9.1). This may account for the greater depth of snow piled on the ground near the wall.

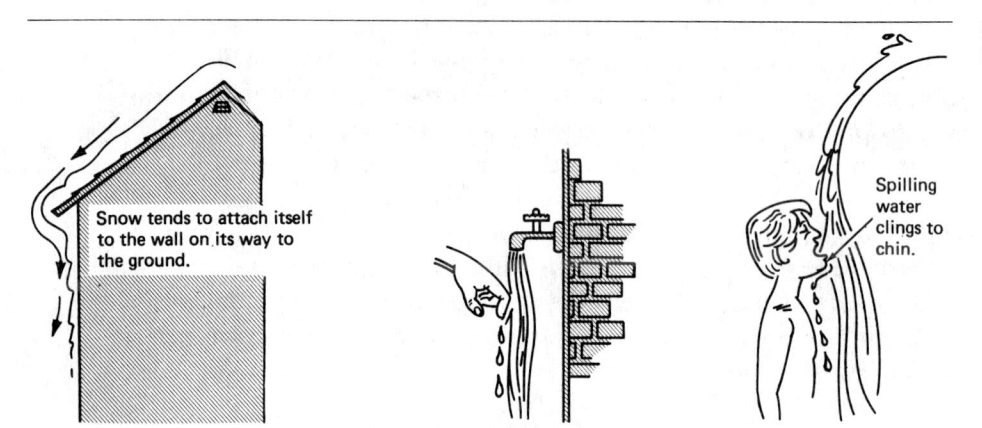

Figure 9.1 Wall-attachment phenomenon.

This principle is also easily grasped by observing water flowing from a faucet. If the velocity profile of the water stream were plotted, it would be noted that velocity causes a pressure difference between the water stream and the ambient pressure. Since pressure tends to equalize, air is entrained in the stream. If you hold a finger near the stream and point away from it, water will follow the finger.

Entrainment

Entrainment is defined as the moving stream of air that tends to "sweep other interface layers of dynamic air currents along."

In Figure 9.2 the stream can entrain air from one side but ultimately the finger creates a wall to entrainment, causing a low-pressure pocket. The higher ambient pressure on the entrainment-free side of the water stream will hold it against the finger or similarly against a wall of a fluidic device.

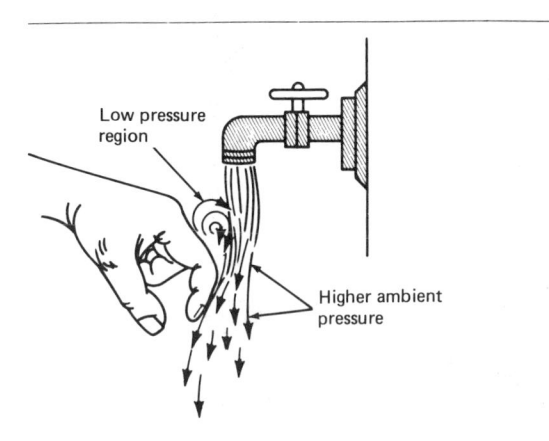

Figure 9.2 Air-entrained fluid flow pattern.

The wall-attachment device was one of the first fluidic elements to be developed and it became the most popular of the digital fluidic amplifier devices (Figure 9.3).

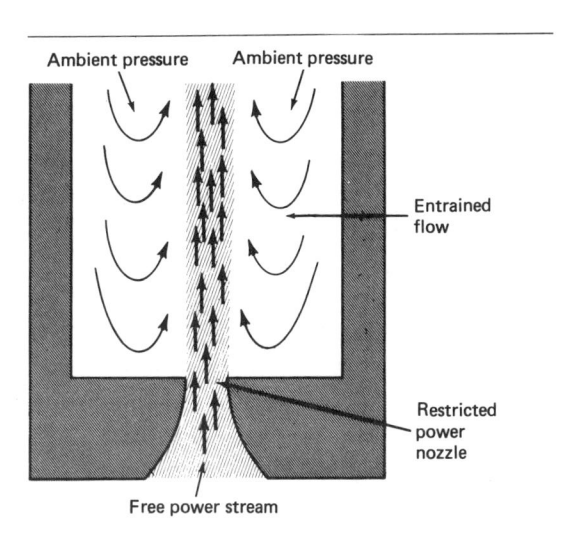

Figure 9.3 As the power stream moves along, it entrains air from both sides, becomes broader, and carries more air toward the open end of the container.

In constructing a fluidic element it can be noted that, when a jet of air is emitted into a confined region from a pressure supply source through a restriction, turbulent flow will occur (Figure 9.4). The turbulent jet entrains air from its surroundings. A turbulent jet is dynamically unstable and will veer in different directions. As it approaches a wall, it will reduce and ultimately interrupt the entrainment flow path. This means that no more air is flowing in to replace the outrushing air.

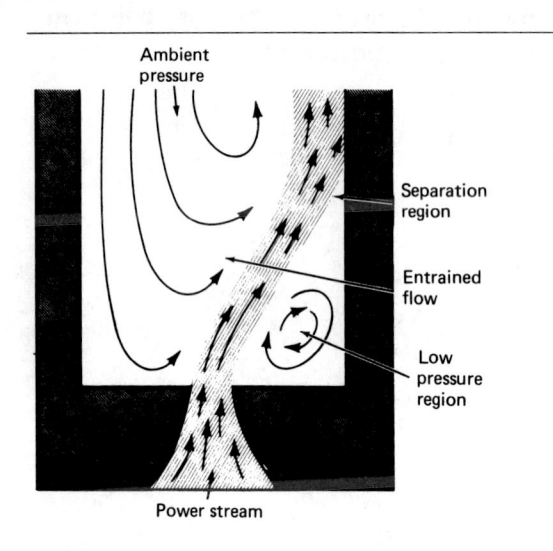

Figure 9.4 The Coanda effect: jet deflection by wall attachment. A jet stream causes pressure to drop in the zones between the stream and the walls, creating an unstable situation.

Jet interaction

The principle involved in the fluid amplifier is not too different from that employed in many electronic applications. For example, in a cathode-ray tube, a weak fluctuating voltage applied to the deflection plate of the tube acts on a stream of electrons flowing through the tube. The resulting modulation of the stream serves to boost the signal to operating levels.

In a basic fluidic control device, a steady stream of liquid or gas (most frequently air) replaces the electrons and is acted on by precisely controlled jets of fluid directed at it from the side. A jet's signal (air under pressure) deflects the mainstream into a preplotted course along the output channel (Figure 9.5). The fluid will continue to flow in this direction until it is acted on by a signal change. The control ports allow the operator to determine at will which wall the stream is to follow, thereby controlling the flow to the desired output. This operating principle of fluidics is known as *jet interaction*.

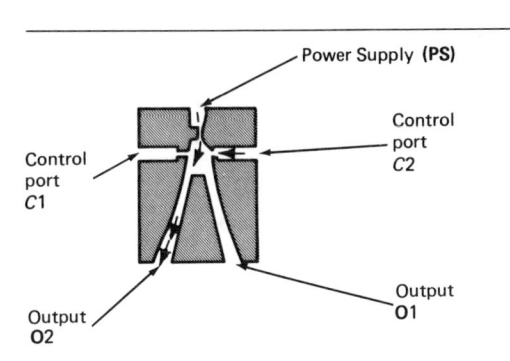

Figure 9.5 Deflection by control of jet interaction. Fluid signal from right side (C2) moves input to left-side output (O2).

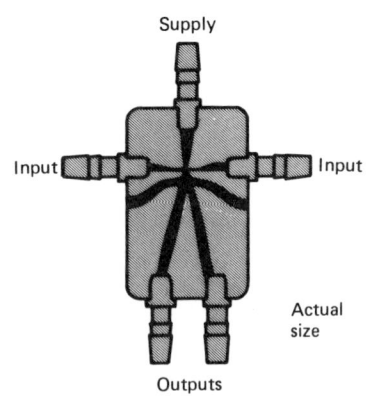

Figure 9.6 Dual-control, bistable Flip-Flop device.

Selection of the output channel that the inlet fluid stream follows is controlled by the signal or no-signal conditions and is also determined by the constructional design of the fluidic element. This type of device uses input control signals to divert the output back and forth between the two output ports (Figure 9.6).

As the air continues to be entrained from the separation region by the power jet, a low-pressure area is established next to the jet. Once this separation region is created, the pressure difference across the stream causes it to become stable and remain attached to the wall (see Figure 9.4).

Three fluid effects are employed in the deflection of the jet: momentum exchange, pressure differential, and the Coanda effect.

Momentum exchange

At the exit of the supply tube, two smaller control tubes are arranged at right angles to the power jet (see Figure 9.5). As a jet of air emerges from either of these control tubes, it impinges on the main power jet. The two jets mix, a momentum interchange occurs, and the main power jet is deflected away from the control jet. The momentum (M) of the fluid streams emerging from the control and power jets is resolved by an angular deflection. After interaction, the single emerging jet must change to a direction represented by momentum ratio (Mr) (Figure 9.7).

To capture and use the power jet efficiently, the jet must be confined within surfaces, so that all or nearly all of it enters the output tube. The device is an amplifier, since a small control flow can switch a much larger power output flow (Figure 9.8), and the action of this

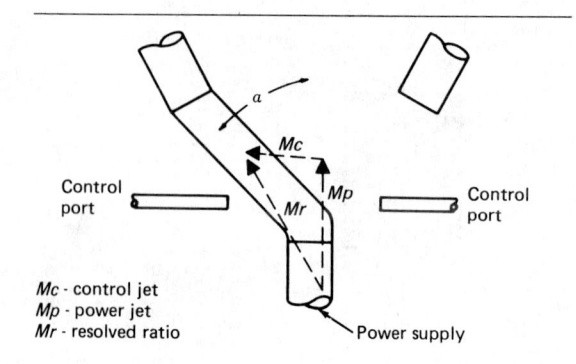

Figure 9.7 Stream interaction.

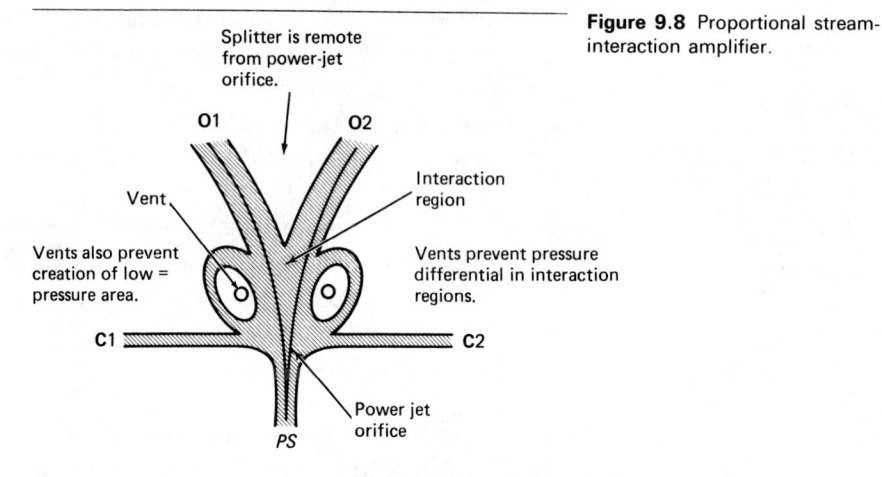

Mc - control jet
Mp - power jet
Mr - resolved ratio

Figure 9.8 Proportional stream-interaction amplifier.

amplifier makes it an analog device. Analog devices or circuits are those whose output is utilized as a continuous function of their input.

Pressure differential

Momentum-exchange devices (also known as jet-interaction devices) are based on the principle that, when two or more fluid streams interact, the effect produced is a function of the momentums of the individual streams and their relationships to each other. Depending on the construction of a fluidic amplifier, both pressure differential and momentum interchange contribute to power-jet deflection.

The proportional stream (jet) interaction is another differential device. For proportional operation, a device must be constructed to avoid the Coanda effect. This is most easily done by omitting the wall (splitter) in the vicinity of the power-jet orifice (Figure 9.8).

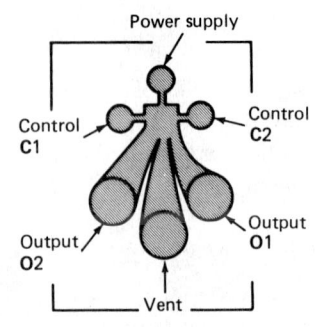

Figure 9.9 Jet-interaction proportional amplifier.

In the absence of a control input, the supply stream exhausts through the vent port (Figure 9.9). When a control signal is applied, the supply stream is deflected through an angle proportional to the ratio of the momentums of the control and supply streams. The difference in pressure between the outputs O2 and O1 in the proportional device is directly proportional to the differential pressure between the control inputs C1 and C2. The recovery, or amount of pressure output, in both the proportional and wall-attachment types varies between 25 and 50% of input pressure.

Monostable jet-interaction amplifier

Another version of the momentum-exchange device is the monostable jet-interaction amplifier. This device uses the same basic principle of momentum exchange as the proportional amplifier. The essential difference, however, is that the device is stable on one side only because of a bias built into the flow channel.

If an aerodynamic bias is formed by an air bleed, the wall-attachment amplifier can be made stable on one side only (monostable). The bleed vent (Figure 9.10) provides the air bleed that biases the power jet stream to the right-hand wall to provide an output signal at O1 with no input control signal. A control signal applied to C1 or C2 will deflect the power stream so that it attaches to the right-hand wall to provide an output at O2.

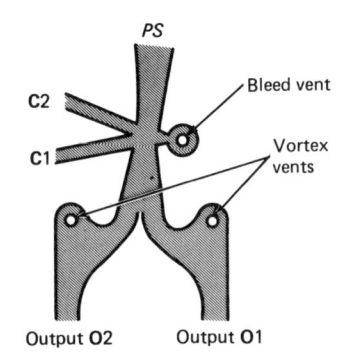

Figure 9.10 Internal layout of monostable Flip-Flop amplifier.

Construction of fluidic devices

Fluidics involves the passage of a fluid—usually air—through mechanical components that have critically designed channels and passages [Figure 9.11(a)]. Physically, a fluidic device is a block of material such as ceramic, plastic, or metal with an internal network of passages and external connectors [Figure 9.11(b)]. Manufacturing of these devices is done by machining, molding, or etching photosensitive glass to create the necessary channels. Flat plates are fused and/or riveted on either side to make one totally enclosed unit [Figure 9.11(c)]. These components are usually interconnected with plastic tubing or they can be internally interconnected as a series of wafers compressed into a monolithic block [Figure 9.11(d)]. For

310 Fluidics

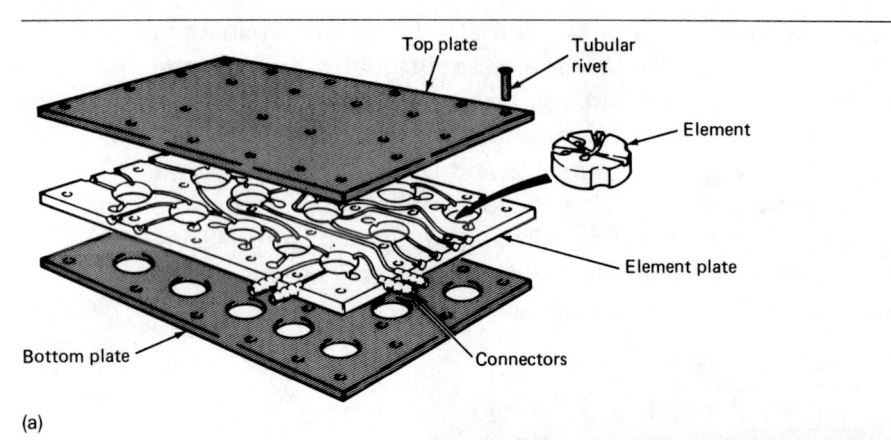

Figure 9.11 (a) Integrated fluidic circuit module. (Courtesy Aviation Electric Ltd.) (b) Etched glass sheet for fluidic elements. (Courtesy Corning Glass Works) (c) Bistable Flip-Flop amplifier device. (Courtesy Aviation Electric Ltd.) (d) Monolithic integrated circuit block. (Courtesy Corning Glass Works) (e) Sample performance characteristics of a bistable fluidic amplifier. (Courtesy Fluidonics)

(a)

(b)

PS = Power supply
C1 = Left control
C2 = Right control
O2 = Left output
O1 = Right output

(Actual size) and construction

(c)

(d)

Figure 9.11 (*cont'd.*)

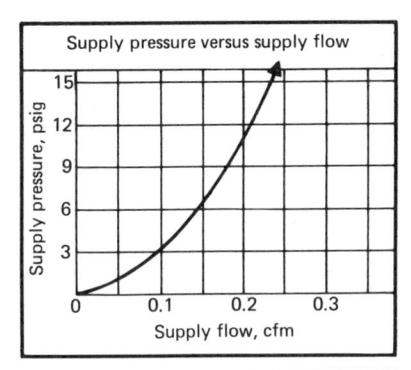

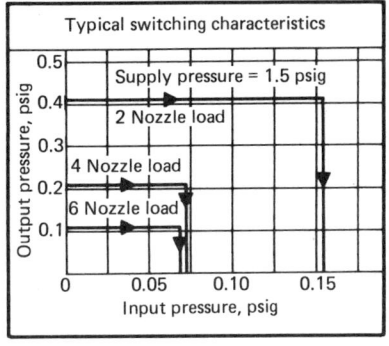

(e)

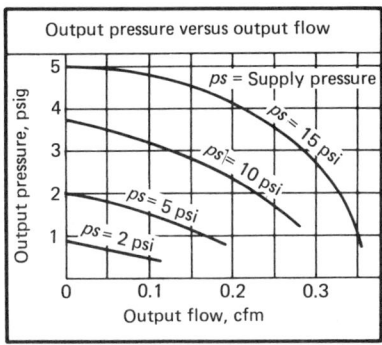

purposes of illustration, the operating characteristics of a bistable Flip-Flop amplifier are shown in Figure 9.11(e) and in Table 9.1.

- **Function** two-input bistable Flip-Flop
- **Operating medium** gaseous fluids
- **Operating principle** wall attachment with vented vortex
- **Temperature range** $-140\ °F$ to $+270\ °F$

Table 9.1 Operating characteristics of Flip-Flop amplifier.

	Maximum	Nominal	Minimum
Input pressure	15 psig	2.5 psig	1.0 psig
Power consumption		1.3 watts	
Pressure recovery (blocked)		45%	
Flow recovery (open)		125%	
Frequency response	800 Hz		
Response time		0.0004 sec	
Switching flow	0.050 SCFM	0.030 SCFM	0.015 SCFM
Switching pressure	0.85 psig	0.35 psig	0.15 psig

It may be observed that, in a jet-deflection fluidic amplifier (see Figure 9.9), the power supply (*PS*) in the form of a gas or liquid under pressure (usually up to 15 psig in gases) is connected to the power jet. The resultant stream that emerges from the power jet is collected in the two output channels as a pressure and flow of fluid. The signal input usually takes the form of a differential pressure of the same fluid applied to the two input ports. Thus two streams of fluid will emerge from the control jets *C*1 and *C*2, perpendicular to the stream from the power jet, and, due to the interchange of momentum occurring at the interaction of these jets, the power stream is deflected toward the control with the lower pressure. The deflection of the power jet stream, which is proportional to the differential pressure across the control input jets, alters the ratio of the output collected in each output channel. The differential pressure recovered across the two output ports is proportional to the differential applied to the two control ports. The recovered output is also larger than the control input, and thus the device acts as an amplifier; hence the term proportional amplifier. The gain is the ratio of the change in output to the corresponding change in input (Figure 9.12). This device will operate just as well with a single control input, when

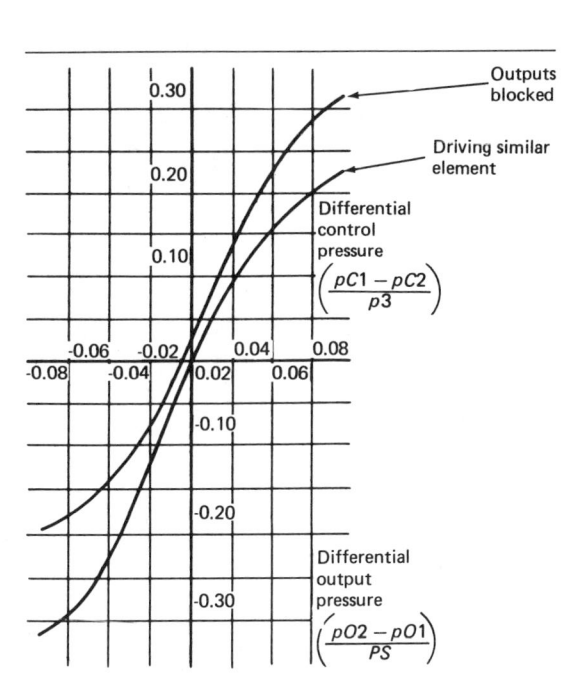

Figure 9.12 Gain characteristics of a bistable amplifier.

a constant bias pressure is applied to the other control port. For increased gain, the output from one amplifier can be applied to the input of a second amplifier, making the gain of the cascaded system the product of the gains of individual amplifiers.

As stated earlier, the first fluidic devices were designed on the basic principle of the Coanda effect. In the beginning, practical applications of fluidics proved to be more difficult than originally envisioned. To make a usable control device utilizing the Coanda effect, it was necessary to find a way to control this wall-attachment phenomenon with precision and at will.

The digital bistable device can be switched from one wall to the other (Flip-Flop) by admitting a pulse of air flow in Control Ports $C1$ or $C2$ (see Figure 9.10). For example, a pulse of air applied at Control Port $C2$ will switch the main stream from the right to the left wall. Either input produces a corresponding output.

By adding a splitter, receivers are created to collect the main flow from the power stream. The basic geometry of Figure 9.13 illustrates that this Flip-Flop element would continue to operate from a single output port $O2$ until a change signal was issued. A Flip-Flop element has symmetrical receivers to retain the bistable quality that gives

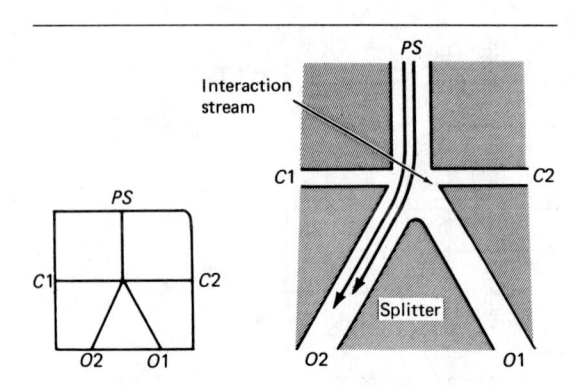

Figure 9.13 A bistable stream-interaction amplifier has two stable positions of jet discharge.

memory. The output will continue through this port even after the command signal has been removed. Therefore, we say the device has "memory."

Active devices

Fluidic units that require a power supply separate from the controls are classified as active devices.

Flip-Flop

Flip-Flop elements serve as the memory bank in a fluidic system. Their characteristics are illustrated by the logic in Table 9.2. An "X" represents the presence of flow and a "O" represents the absence of flow. The Flip-Flop gate can be used as a memory device for any set-and-reset function or where it is necessary to hold a signal and retain memory for some time.

C1	C2	O1	O2
X	0	X	0
0	0	X	0
0	X	0	X
0	0	0	X

Table 9.2 Truth table for bistable Flip-Flop.

The memory or bistable function has its limitations, too. There must also be elements that have a spring return or monostable function—in other words, a favored receiver by geometrical design. There are two basic monostable elements used in fluidics technology: the OR-NOR element (Figure 9.14) and the AND-NAND element.

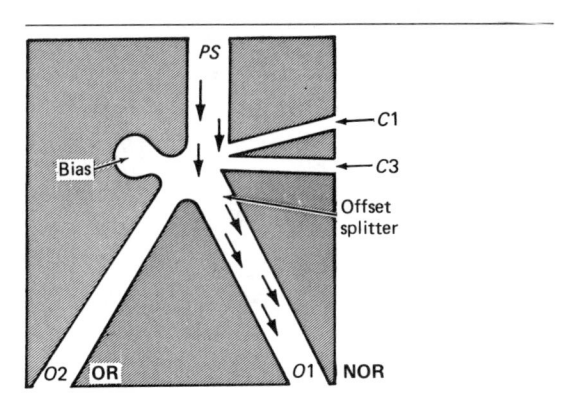

Figure 9.14 An OR/NOR device employing an asymmetrical configuration.

OR-NOR

The OR-NOR element is biased either by geometrical asymmetry or by venting. This device favors one leg. A control signal will switch the power stream over, but the control must be maintained to keep it switched. If the control is removed, the device will switch back to the preferred output channel (Figure 9.15). In the OR position, out-

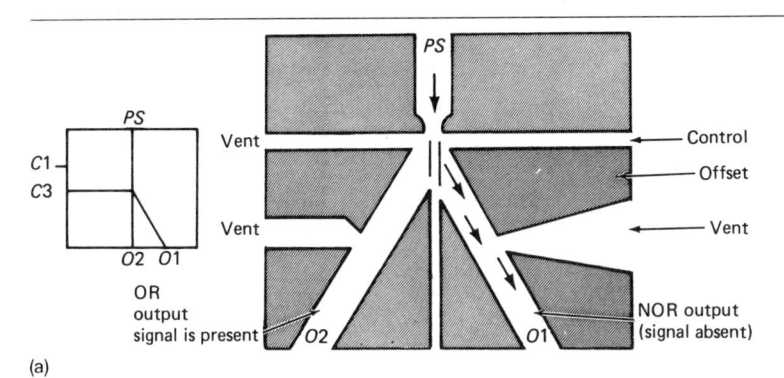

Figure 9.15 (a) A monostable OR/NOR gate. When a control signal is removed, the device switches back to the preferred output. (b) OR/NOR gate, symmetrical design with vent.

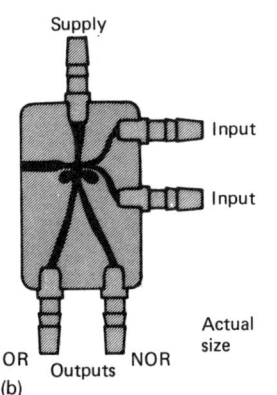

put will occur if either input is on, but not if both inputs are on. In the NOR position, output occurs only if all inputs are off.

The device in Figure 9.12 is an OR-NOR gate. If a control signal is introduced through either one of the two control ports, the stream will exit by the OR output leg. The stream switches to the NOR output leg when both of the control inputs are off (Table 9.3).

Table 9.3 Truth table for OR-NOR gate function.

C1	C3	O1 (NOR)	O2 (OR)
0	0	X	0
0	X	0	X
0	0	X	0
X	0	0	X
0	0	X	0
X	X	0	X

AND-NAND

The AND device has two or more inputs and a single output; the output is on only when all inputs are on, and off when any one or all inputs are off (Figure 9.16). It may be observed that, while the NOR gate is used to detect when none of the control signals are present, the AND gate is used to detect when all the control signals are present. Figure 9.16 indicates that, when C1 and C3 signals are present, an output at O1 is also present. Absence of C1 or C3 will result in an output at O2 port.

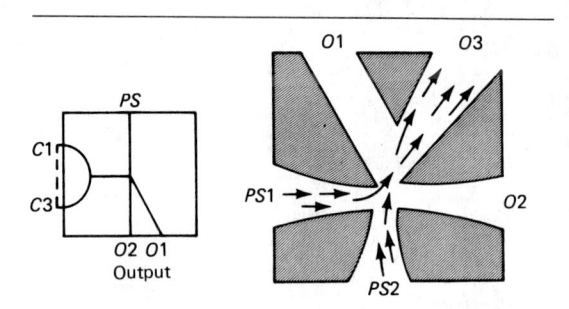

Figure 9.16 The AND gate requires at least two inputs. Activating jet PS1 creates output at O2; activating only PS2 creates output at O1; and simultaneously and continuously activating jets PS1 "AND" PS2 creates output at O3.

The NAND device has two or more inputs and a single output; the output is off only if all the inputs are on (Figure 9.17). The term NAND is a contraction for NOT-AND.

The AND-NAND device is basically a monostable element. A control signal applied to C1 will go through the network and return to atmos-

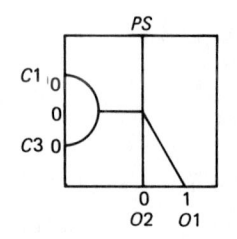

Figure 9.17 NAND gate.

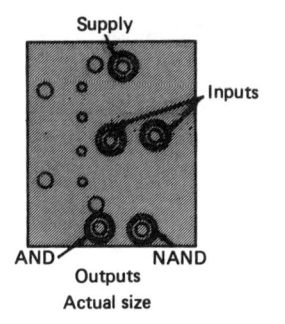

Figure 9.18 AND-NAND gate. (Courtesy Fluidonics)

phere. A control signal at C3 will also go straight through the network and return to atmosphere. If signals at C1 and C3 are applied simultaneously, they will meet at a point, each deflecting the other into the control inlet of the monostable element. Logic functions can be performed by AND-NAND devices (Figure 9.18) for interlocks in control circuitry. The basic logic functions for the AND and NAND gates are expressed as follows:

$$O1 = (C1) \cdot (C3) = \text{AND}$$

$$O2 = \overline{(C1)} \cdot \overline{(C3)} = \text{NAND} \quad \text{(not AND)}$$

Table 9.4 gives a truth table for the AND-NAND device.

Table 9.4 Truth table for AND-NAND gate.

C1	C3	O1 (AND)	O2 (NAND)
0	0	0	X
X	0	0	X
0	0	0	X
0	X	0	X
0	0	0	X
X	X	X	0
0	0	0	X

Schmitt trigger

The Schmitt trigger is another active fluid logic device, which provides adjustable triggering settings and sensitivity to low-level input signals. It is adapted particularly for applications in conjunction with back-pressure sensors, because the switching points can be adjusted over a wide range by varying the bias pressure. The outputs of the back-pressure sensors are adjusted to switch the Schmitt trigger (Figure 9.19) at specific settings. For example, it may be necessary

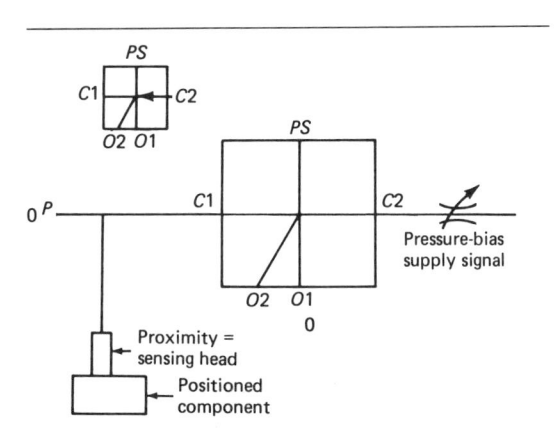

Figure 9.19 Component causes the Schmitt trigger to switch to output at O2.

for a Schmitt trigger to have an output signal at O1 when the back-pressure signal at C1 is greater than 1.5 psi and to switch over to O2 output when the control signal at C1 drops below 1.5 psi. The triggering set-point is determined by the adjustment of the bias supply signal. Where the differential pressure being sensed is very small, it may be necessary to amplify the signal. The supply input and output points are indicated in the function symbol in Figure 9.19. Table 9.5 contains the truth table for the Schmitt trigger device.

Table 9.5 Truth table for Schmitt trigger.

C1	C2	O1	O2
X	X	X	0
X	0	0	X

Binary counter

The Flip-Flop, OR, NOR, AND, NAND, and Schmitt trigger are active fluidic devices; that is, they require a power supply. In many applications it is necessary to count the number of components, operations, motions, functions, or other discrete quantities that have been presented.

The *binary counter* device illustrated by the fluid symbol in Figure 9.20 is used as a single unit or as a binary module including several units in a circuit to provide the count function. The operating principle of a binary device is characterized by two states, 1 and 0. The states of fluidic devices are characterized by the 1 state when a particular port is under pressure and by the 0 state when a port is connected to exhaust or atmosphere.

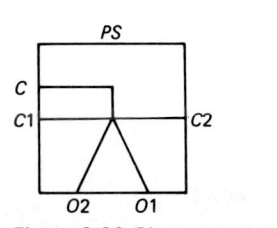

Figure 9.20 Binary counter.

Operationally, the binary counter has three control inputs and two outputs (Figure 9.21). A signal at Pin 3 is called a set signal and

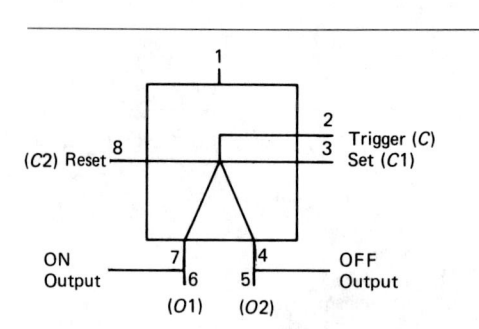

Figure 9.21 Binary counter module symbol.

always produces an output at Pins 6 and 7. An input at Pin 8 is called a reset signal and always produces an output at Pins 4 and 5 A signal at Pin 2 is called a trigger signal and always changes the output, except when set or reset is applied. In a binary counter, the output is switched from one output port to the other with each succeeding input signal. The truth table for the binary counter is provided in Table 9.6.

Table 9.6 Truth table for binary counter.

C Count	C1 Set	C2 Reset	O1 One	O2 Zero
0	X	0	X	0
0	0	0	X	0
0	0	X	0	X
0	0	0	0	X
X	0	0	X	0
0	0	0	X	0
X	0	0	0	X
0	0	0	0	X
X	0	0	X	0
0	0	0	X	0

The binary-up counter counts from 0 upward in the binary number system (Figure 9.22). To build up counters, connect the 0 state output of each stage to the toggle input of the next higher stage. The count signal is applied to the trigger input of the lowest stage. The counter can be extended to several stages.

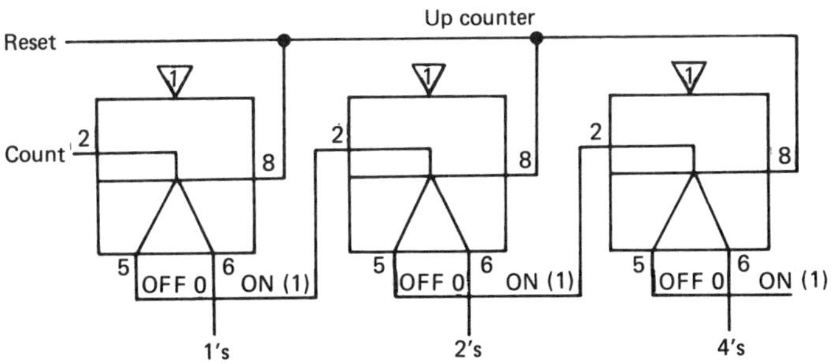

Figure 9.22 Binary-up counter.

A three-stage binary counter will assume the states shown in Table 9.7. Stage 1 changes state on the trail edge of each trigger signal; Stage 2 changes state on the trail edge of every second trigger signal; Stage 4 changes state on the trail edge of every fourth trigger signal. There are eight unique states for the three-stage counter of Table 9.7. The number of unique states for any binary counter is 2^n, where n is the number of stages.

Trigger signal number	Stage 4	Stage 2	Stage 1	Table 9.7 States of a three-stage binary counter.
0	0	0	0	
1	0	0	1	
2	0	1	0	
3	0	1	1	
4	1	0	0	
5	1	0	1	
6	1	1	0	
7	1	1	1	

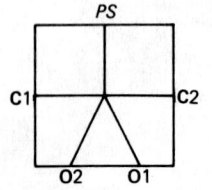

Figure 9.23 Memory device symbol.

Memory logic

The digital logic devices can also serve to create the memory function (Figure 9.23). This particular single memory device would lose state—that is, it would revert to 0—when an interruption in the supply to the device occurred. Table 9.8 provides the truth table for the memory device.

Another composite memory device can be assembled using a bistable Flip-Flop made of two NOR gates (Figure 9.24). The memory-function module employing the NOR logic devices is created with holding circuits. The basic interconnections illustrated in Figure 9.24 produce the same input-output relationships as are present in the wall-attachment Flip-Flop digital device. An input at Pin 1, 2, or 3 produces an output at *A*. Similarly, an input at Pin 4, 5, or 6 produces an output at *B*.

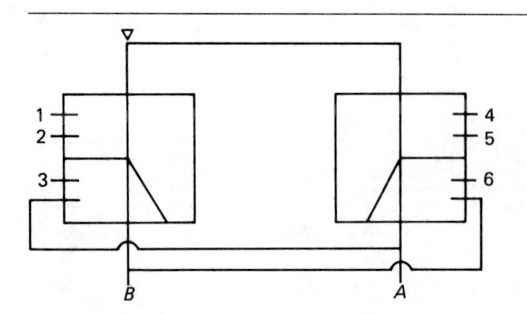

Figure 9.24 Bistable Flip-Flop memory module.

There is an important difference, however, between this interconnected module and the wall-attachment Flip-Flop gate. If an input signal is present simultaneously at Pins 1 and 4, the wall-attachment device may become unstable, or instability can be caused by elevated supply pressure. In such a case, it is unpredictable whether the output will occur at Port *A*, Port *B*, be proportional, or oscillate back and forth. With the interconnecting NOR gates, this module has no

C1	C2	O2	O1
1	0	0	1
0	0	0	1
0	1	1	0
0	0	1	0

Table 9.8 Truth table for memory device.

output when two opposing input signals are present. This function is maintained. It can be observed that output at Port A feeds back to the input of the NOR gate that has a potential Port B output and maintains the OR condition, that is, no output at Port B. However, when the signal stops the output at Port A, there is no longer a signal to maintain the OR condition, and B is allowed to function. If it happened that there occurred simultaneously an input at Pins 1 and 4, Input 1 would stop the output at Port B and Input 4 would stop the output at Port A. Consequently, there would be no output until one of the inputs was stopped. There are many varieties of fluid logic memory devices, particularly as these are interfaced with electronic devices.

Fluid amplifiers

Fluid amplifiers are responsible for the technological breakthroughs and capabilities of fluidic systems. Amplifiers are elements in which changes in a low-energy control signal are used to cause greater changes of output. Because the larger changes of output are amplified changes of the input signal, the designation "amplifier" is given to these devices. In reality, the recovered output is greater than the control input signal. Active fluidic elements are usually classified as either digital or proportional devices, depending on the output change that results from a change of the input signal. Digital devices have distinct output states. Wall-attachment devices are considered digital elements. Digital devices perform on-off functions because of the wall-attachment principle.

Proportional, or analog, elements have an output that can assume an infinite number of different states in response to minute changes of the input signal. To construct an analog device—one that allows some output through each leg—the output channels are not connected by a common wall (see Figure 9.23). The amount of air flow exiting through each leg is determined by the difference in pressure being introduced through the two control ports. For example, if in Figure 9.25(a) a stronger signal is applied through Control Port C2, a greater proportion of the power-port air will be diverted to Output O2, rather than Output O1. The action is proportional and amplification is obtained. Typical output curves are shown in Figure 9.25(b). Fluid amplifiers produce output signals that are amplified replicas of fluid input signals. Switching from one output to the other can be accomplished without shock.

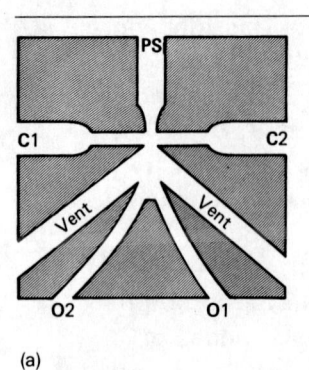

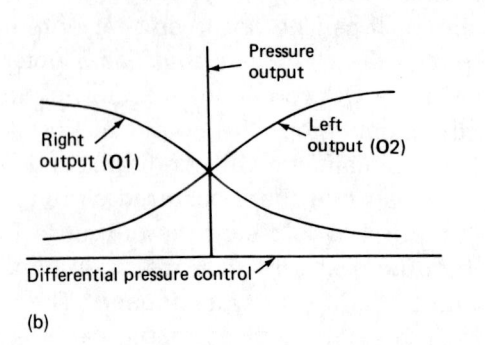

Figure 9.25 (a) Proportional fluidic device. (b) Typical output curves for stream-interaction amplifier.

Vortex

The vortex amplifier operates on the principle of conservation of momentum. The power flow is confined within a flattened, cylindrical chamber (Figure 9.26). Usually, the supply fluid (air) is uniformly fed into the chamber around the periphery of the outer cylindrical wall. If no interfering impedance signals are present, the fluid flows along a radius toward the center of the device and ultimately out through the output tube. The flow pattern in the chamber can best be described as radial. To introduce a vortex fluid motion in the chamber, a tangential input signal is applied through the porous ring (Figure 9.26). The vortex amplifier is intended for applications in which the input or control signal pressure is higher than the supply pressure. Input pressure will range from 20 to 50% higher than the supply pressure, depending on the amplifier design for a particular amplification.

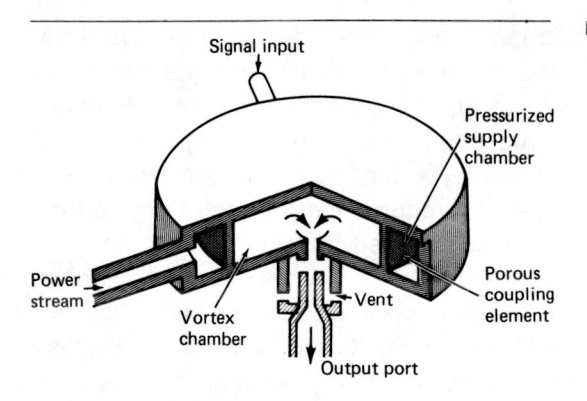

Figure 9.26 Vortex amplifier.

It should be observed that, as the power stream enters the vortex chamber through the porous coupling element (Figure 9.26) and flows toward the output port, its radial velocity must increase. Figure 9.27 shows what happens as this power stream enters through the porous element, with tangential or angular signal velocity imparted to the stream as it is about to exit from the device. To conserve angular momentum, the angular velocity of the fluid must increase as it approaches the output port. Thus there are two amplifying properties of the vortex amplifier: (1) increase in radial velocity and (2) increase in angular velocity as the combined fluid flows toward the output port. When the component is used as a power-amplification device, vents are employed to make the chamber insensitive to back pressure. These vents permit the output power to be modulated from full to zero output.

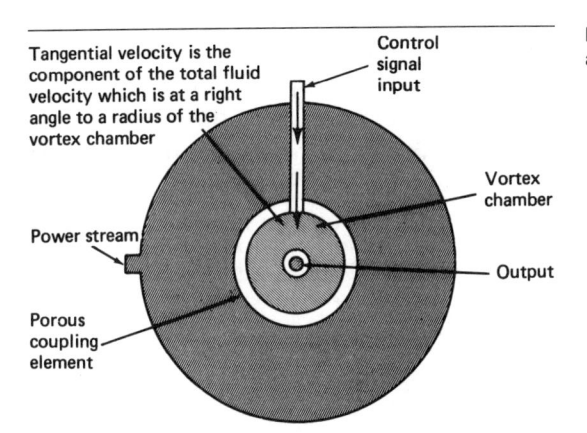

Figure 9.27 Vortex pressure amplifier.

It should be understood that, if a control signal is introduced, the power stream is deflected and follows a vortex path toward the output port. Since the output flow is inversely proportional (within limits) to the control flow, the action of this type of amplifier is proportional, thus providing another analog device.

Turbulence

The operation of a turbulence amplifier depends on the change in flow conditions that occur during a change from laminar to turbulent flow in a fluid stream. The air stream flowing between the supply nozzle and the receiver tube is laminar if there is no control signal (Figure 9.28). This flow condition produces maximum pressure recovery.

When a control flow is applied, the power supply stream becomes

Figure 9.28 Turbulence amplifier.

turbulent and pressure recovery is reduced (Figure 9.29). One or more input signals can be built into the device to upset the laminar flow. Supply pressure is adjusted so that the jet becomes turbulent immediately before entering the output tube. The greater the distance between the supply tube and the output tube, the greater the sensitivity of the device, that is, the greater the effect of jet disturbance. Although this device can theoretically be operated proportionally, it is used primarily as a bistable digital device in logic circuitry. The turbulence amplifier performs a logical NOR function.

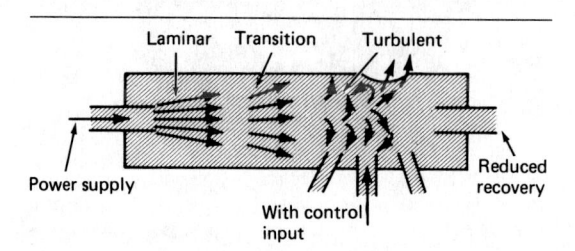

Figure 9.29 Turbulence amplifier.

Focused-jet

The focused-jet amplifier operating without a control signal permits the power stream to attach itself to the wall and flow toward the output area, as shown by the solid lines in Figure 9.30. The resulting wall attachment focuses the supply jet into the control output port. When a control signal is applied, the power stream is deflected from the output port to the outer vent region. This device performs the basic NOR logic function, producing a high output when the control signal input approximates zero and a low output when the control signal input is present. The focused-jet amplifier is a bistable logic device.

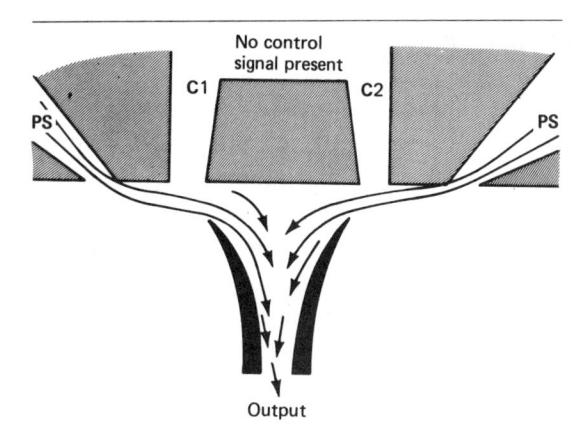

Figure 9.30 Focused-jet amplifier.

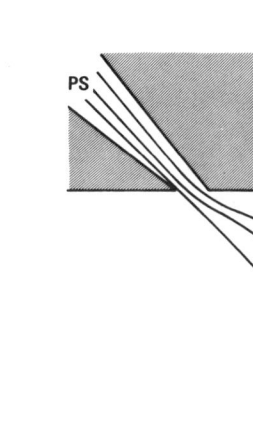

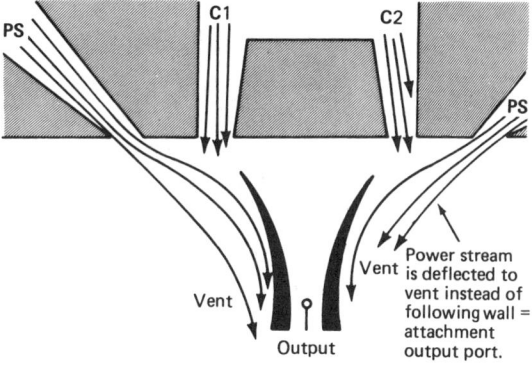

Diaphragm

The diaphragm amplifier (Figure 9.31) is an active device that requires a supply connection. When it is used in conjunction with fluidic elements, the supply pressure can be the same as for fluidic elements, 8 to 15 psi. The signal input port is in the diaphragm chamber cover. The power stream supply is the lower port. The metered supply flow is passed through the filter's orifice, past the output connection, and through the larger orifice to atmosphere. When a control signal is impressed on the diaphragm, the lower portion of the diaphragm closes off the secondary orifice. This closed condition diverts the supply flow to the output port. This miniature device can accept very low pressure signals and boost them to a pressure level sufficient to operate fluidic devices. Diaphragm amplifiers are used in conjunction with noncontact fluidic sensors [Figure 9.31(b)] and other fluidic circuitry systems that require a boost of very small signals.

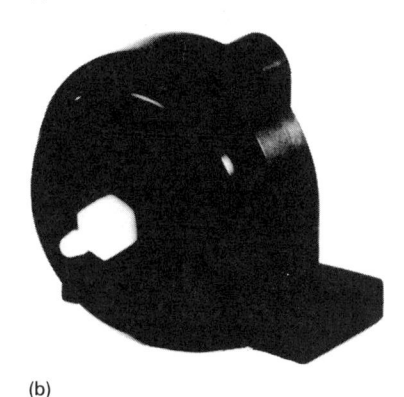

Figure 9.31 Diaphragm amplifier. (Courtesy Norgren Company)

Impact modulator

Impact modulators are active devices that utilize two impacting fluid jets. The principle of operation of the impact modulator is momentum exchange. Two axially aligned power jets impact to create a pressure-balance condition that is identified by a resulting radial jet (Figure 9.32). As the jets leave their respective orifices, each stream has a momentum. When the jets collide, the momentums cancel each other out. The location of the radial jet is determined by the momentum delivered by each of the two impinging jets in the impact zone. Increasing the momentum of the supply from the left emitter will move the radial jet to the right, partially beyond the annular collector. The result is a decrease in the output of the impact modulator.

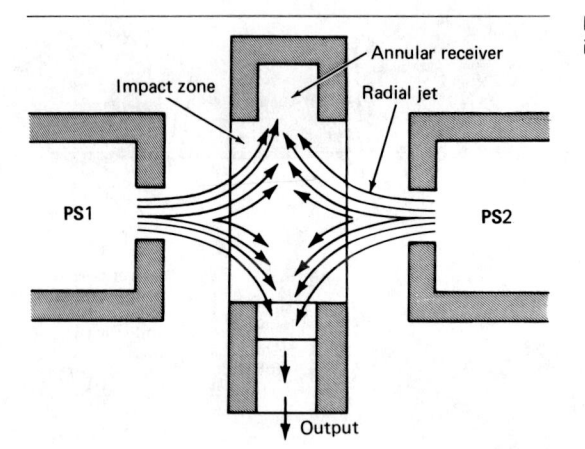

Figure 9.32 Operation of an impact modulator.

A second method of controlling the impact modulator, shown in Figure 9.33, uses a transverse control signal. Flow from the control port deflects the first supply power stream $PS1$, thus reducing the momentum density delivered to the impact zone. Consequently, the radial jet shifts to the left, and output pressure in the output receiver

Figure 9.33 Transverse impact modulator.

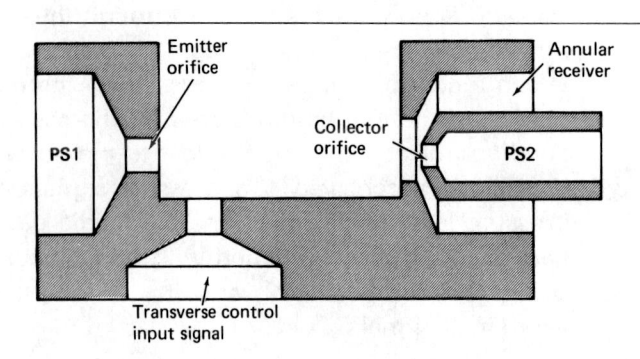

decreases. Typical pressure gain for the transverse impact modulator (TIM) is in the range of 20 to 60%.

The collector-jet pressure is somewhat lower than the emitter-jet pressure. For example, with an initial emitter-jet pressure of 14 psi and a fixed collector-jet pressure of only 8 psi, the radial jet is shifted well toward the output chamber. Output pressure would be approximately 8.4 psi.

Some applications require a differential amplifier, which can be acquired by using both the emitter and collector jets as inputs. Output pressure is a function of the difference between the pressures applied to the emitter and the collector jets.

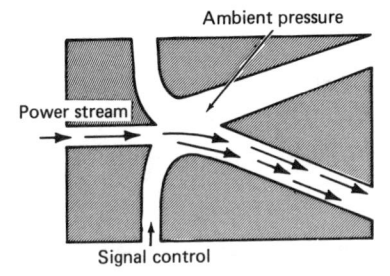

Figure 9.34 Induction amplifier.

Induction

The induction amplifier uses boundary-layer control to produce a bistable logic device (Figure 9.34). The curved-surface amplifiers are based on the principle that fluid flow has a tendency to follow or break away from a curved surface. The characteristics of this amplifier are similar to those of the bistable stream-interaction device, but the switching principle is different. In the bistable stream-interaction amplifier, the output stream is locked onto the wall by the Coanda effect. In the induction amplifier, the surface is curved like an airfoil and the power stream has a tendency to follow it. The fluid stream is detached from the airfoil and switched to the other output port by a control fluid signal injected into the boundary layer of flowing fluid. Thus the power stream is able to unlock and return the amplifier to the off position. The bistable logic functions of on-off conditions in this amplifier are made possible by the airfoil surface design and the location of control ports.

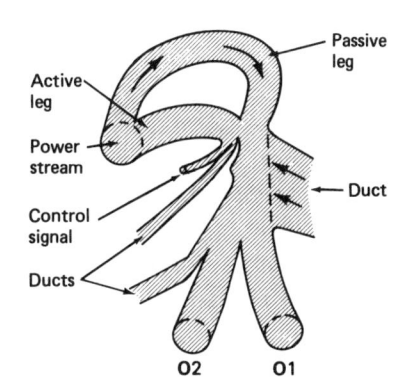

Figure 9.35 Double-leg elbow amplifier.

Double-leg elbow

The double-leg elbow amplifier is another digital device that operates on the basis of controlled boundary-layer separation. The power stream divides into two branches or legs (Figure 9.35). The control nozzle is placed near the boundary-layer separation point of one branch, the active leg of the amplifier. The dual-output receiver is placed to receive the combined supply flow from both the active and passive legs of the device. Changes in control flow vary the separation point in the active leg of the amplifier and deflect the total flow from the active and passive ducts in various amounts to the output ports of the amplifier. An increase in the control signal gives a proportionally increased output at the O1 output. This device produces flow gains of approximately 75% and pressure gains of 3.2 psi.

Many of these fluid amplifiers are available in single or cascaded units with specific configurations, capacities, and operational characteristics. It is helpful to get the manufacturer's technical data charts and tables for particular applications. Product and system designers have not only to understand the fundamentals of the basic operation of these devices (as provided in this chapter), but also be familiar with more of the performance characteristics, mathematical equations, geometry, and logic-function capabilities of fluid amplifiers.

Passive circuit elements

Passive logic devices create an output signal directly from the input signal without amplification of the input signals. There is no power supply in a passive device. Passive circuit elements, in addition to basic active elements, are usually necessary to assemble a functional circuit. The most frequently used passive elements are capacitors, resistors, and oscillators.

Capacitors

Fluidic capacitors are small-volume chamber devices or accumulators (Figure 9.36). When an air signal is fed into a capacitor, the volume begins to fill. Within a short interval determined by signal pressure and flow rate, the capacitor charges to a pressure high enough to constitute a signal output. Operationally, only compressible fluids retain capacitive effects. It should be noted that, as fluid flows into the volume, the additional energy to the fluid already within the volume is available for pulse and time-delay functions in control applications. Fluid capacitors generally have small volumes, 0.6 to 2.4 in^3. Capacitors are designed for pulse-control applications such as the time-delay function. Fluid capacity may be likened to electric capacity.

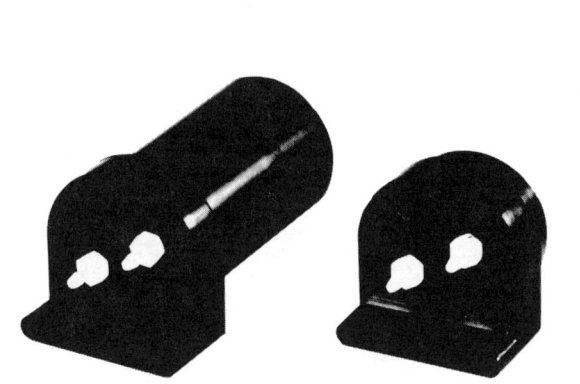

Figure 9.36 Fluidic capacitors. A capacitor is a small tank of a fixed calibrated volume to provide time-delayed impulses or provide dampening within both test and completed circuits. (Courtesy Norgren Company)

Fluid resistors

Restrictors to provide fluid resistance to flow are occasionally necessary in connecting fluidic circuits. The primary types of fixed resistors used in fluidic control systems are orifices and small tubes (Figure 9.37).

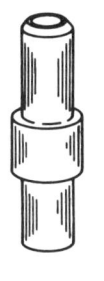

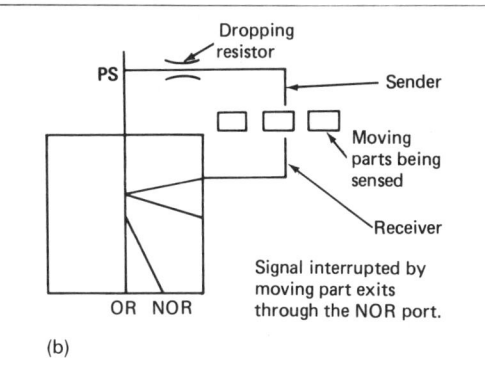

Figure 9.37 (a) Fluidic resistors. Synthetic sapphire orifices are mounted in ⅜-in. brass housings to provide precision restrictors for fluidic use. Fluidic resistors are available in sizes of 0.0028-in. to 0.0252-in. diameter. (b) Resistor in fluidic eye-sensor circuit. Resistor helps to create necessary back pressure, causing the device to switch.

(a)   (b)

When orifices are used with low-pressure gases, their pressure-flow characteristics are nonlinear. The tube or capillary can provide essentially linear restriction. In contrast to the fixed resistor, the variable fluid resistor is simply a valve. The needle valve is the most widely used type in fluidic circuits. A variable resistor of needle-type design offers the convenience of fine adjustment in flow over wide ranges. Needle resistors permit the proper values of pressure and flow in a fluidic circuit. Resistors are frequently used in sensor circuits to keep pressure below the level necessary to switch back-pressure devices (Figure 9.38).

Figure 9.38 Variable resistor. (Courtesy Norgren Company)

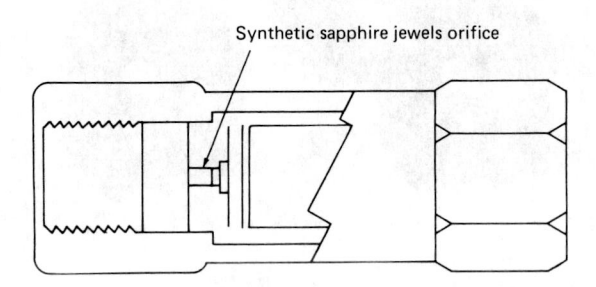

Fluidic diode

The fluidic diode (Figure 9.39) is a miniature duckbill check valve. This valve provides low flow resistance in one direction and nearly absolute shutoff to reverse pressures. A fluidic diode is especially valuable for large fan-in (collecting input from numerous similar devices) to a single input and for performing passive logic functions in fluidic circuits.

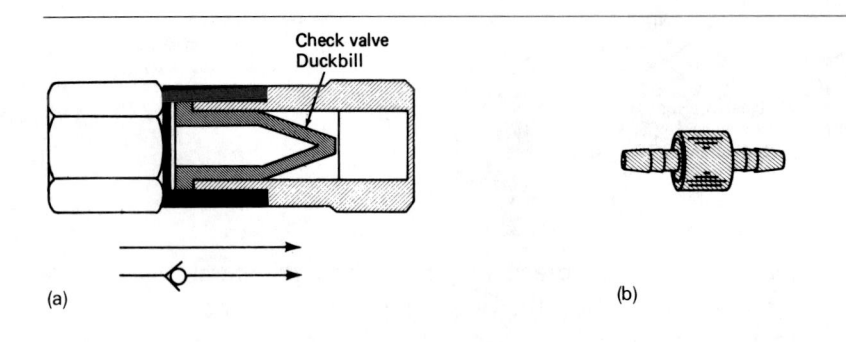

Figure 9.39 The fluidic diode is a miniature duckbill check valve. This valve provides low flow resistance in one direction and nearly absolute shut-off to reverse pressures. A fluid diode is especially valuable for allowing large fan-in to a single input and for performing passive logic in fluidic circuits. (Courtesy Fluidonics)

Fluid oscillators

An oscillator in its simplest form is constructed with a power supply set into a feedback ring supplying either of two outlets, as illustrated in Figure 9.40(a). When the power stream switches to outlet O1, ambient air flows counterclockwise within the feedback ring, moving an acoustic shock wave ahead of it. This shock wave eventually collides with the power stream and provides sufficient energy to switch the power stream to O2, and the action reverses. Thus an oscillator serves the function of delivering alternate pressure pulses. A variable oscillator [Figure 9.40(b)] has provisions for altering the flow of ambient air, thus reducing or increasing the frequency pulse cycle. The longest fluidic time delays are achieved by using the accumulator in reverse—that is, filling the volume rapidly and allowing it to vent through the control port of an element.

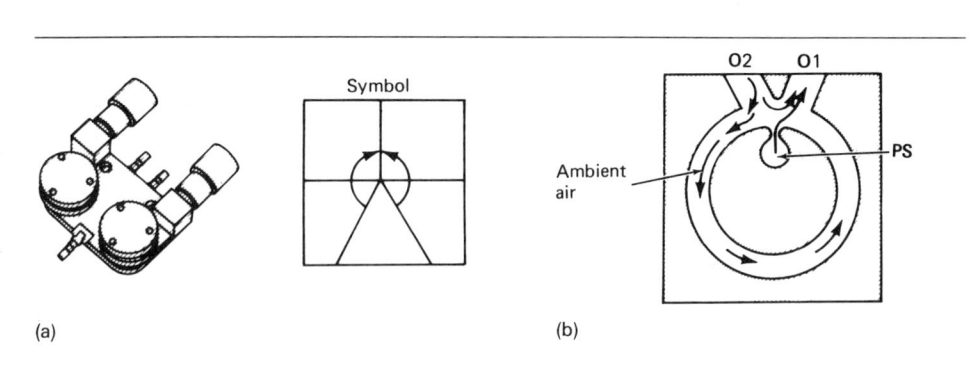

Figure 9.40 (a) Fluidic oscillator. (b) Variable oscillator timer delivers alternating pressure pulses. (Courtesy Fluidonics)

Digital logic functions

Binary arithmetic, which considers only two values—1 and 0—is the most efficient tool for systemizing circuit design and minimizing the possibility of errors (Figure 9.41). Among the reasons for the widespread use of the binary system are (1) the simplicity of arithmetical manipulations such as addition or multiplication and (2) the ease with which the desired function can be implemented with any two-state device such as a switch for controlling on-off or stop-go functions. Digital controls also consider only two values. Logic functions are the basic operations of Boolean algebra and they serve also as the basis for the creation of fluid logic devices. Using binary arithmetic permits a technical specialist to design, understand, and troubleshoot rather complex fluid circuits. A wide array of mechanical and fluid logic devices is available today.

In digital control logic, as derived from Boolean algebra principles, only binary logic devices can be used, that is, only devices capable of creating two discrete signal states, 1 and 0. Designers working

Figure 9.41 Equivalent circuits and basic logic functions.

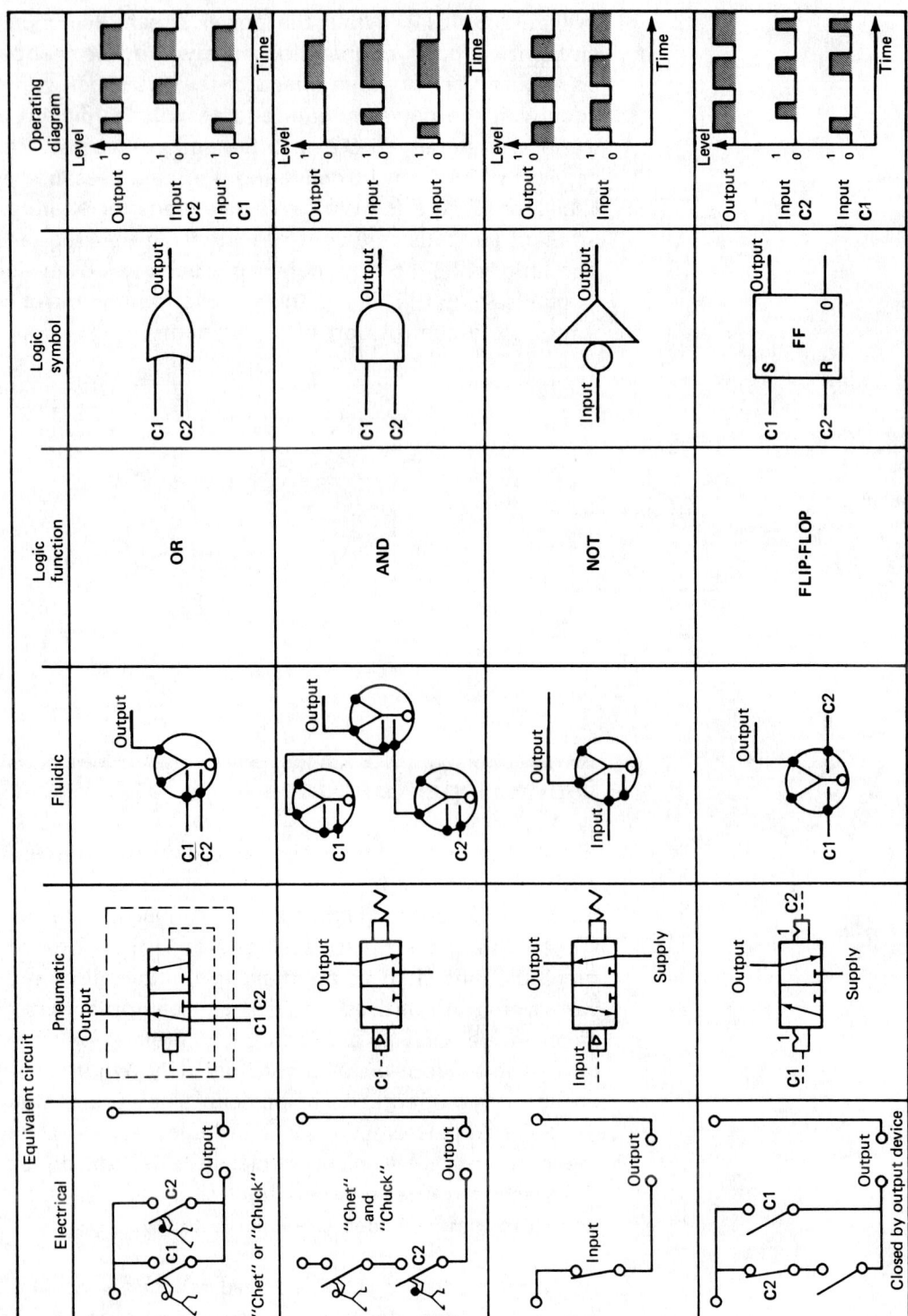

in fluidic technology learn that in fluid controls these two signal states include (1) air compressed to levels higher than atmospheric air pressure for State 1 and (2) compressed air exhausted back at the atmospheric pressure for State 0. Note that fluid logic functions are always related to the binary values of pressures acting on the various devices within the control circuit. It is also true that fluid logic devices are designed to respond to specific logic functions.

The use of two-state devices in fluidic circuits suggests the importance of understanding the binary system. It has been mentioned that the binary system uses the concept of absolute value and position value in the same way as the decimal system. The difference is that the binary system uses only two absolute values, 0 and 1, and the positional values are powers of 2. Binary-number equivalents of decimal numbers up to 10 are represented in Table 9.9. The term "bit" is an abbreviation of a binary digit. Table 9.10 illustrates the positional values in a binary system.

Decimal	Pure binary
0	0000
1	0001
2	0010
3	0011
4	0100
5	0101
6	0110
7	0111
8	1000
9	1001
10	1010

Table 9.9 Decimal numbers and binary-number equivalents

Position	4	3	2	1	0	Binary point	-1	-2	-3	-4
Position value	2^4	2^3	2^2	2^1	2^0		2^{-1}	2^{-2}	2^{-3}	2^{-4}
Quantity Represented by position value	16	8	4	2	1		$\frac{1}{2}$	$\frac{1}{4}$	$\frac{1}{8}$	$\frac{1}{16}$

Table 9.10 Some of the positional values in the binary system

The progression of numbers is by powers of 2, as follows:

Binary		Decimal
1	=	$1 = 2^0$
10	=	$2 = 2^1$
100	=	$4 = 2^2$
1 000	=	$8 = 2^3$
10 000	=	$16 = 2^4$
100 000	=	$32 = 2^5$

Thus the binary number

$$10101 = 16 + 4 + 1 = 21$$

Similarly, numbers to the right of a decimal point can be found from negative powers of 2:

$10^{-1} = 0.1 \quad\quad 2^{-1} = \frac{1}{2}$
$10^{-2} = 0.01 \quad\quad 2^{-2} = \frac{1}{4}$
$10^{-3} = 0.001 \quad\quad 2^{-3} = \frac{1}{8}$

Any decimal number can be expressed in binary form, and vice versa. Observe that in the binary system all numbers are made up of only the two chosen marks. Numbers may appear like this: 11101, 111011, 1101101, 1010111.

Addition with binary numbers

Binary numbers are the easiest set in which to perform arithmetic; this is why the binary system is used in computers and logic networks. In decimal addition, you are required to remember that 9 plus 8 are 17, that 7 plus 8 are 15, and so forth. In binary addition, you need remember only the following two simple rules:

- **Rule 1** 0 plus 1 is 1.
- **Rule 2** 1 plus 1 is 0 and carry a 1 to the next column to the left.

With these two rules it is possible to perform addition. For example, to add decimal 2 and 3 (binary 10 and 11)

```
  10 = 2
  11 = 3
   1
   0   carry 1 to the left
   1   the carry
 101 = 5
```

The next step is to add two "bits" (binary digits), A and B. As each bit must be either a 0 or a 1, there are four possible combinations of A and B, as shown in the Input column in Table 9.11. Note that a sum (S) exists when either A or B is 1; a carry (C) exists only when

Table 9.11 Truth table for adding A and B

Input	Output	S exists when	C exists when
A B	S C		
0 0	0 0		
0 1	1 0	$\overline{A}B$	
1 0	1 0	$A\overline{B}$	
1 1	0 1		AB

both A and B are 1. These situations are expressed in the language of logic in the last two columns of Table 9.11.

Symbolic logic notation

Applying the language of logic, each bit is called A, B, etc., if it is a 1; it is called $\overline{A}$, $\overline{B}$, etc., if it is a 0, with the bar representing the word NOT. That is,

$\overline{A}$ = NOT A

$\overline{B}$ = NOT B

From Table 9.11, if the inputs are A = 0 and B = 1, then a sum exists. This (NOT A and B) is written

$S = \overline{A}B$

Similarly, if A = 1 and B = 0, a sum exists. This (A and NOT B) is written

$S = A\overline{B}$

To conclude, an adder is a device that develops an output signal when

S = (A and NOT B)   or   (B and NOT A)
   $= A\overline{B}$ or $B\overline{A}$

Logic notations have been developed for other available functions such as the AND, OR, and NOR gates.

Applications of fluidic devices

To retain the efficiency and reliability of fluidic control devices, it is necessary to take a number of precautions. Several practical suggestions are offered:

- Enclosures containing fluidic components should be vented to atmosphere, or pressure will build up to delay or even shift components.
- Some circuits using fluidic devices use low pressures and thus require precise regulation.
- Some fluidic components create a vacuum through some ports and can become contaminated and malfunctioning if the environment contains suspended contaminants.
- As with other controls—electric, solid state, or compressed air—these devices should be installed properly according to job specifications within the circuit.
- Fluidic devices are applied, in some cases, to generate ultrasonic waves, which other devices can receive, and amplify the signals to conventional values for full-line pressures.
- Filtration remains critical in low-pressure-actuated devices. Filters in the range of 3 to 5 microns are commonly used for fluidic circuits.
- Fluidic elements can be washed with liquid detergents if the passages become clogged by contaminants.

Basic fluidic circuit applications

Fluidic devices often serve similar or parallel functions to electric or electronic transmission and control applications. Fluidic circuits for sensing and controlling liquid levels, as in a clothes washer, can be accomplished with fluidic devices. In Figure 9.42(a) a Flip-Flop ampli-

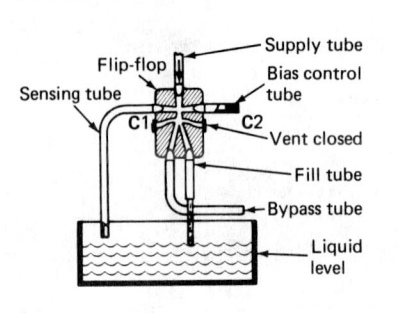

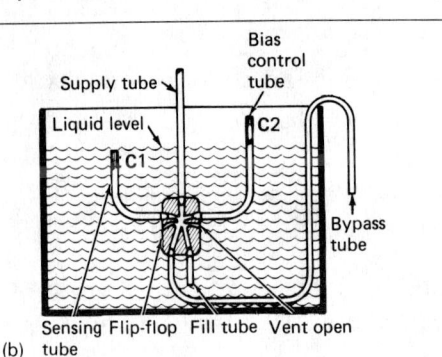

Figure 9.42 (a) Liquid-level control with a nonsubmerged amplifier. (b) Liquid-level control with a submerged amplifier.

fier is mounted above the surface to sense and control the liquid level. Figure 9.42(b) shows a submerged amplifier performing the same function.

The nonsubmerged bistable amplifier circuit can be placed remotely and have no direct contact with the liquid. The end of the sensing tube is placed at the desired liquid level while the amplifier vents are closed. The biased control tube is connected to the pressurized air supply through a variable resistor. As the supply liquid passes through the amplifier, it forms a jet and entrains some of the surrounding air in the interaction region. The air can only enter through the two control ports ($C1$ and $C2$). If the liquid level drops, the end of the sensing tube becomes open and, as the resistor in the bias supply restricts the flow of air into the interaction region to a greater degree than does the open sensing tube, the pressure on the bias supply drops to a lower value than the pressure on the sensing tube. When this happens, the liquid stream switches and flows out on the fill-tube side until the liquid reaches the appropriate level to close the sensing tube. Air flow from the sensing tube is then shut off completely, while air continues to flow in from the bias supply. The resulting pressure changes in the interaction region cause the stream jet to switch to the bypass side of the amplifier.

Sometimes it is necessary to have the amplifier submerged in the liquid rather than remotely located. The principles of operation of a submerged amplifier are basically the same except that it is not necessary to have vent closures, since the device is covered with a liquid. Frequently, noncontact sensors are used in packaging, counting, and detection applications (Figure 9.43).

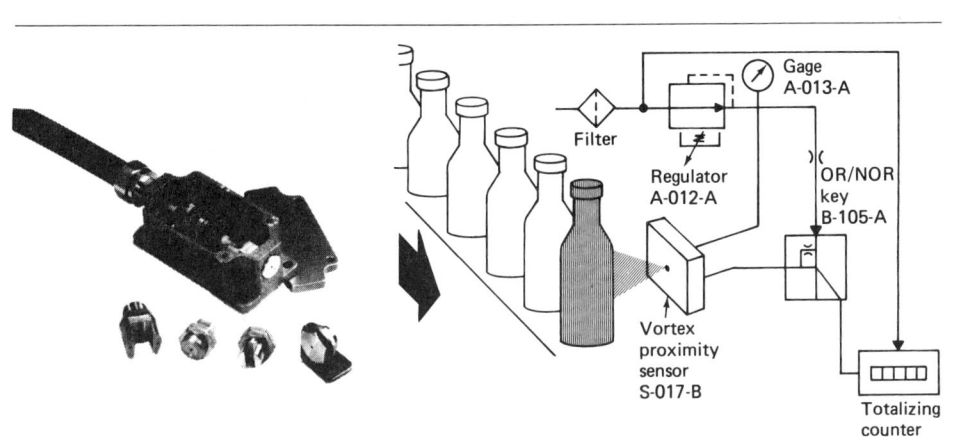

Figure 9.43 (a) Noncontact sensor. (Courtesy Norgren Company) (b) Vortex proximity sensor. High-cycle counting operations are often hard on contact-type electric or pneumatic limit switches. Contacting the object with the sensor is not practical in many cases. The totalizing system diagrammed here can be assembled from inexpensive, off-the-shelf fluidic components. (Courtesy Bowles Fluidics Corporation)

(a)   (b)

Two experiments are provided for the student to observe the principles of fluid logic devices. Both are easy to construct, assemble, and demonstrate. Figure 9.44 shows a fluidic pneumatic servomechanism designed to follow a positioned object such as the transmitting and receiving array shown. As the array is moved, selective blocking of flow to one of the stream receivers causes a differential input to the proportional amplifier. Its resulting differential output then causes cylinder motion in the direction that will equalize the differential signal; this constitutes the follower action.

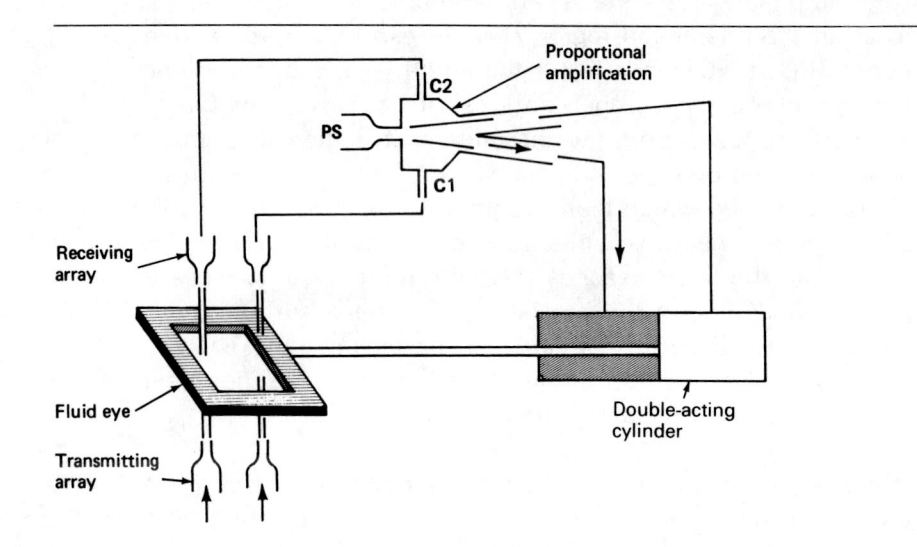

Figure 9.44 Fluidic servomechanism.

The next example, in Figure 9.45, typifies the use of the medium being controlled as its own information carrier in the decision-making process. Assume that you wish to pressure regulate the flow of fluid from a vessel. Regulating the fluid depth would accomplish this quite well. The monostable element in the bottom of the tank has an

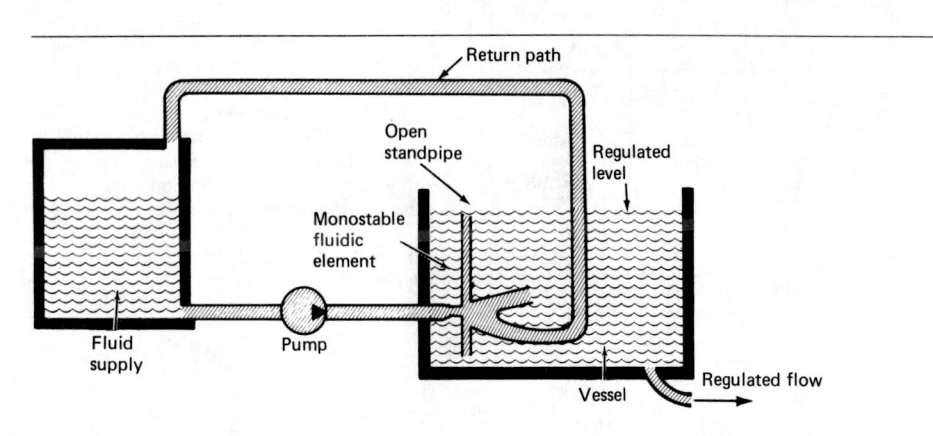

Figure 9.45 Submerged monostable element controls liquid level.

open standpipe connected to the control port of its preferred-flow side. Fluid pumped from the supply emerges into the tank until the level rises and liquid begins flowing into the open standpipe. Gravity-supported standpipe flow then pushes the stream to the opposite outlet and the excess liquid returns to the supply tank as a result. As soon as the level falls below the standpipe height, the original condition returns. The system oscillates back and forth and maintains the depth in response to the signals.

Another simple fluidic application circuit can be constructed to detect and reject uncapped containers (Figure 9.46). As a bottle approaches the detector station, Sensor 1 sets Flip-Flop 1, thus removing one input ($C1$) from the AND gate (an OR-NOR amplifier). If a cap is present, Sensor 2 resets Flip-Flop 1, and the bottle proceeds along the assembly line. If the cap is absent, Sensor 2 is not actuated, and Sensor 3's output removes the other input ($C3$) to the AND gate. The output from the AND gate ($O2$) switches Flip-Flop 2, which shifts the control valve to advance the cylinder, thus rejecting the uncapped bottle. As the rod advances, it strikes Sensor 4, which in turn resets Flip-Flop 2, and the cylinder is retracted, allowing the cycle to continue until it is uninterrupted by an uncapped bottle.

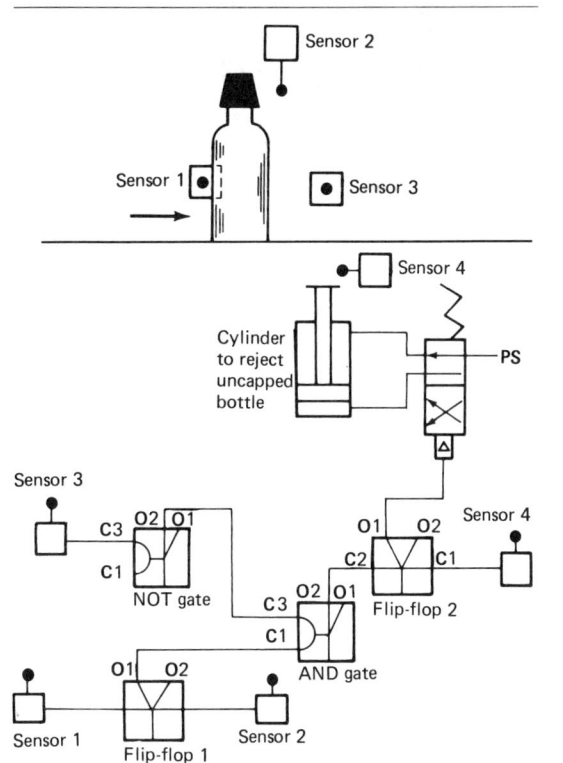

Figure 9.46 A detection and rejection fluid control system.

It should be remembered that fluidics is a control technique in which the control is attained primarily through the use of fluid dynamic phenomena. Fluidic circuits can and do control appliances, flight guidance autopilots, marine guidance systems, machine tools, air-conditioning units, health-monitoring equipment, and many other automated machines [Figure 9.47(a)]. Fluidic controls [Figure 9.47(b) and (c)] will continue to compete with other control techniques or be interfaced to provide economical, reliable, and easily maintained control functions.

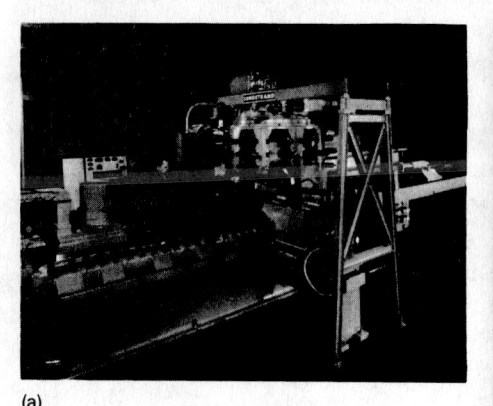

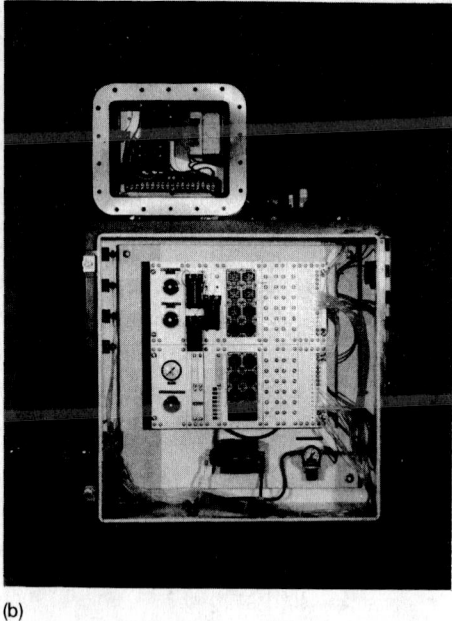

Figure 9.47 (a) Application of fluidics in manufacturing. Approximately 100 gates control 20 tool drum positions in this "Parts Maker." Fluidic controls are interfaced with the master N/C unit through solenoid valves. (Courtesy Corning Glass Works) (b) An industrial fluidic control panel installation for an explosion-proof environment. (Courtesy Norgren Company) (c) Interface between fluid logic circuitry and a two-position, four-way, air directional-control valve operating at working-level pressure. (Courtesy Double A Products)

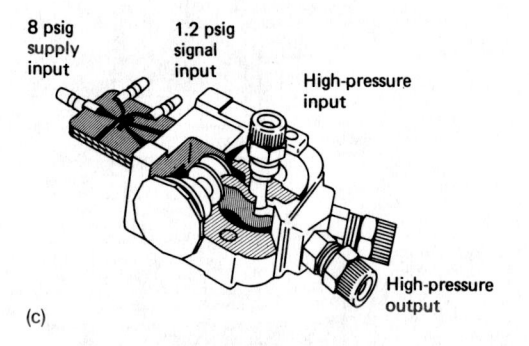

Moving-parts fluid controls

In Figure 9.48 *A* and *B* show the normal and shifted positions of

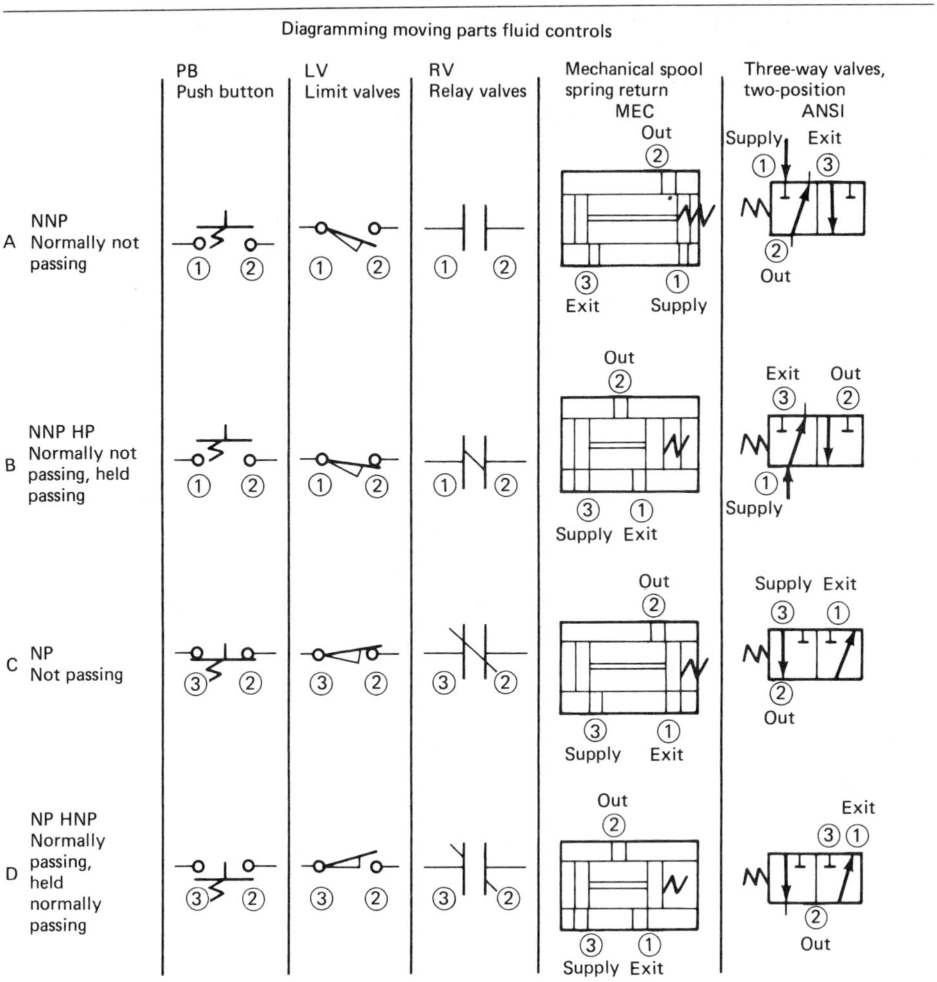

Figure 9.48 Symbology for diagramming moving-parts fluid controls.

the same valve; C and D also show the two positions for the same valve. The difference between $\overline{AB}$ and $\overline{CD}$ is the piping arrangement. Since four-way valves are a combination of 2 three-way valves, the four-way functions can be achieved by combining the appropriate symbols to represent the normal and shifted positions of the valves.

Pneumatic moving-parts systems

Pneumatic logic is a moving-parts system, whereas fluidic devices have no moving parts. In recent years there has been a considerable effort made by persons in industry to develop and standardize pneumatic symbols for purposes of diagramming pneumatic logic circuits.

Today, new symbology is being applied to air logic, since this technology has experienced enormous expansion of applications during the past decade. Designers, manufacturers, salespersons, and maintenance personnel need a common symbology to communicate. Symbols have been borrowed essentially from the electric contact systems and applied to air logic. Since the flow capabilities of electricity and air are somewhat similar, the majority of symbols could be readily adapted, although others had to be created where dissimilar functions were performed by air. To a large extent, any functions of logic performed by electric devices can be performed with air controls as long as they do not involve heat or magnetic fields.

Control functions such as monitoring pressure, sequence, and area detection can be readily performed with air. Compressed-air applications are becoming more widespread in industry, and there is a corresponding increase in the acceptance of air controls on air-powered equipment. Compatible control components, duplicating the functions of electric controls, are now readily available and electric, hydraulic, and pneumatic controls can reinforce and complement each other, often resulting in superior control systems. Currently, there is a need for individuals in industry who have a working knowledge of all mechanical control systems, including electric, hydraulic, and pneumatic applications. Production equipment is only as reliable and efficient as the logic circuits that control the numerous functions (Figure 9.49). Conversely, even well-designed logic circuits cannot correct faulty mechanical equipment.

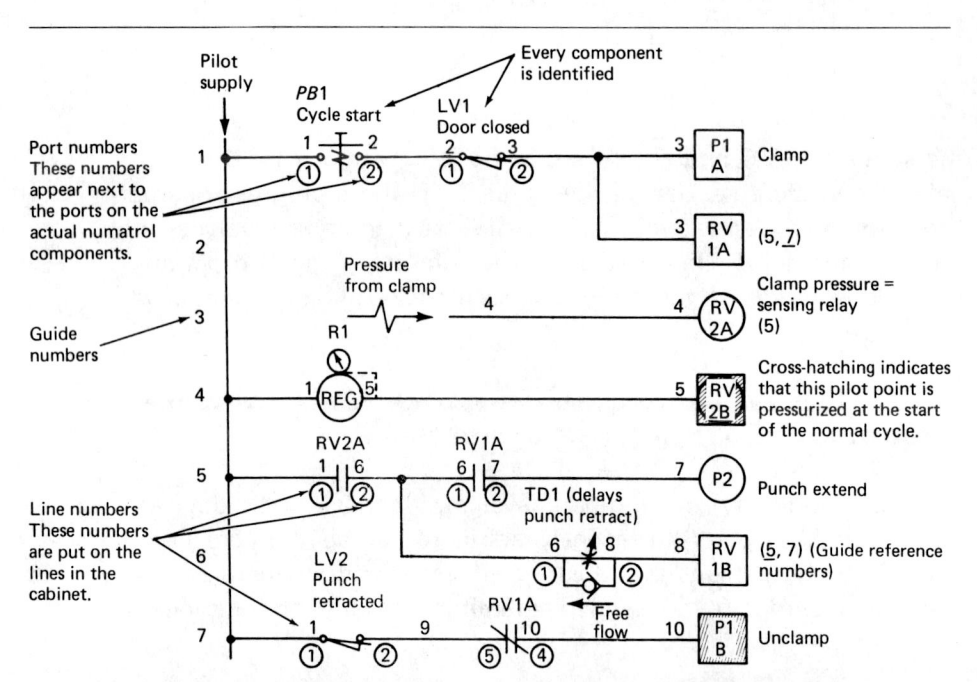

Figure 9.49 Typical moving-parts fluid logic diagram. (Courtesy Numatrol Control Division, Numatics)

Summary

Fluidics is a relatively new method of control that is gaining widespread application in automated equipment. Any fluidic control system performs the following functions: (1) collects information on system conditions (inputs), (2) uses this information to decide what the controlled equipment should do (logic), and (3) commands the controlled equipment to function as directed (output). Logic is simply a way to process information so that something will happen as pre-planned. Symbolic logic notation is the language used to express binary arithmetic functions and also logical decision-making functions. It is helpful to remember that each fluid logic device is designed to respond to specific logic functions. Each gate has only two possible output states—"go" or "no go"—determined by input.

In sequencing control systems the functions performed by fluidic devices are the typical logic operations OR, NOR, AND, NAND, Memory, etc. Basically, a digital device need only discriminate between the absence or presence of signals, while an analog device must identify and yield instantaneous signals. In digital amplifier devices, the stream wall-attachment, or Coanda, effect is used to yield the switching action. On the other hand, in the proportional amplifier devices, this principle is avoided to permit proportional splitting of the power stream. The degree of the deflection of the power stream is proportional to the momentum of the control jet. From a very practical point of view, fluidic devices have found numerous applications because of their many advantages and their compatibility for interfacing with other control systems, particularly pneumatics and electronics.

Opportunities in fluidic technology will continue for those who master the major concepts of fluid mechanics, specialists in material processes, specialists in design, specialists in logic and circuits, and those capable of mathematical and system synthesis. Even with the simplicity and high reliability of fluidic circuits and systems, there will always be a need for technicians to troubleshoot, service, and maintain such systems.

The future potential of fluidic technology has aroused the interest of numerous large industries, research centers, and scientists and engineers. Space, oceanographic, military, medical, chemical, and numerous other agencies are constantly alert to maximizing control capabilities and reliability. Fluidic logic is very likely to expand in sensing and programming functions.

Review questions and problems

1. List four advantages of fluidic devices that make them more effective in hostile environments than other types of controls.
2. Explain the Coanda effect.
3. What is meant by entrainment?
4. How is amplification of the input signal created in a fluid amplifier?
5. What functions does a splitter serve in a bistable amplifier?
6. What function does a vent serve in a proportional amplifier?
7. What is the upper frequency response of fluidic devices?
8. What are the two signal states of digital fluidic devices?
9. Draw the symbols for the following logic devices: OR-NOR, NOR, and Flip-Flop.
10. Write out the truth table for the bistable Flip-Flop gate.
11. What is meant by a truth table?
12. What is the essential difference between an active and a passive fluid logic device?
13. Name three active and two passive fluidic devices.
14. What is a variable fluid resistor?
15. (a) In inputs to logic elements, the complement of the signal A is designated as what? (b) If $A = 0$, $\overline{A} = 1$ or if $A = $ ____, $\overline{A} = $ ____?

Selected readings

- Bouteille, Daniel. *Fluid Logic Controls and Industrial Automation,* translated and edited by Stuart North and Leonard P. Gau. John Wiley & Sons, New York, 1973.
- *Fluid Fundamentals—Fluidics.* Programmed Learning Series, Society of Manufacturing Engineers, Dearborn, Michigan, 1971.
- Interface Workshop Subcommittee. *An Introduction to Fluid Input and Output Interfaces.* National Fluid Power Association Marketing Board, 1969.
- Kirshner, J. *Fluid Amplifiers.* McGraw-Hill, New York, 1966.
- Letham, Daryl L. *Fluidic System Design,* Vol. 1, Parts 1–4. Penton Publishing, Cleveland, 1966.

- Miller, C. J. "Fluidic Systems," *Electronics World,* Vol. 77, No. 6, (June 1967), pp. 23-25 and 78-79.
- Pease, Dudley A. *Basic Fluid Power.* Prentice-Hall, Englewood Cliffs, New Jersey, 1967.
- *Progress Report on Formulation of Proposed NFPA Recommended Standards for Fluidic Devices.* National Fluid Power Association, Thiensville, Wisconsin, 1967.
- Stewart, Harry L. *Hydraulic and Pneumatic Power for Production,* 3rd ed. Industrial Press, New York, 1970.
- Taplin, L. B., and W. F. Datwyler. *Fluidics for Computation and Control.* The Bendix Corporation, Bendix Research Laboratories, Southfield, Michigan, 1968.

———. *The Study of Fluidics Laboratory Manual.* Digiac Corporation, New York, 1970.

———. *Norgren Fluidics Systems Manual.* Norgren Corporation, Littleton, Colorado, 1968.

Appendixes

Appendix A Conversion of English gravitational unit system to international
Appendix B The nature of fluids
Appendix C Important fluid power facts and formulas
Appendix D Abridged graphic symbols for fluid power diagramming
Appendix E Symbols and abbreviations used in this text

Appendix A
Conversion of English gravitational unit system to international

Quantity	English unit	SI unit	Metric symbol	Equivalent unit
Length	1 foot (ft)	0.3048 meter	m	—
Mass	1 slug	14.59 kilograms	kg	—
Time	1 second (sec)	1.0 second	s	—
Force	1 pound (lb)	4.448 newtons	N	kg · m/s
Pressure	1 lb/in² (psi)	0.3325 pascals	Pa	N/m² or kg/(m · s²)
Temperature	Fahrenheit	$\frac{°R}{1.80} = °K$	°K	—
Energy	1 ft-lb	1.356 joule	J	kg · m²/s or m · N
Power	1 ft-lb/sec	1.356 watt	W	J/s or N · m/s

$T_R = T_F + 459.67 = 1.80(T_C) + 491.67$

Supplementary conversion units

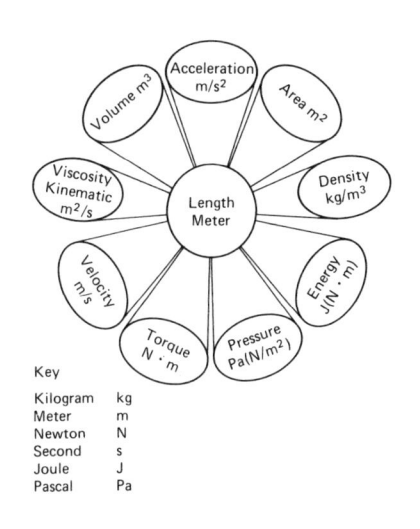

English unit	SI unit	SI symbol	Equivalent unit
Length			
1 inch (in.)	2.540 000 centimeters	cm	25.400 000 mm
1 foot (ft)	0.304 800 meter	m	30.4800 cm
1 yard (yd)	0.914 400 meter	m	91.4400 cm
1 mile (mi)	1.609 344 kilometer	km	1609.344 m
Area			
1 in²	6.451 600 square centimeters	cm²	
1 ft²	0.092 903 square meter	m²	
Volume			
1 in³	16.387 064 cubic centimeters	cm³	
1 ft³	28.316 850 cubic decimeters	dm³	
1 ft³	0.028 317 cubic meter	m³	

Key
Kilogram kg
Meter   m
Newton   N
Second   s
Joule   J
Pascal   Pa

Conversion of English gravitational unit system to international

English unit	SI unit	SI symbol	Equivalent unit
Velocity			
1 ft/sec	30.480 centimeters/second	cm/s	
1 ft/sec	0.304 800 meter/second	m/s (preferred)	
1 mile/hour	1.609 344 kilometer/hour	km/h	
Volume flow rate			
1 in^3/sec	16.3871 cubic centimeters/second	cm^3/s	
1 in^3/min	983.224 cubic centimeters/second	cm^3/s	
1 ft^3/sec	28.316.85 liters/second or cubic decimeters/second	or dm^3/s	
1 ft^3/min	1699.011 liters/second or cubic decimeters/second	l/s or dm^3/s	
1 U.S. gal/sec	3.785 412 liters/second or cubic decimeters/second	l/s or dm^3/s	
1 U.S. gal/min	227.1247 liters/second or cubic decimeters/second	l/s or dm^3/s	
Mass			
1 lb	453.592 37 grams	g	
1 lb	0.453 592 kilogram	kg	
Pressure			
Standard atmosphere (14.7 psi)	101.352 kilopascals	kPa	
1 inch of water (at 4 °C or 39.2 °F)	249.082 pascals	Pa	
1 lb force per square inch (psi)	6.894 757 kilopascals	kPa	
Energy			
1 foot-pound (ft-lb)	1.355 818 joule	J = N · m	
Force			
1 lb-(lb$_f$)	4.448 222 newton	N	
1000 lb	4.448 222 kilonewton	kN	
Power			
1 ft-lb/sec	1.355 818 watt	W	
1 ft-lb/min	0.022 597 watt	W	
1 U.S. hp (550 ft-lb/sec)	745.700 watts	W = 1 J/s or = 1 N · m/s	
Kilowatts	$\dfrac{\text{torque(newton-meters)} \times \text{rpm}}{0.000955}$	kW	
Torque			
1 lb force-inch (lb-in.)	0.112 985 newton-meter	N · m	
1 lb force-foot (lb-ft)	1.355 818 newton-meter	N · m	

English unit	SI unit	SI symbol	Equivalent unit
Viscosity			
pound force sec/foot (lb-sec/ft)	47.880 26 pascal-seconds	Pa · s	
square foot/second (kinematic); ft^2/sec	0.092 903 square meter/second	m^2/s	

Prefixes for decimal multiples and submultiples

The prefix scheme used in the SI metric system is the scientific notation system. This system uses a base-10 function. There are six prefixes for multiples of the base units (Table A.1) and eight prefixes for the submultiples of the base units (Table A.2).

Table A.1 Multiples

Prefix	Symbol added to base unit	Value	Power of 10
deca	da*	10 units	10
hecto	h*	100 units	10^2
kilo	k	1 000 units	10^3
mega	M	1 000 000 units	10^6
giga	G	1 000 000 000 units	10^9
tera	T	1 000 000 000 000 units	10^{12}

*To be avoided.

Table A.2 Submultiples

Prefix	Symbol added to base unit	Value	Power of 10
deci	d	0.1 unit	10^{-1}
centi	c	0.01 unit	10^{-2}
milli	m	0.001 unit	10^{-3}
micro	µ	0.000 001 unit	10^{-6}
nano	n	0.000 000 001 unit	10^{-9}
pico	p	0.000 000 000 001 unit	10^{-12}
femto	f	0.000 000 000 000 001 unit	10^{-15}
atto	a	0.000 000 000 000 000 001 unit	10^{-18}

Appendix B
The nature of fluids

Fluid	Advantages	Disadvantages
Water	Inexpensive Readily available Fire-resistant	No lubricity Corrosion-inducing Temperature limitations
Water-oil emulsion	Good fire resistance Inexpensive Compatible with most seals	Sometimes difficult to maintain Low lubricity
Water-glycol	Good fire resistance Inexpensive Compatible with most pipe compounds and seals	Not good for radial-piston pumps Poor corrosion resistance May cause excessive pump wear at high pressures
Phosphate esters	Excellent fire resistance Good lubricity Noncorrosive	Not compatible with some pipe compounds and some seals Fairly expensive No rust protection
Petroleum oils	Excellent lubricity Reasonable cost Noncorrosive	Tendency to oxidize rapidly No fire resistance
Air	Available universally Relatively clean Easily harnessed Can be exhausted to atmosphere Clean leakage	Expensive Limited to low pressure Subject to moisture
Dry nitrogen	Clean Safe Inert	Expensive Not readily available Sponginess (sometimes)

Appendix C
Important fluid power facts and formulas

Theory

Bernoulli's equation:

$$\frac{p}{w} + \frac{v^2}{2g} + h = \text{constant}$$

Boyle's law: $p_1 V_1 = p_2 V_2$
Charles' law: $V_2 T_1 = T_2 V_1$
Gay-Lussac's law: $p_2 T_1 = T_2 p_1$
Reynolds number:

$$N_R = \frac{VD}{\nu}$$

Darcy-Weisback equation:

$$h = f\frac{Lv^2}{2gD}$$

Torricelli's equation: $v = \sqrt{2gh}$

Fluids

Oil is the most commonly used hydraulic fluid because it serves as a lubricant for hydraulic components and is virtually incompressible (0.004% at 1000 psi).

A fluid is pushed into a pump. Atmospheric pressure supplies this push when the pump element reduces the pressure sufficiently within the pump cavity to create a flow condition.

The weight of oil varies considerably with changes in viscosity. However, the viscosity range of common hydraulic oils is in the range of 55 to 58 lb/ft³.

A 1-in. square column of air extending from sea level all the way to the end of the atmosphere would weigh 14.696 lb.

One standard cubic foot of air per minute (SCFM) is 1 ft³ of air per minute at the standard conditions of 68 °F, 14.7 psi, and 36% relative humidity.

Pressure at the bottom of a 1-ft column of oil is approximately 0.4 psi (58/144). To find the approximate pressure at the bottom of an oil column, multiply the height in feet by 0.4.

Area

Area of a circle $= \pi r^2 = \dfrac{\pi D^2}{4} = 0.7854\, D^2$

Volume

1 U.S. gallon = 231 in³

1 ft³ = 1728 in³

1 ft³ = 7.48 gal

Cylinders

Area = $D^2 \times 0.7854$

Force output (lb) = psi × effective area (in²)

$\qquad\qquad\qquad = pA$

The speed of a cylinder is dependent on its effective piston area and the rate of fluid flow.

Piston speed (in./min) = $\dfrac{\text{pump output (in}^3/\text{min)}}{\text{effective area of piston (in}^2)}$

353 Important fluid power facts and formulas

$$\text{or} \quad \frac{231 \times \text{gpm}}{\text{area}}$$

Displacement of a hydraulic cylinder (in³) = effective area (in²) × stroke (in.)

$$\text{Time to extend piston (min)} = \frac{\text{cylinder displacement (in}^3\text{)}}{\text{pump output (in}^3/\text{min)}}$$

Conductors

Flow velocity through a pipe varies inversely as the square of the inside diameter. Doubling the size of the D_i will increase the area four times.

$$\text{Velocity (ft/sec)} = \frac{231 \times \text{gpm}}{12 \times 60 \times \text{area}} \quad \text{or} \quad \frac{0.3208 \times \text{gpm}}{\text{area}}$$

Rate of flow (Q) = area × velocity

To find the area of a pipe needed to handle a given flow, use this formula:

$$\text{area} = \frac{\text{gpm} \times 0.3208}{\text{velocity (ft/sec)}}$$

Power

1 hp = 550 ft-lb/sec

7.48 gal = 1 ft³

Then

$$1 \text{ gpm} = \frac{1}{7.48 \times 60} \text{ ft}^3/\text{sec}$$

$$1 \text{ psi} = 1 \text{ lb/in}^2 \times \frac{144 \text{ in}^2}{\text{ft}^2} = 144 \text{ lb/ft}^2$$

Therefore, the energy required to pump 1 gal at 1 psi in 1 sec is as follows:

$$\text{Energy} = \frac{1}{7.48 \times 60} \times 144 = 0.321 \text{ ft-lb/sec}$$

Converting the above quantity of energy to horsepower:

$$\text{Energy} = \frac{0.3208}{550} = 0.000583 \text{ hp}$$

$$\text{hp} = \text{gpm} \times \text{psi} \times 0.000583$$

Horsepower input to pump

$$\text{hp}_{in} = \frac{\text{gpm} \times \text{psi} \times 0.000583}{e_o}$$

e_o presents the overall efficiency of a unit

$$\frac{\text{hp}_{out}}{\text{hp}_{in}}$$

Motors

$$\text{Torque (lb-in.)} = \frac{\text{in}^3/\text{rev.} \times \text{psi}}{2\pi}$$

Theoretical motor speed =
$$\frac{\text{theoretical displacement of pump (in}^3/\text{rev.)} \times \text{pump speed (rpm)}}{\text{theoretical displacement of motor (in}^3/\text{rev.)}}$$

$$= \text{rev./min of motor}$$

The actual speed will vary with the volumetric efficiency of both the pump and the motor

Appendix D
ANSI graphic symbols for fluid power diagramming

Solid line — Main line conductor, outline, and shaft

Dash line — Pilot line for control

Dotted line — Exhaust or drain line

Center line — Enclosure outline

Lines crossing The intersection is not necessarily at a 90° angle.

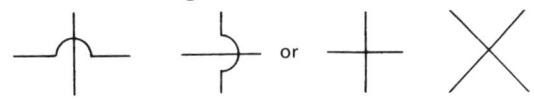

Lines joining Basic symbols may be shown any suitable size. Size may be varied for emphasis or clarity. Relative sizes should be maintained, as in the following example.

Circle and semicircle

Large and small circles may be used to signify that the component is the "main" and the other the auxiliary.

Any part of this standard may be quoted. Credit lines should read: "Extracted from USA Standard Graphic Symbols for Fluid Power Diagrams (USASY32.10—1967) with the permission of the publisher. The American Society of Mechanical Engineers, United Engineering Center, 345 East 47th Street, New York, NY 10017."

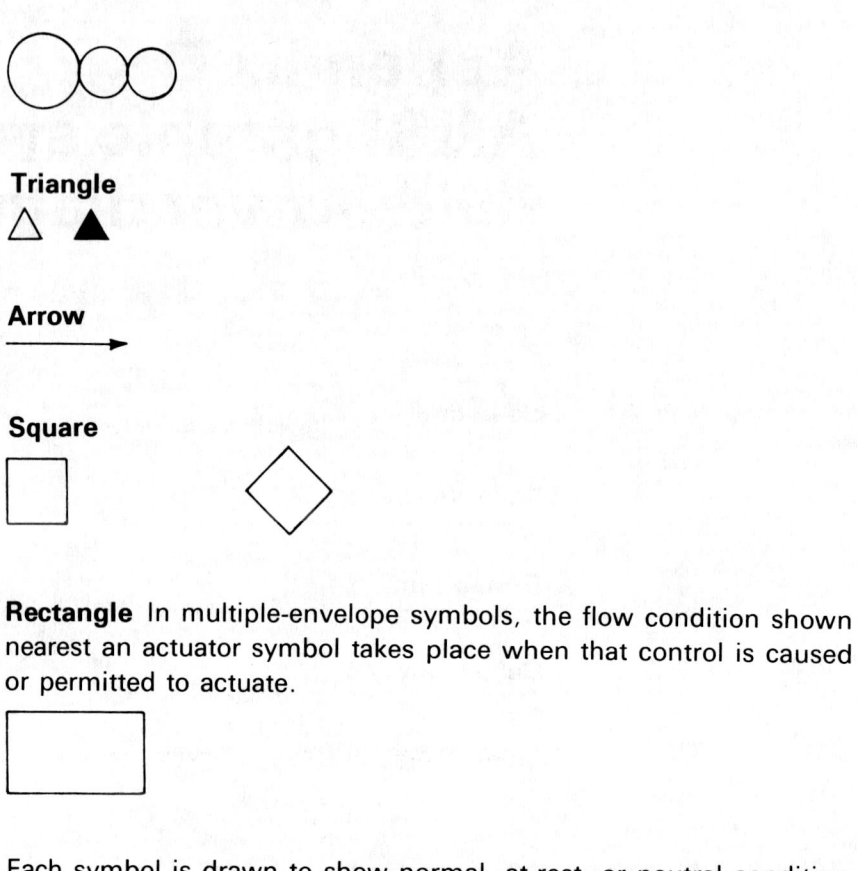

356 Abridged graphic symbols for fluid power diagramming

Triangle

Arrow

Square

Rectangle In multiple-envelope symbols, the flow condition shown nearest an actuator symbol takes place when that control is caused or permitted to actuate.

Each symbol is drawn to show normal, at-rest, or neutral condition of component unless multiple diagrams are furnished showing various phases of circuit operation. Show an actuator symbol for each flow path condition possessed by the component.

An arrow through a symbol at approximately 40° indicates that the component can be adjusted or varied.

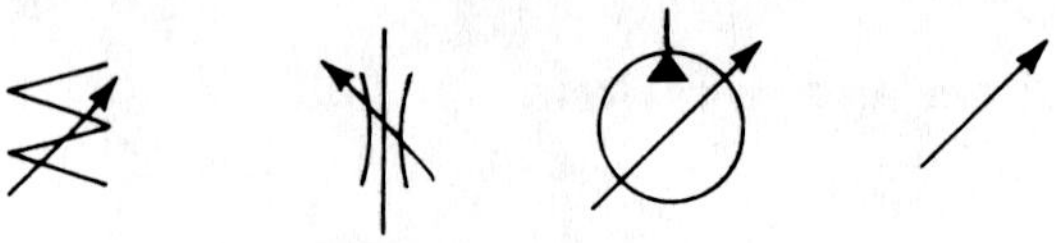

An arrow parallel to the short side of a symbol within the symbol indicates that the component is pressure compensated.

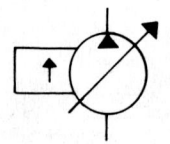

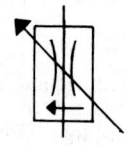

A line terminating in a dot to represent a thermometer is the symbol for temperature cause or effect. (See Temperature Controls, Temperature Indicators and Recorders, and Temperature compensation.)

External ports are located where flow lines connect to basic symbol except where component enclosure symbol is used.

External ports are located at intersections of flow lines and component enclosure symbol when enclosure is used.

Rotating shafts are symbolized by an arrow which indicates direction of rotation (assume arrow on near side of shaft).

Conductor, fluid

——————————— Line, working (main)

— — — — — — Line, pilot (for control)

Line, sensing, etc., such as gauge lines shall be drawn the same as the line to which it connects.

Flow, direction of

——▷— —◁—▷— Pneumatic

——▶— —◀—▶— Hydraulic

Line, pneumatic outlet to atmosphere

———————— Line, liquid drain

Plain orifice, unconnectable

Connectable orifice (e.g., thread)

Line with fixed restriction

Line, flexible

Station, testing, measurement, or power takeoff
Plugged port

Quick disconnect

Without checks

Connected

359 Abridged graphic symbols for fluid power diagramming

———⇥⊣ ⊢⇤——— Disconnected

With two checks

———⊂⇥⊣ ⊢⇤⊃——— Connected

———⊂⇥⊣ ⊢⇤⊃——— Disconnected

With one check

———⇥⊣ ⊢⇤⊃——— Connected

———⇥⊣ ⊢⇤⊃——— Disconnected

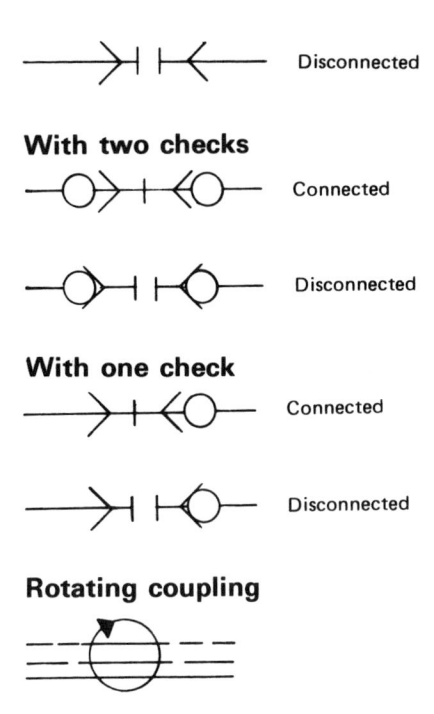

Rotating coupling

Energy storage and fluid storage

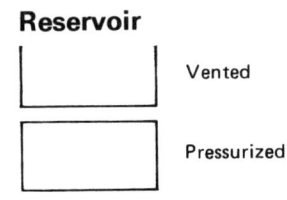

Reservoir

Vented

Pressurized

Note: Reservoirs are conventionally drawn in the horizontal plane. All lines enter and leave from above.

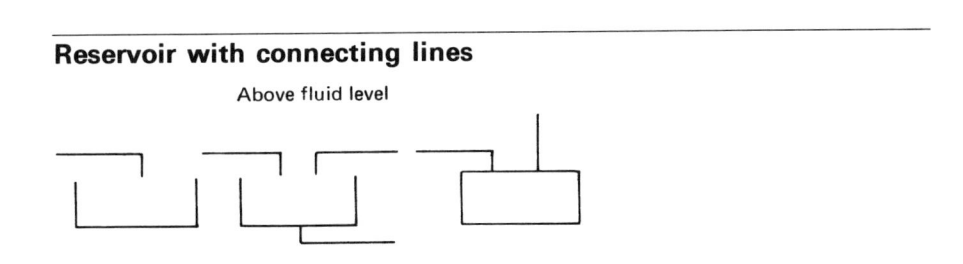

Reservoir with connecting lines

Above fluid level

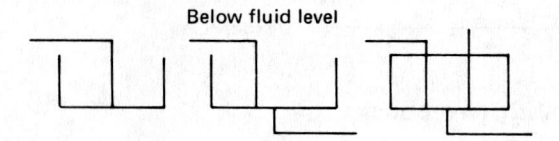

360 Abridged graphic symbols for fluid power diagramming

Below fluid level

*Show line entering or leaving below reservoir only when such bottom connection is essential to circuit function.

Simplified symbol The symbols are used as part of a complete circuit. They are analogous to the ground symbol of electric diagrams —||ı **IEC**. Several such symbols may be used in one diagram to represent the same reservoir.

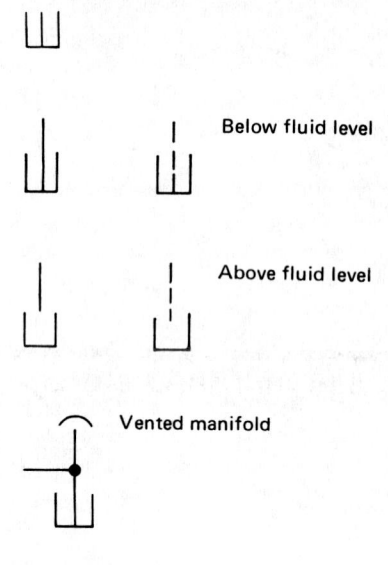

Below fluid level

Above fluid level

Vented manifold

Accumulator

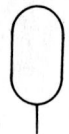

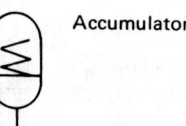

Accumulator, spring loaded

361 Abridged graphic symbols for fluid power diagramming

Accumulator, gas charged

Accumulator, weighted

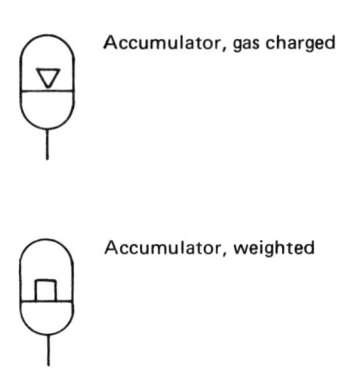

Receiver, for air or other gases

Energy source

(Pump, compressor, accumulator, etc.) This symbol may be used to represent a fluid power source which may be a pump, compressor, or another associated system.

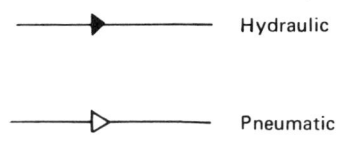

———▶——— Hydraulic

———▷——— Pneumatic

Simplified symbols Example:

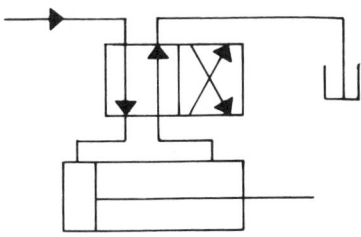

Fluid conditioners

Devices that control the physical characteristics of the fluid.

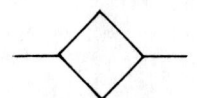

Heat exchanger

Heater Inside triangles indicate the introduction of heat.

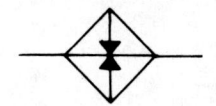

Outside triangles show the heating medium is liquid.

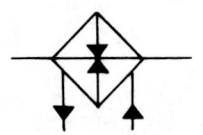

Outside triangles show the heating medium is gaseous.

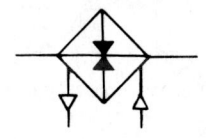

Cooler

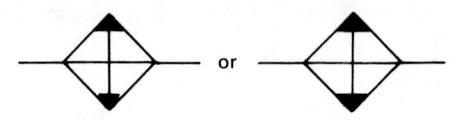

Inside triangles indicate heat dissipation. (Corners may be filled to represent triangles.)

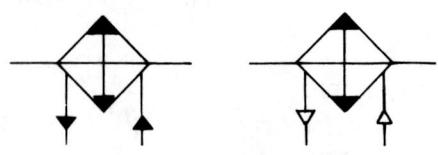

Temperature controller

(The temperature is to be maintained between two predetermined limits.)

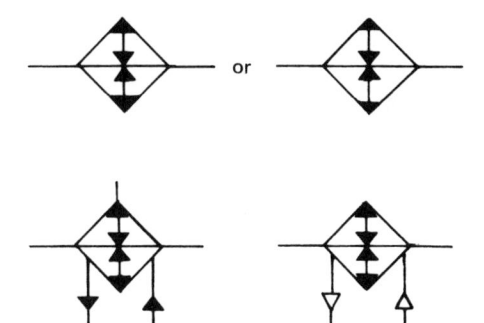

363 Abridged graphic symbols for fluid power diagramming

Filter-strainer

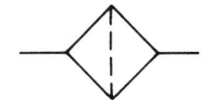

Separator

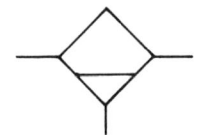

With manual drain

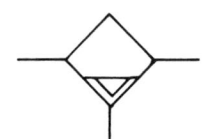

With automatic drain

Filter-separator

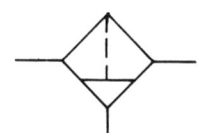

With manual drain

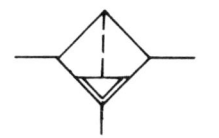

With automatic drain

Desiccator (chemical dryer)

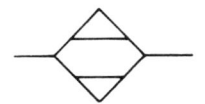

Lubricator

Less drain

With manual drain

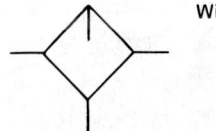

Linear devices

Cylinders, hydraulic, and pneumatic

Single acting

Double acting *Single end rod*

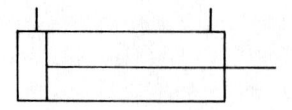

Double end rod

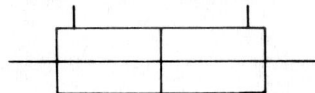

Fixed cushion, advance, and retract

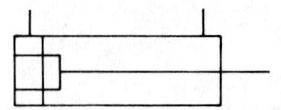

Adjustable cushion, advance only

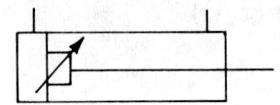

365 Abridged graphic symbols for fluid power diagramming

Use these symbols when diameter of rod compared to diameter of bore is significant to circuit function.

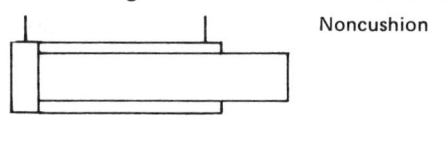

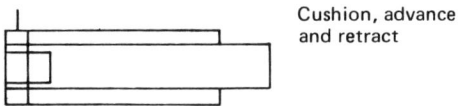

Noncushion

Cushion, advance and retract

Pressure intensifier

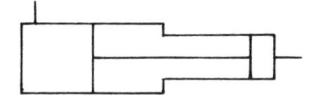

Servo positioner (simplified)

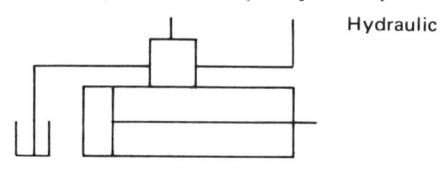

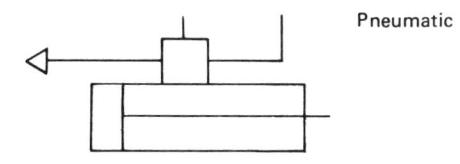

Hydraulic

Pneumatic

Discrete positioner Combine two or more basic cylinder symbols.

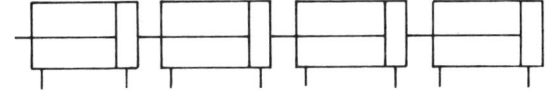

Actuators and controls

Spring

Manual

Use as general symbol without indication of specific type: i.e., foot, hand, leg, arm.

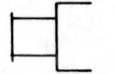

Push button

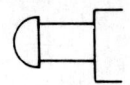

Lever

Pedal or treadle

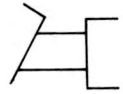

Mechanical

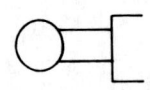

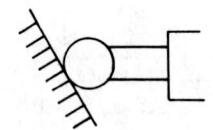

Detent

Show a notch for each detent in the actual component being symbolized. A short line indicates which detent is in use. Detent may, for convenience, be positioned on either end of symbol.

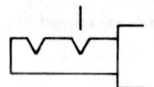

Pressure compensated

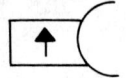

367 Abridged graphic symbols for fluid power diagramming

Electric *Solenoid (single winding)*

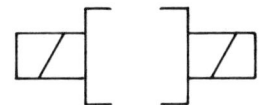

Reversing motor

Pilot pressure

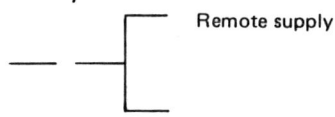

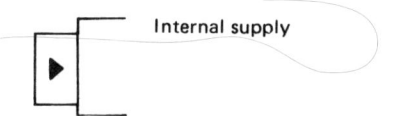

    Remote supply

    Internal supply

Actuation by released pressure

    Remote exhaust

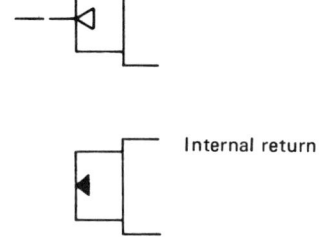

    Internal return

Pilot controlled, spring centered

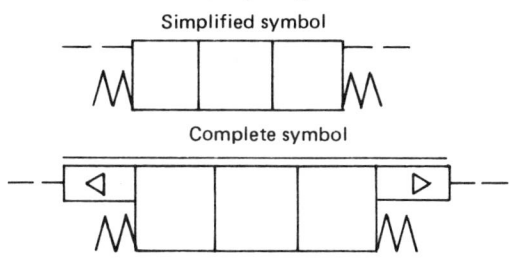

Simplified symbol

Complete symbol

Pilot differential

Simplified symbol

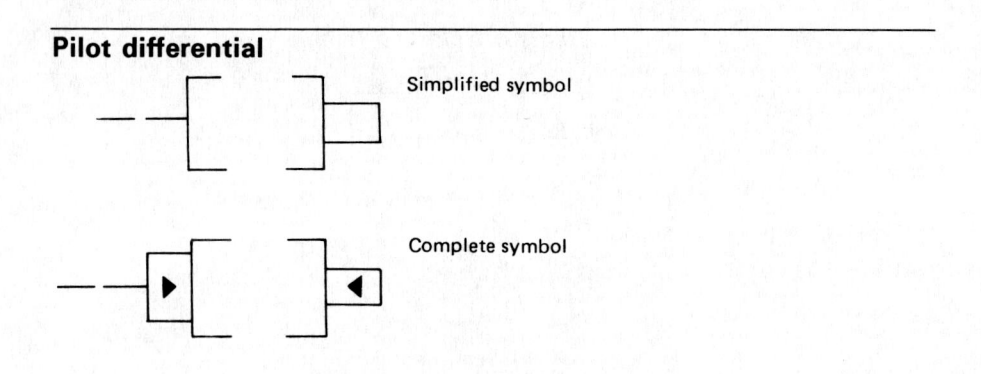

Complete symbol

Solenoid pilot

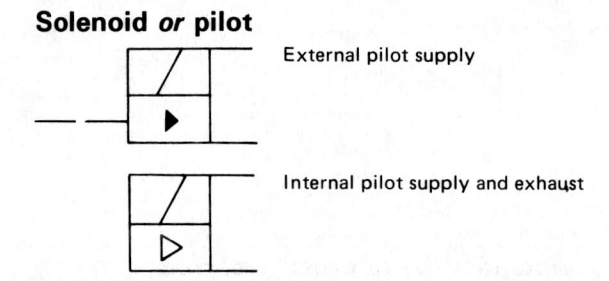

Solenoid *or* pilot

External pilot supply

Internal pilot supply and exhaust

Solenoid *and* pilot

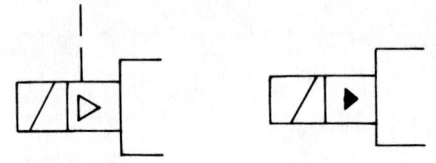

Thermal A mechanical device responding to thermal change.
Local Sensing

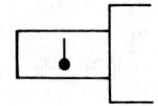

With bulb for remote sensing

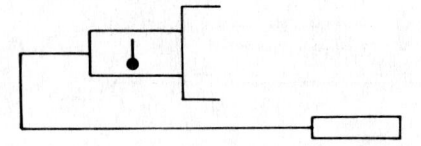

369 Abridged graphic symbols for fluid power diagramming

This symbol contains representation for energy input, command input, and resultant output.

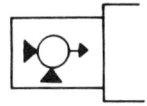

Composite actuators (*and, or, and/or*)

One signal only causes the device to operate.

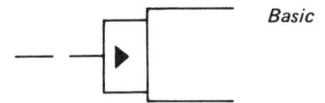

Basic

One signal *and* a second signal both cause the device to operate.

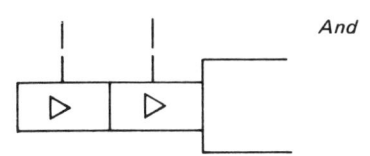

And

One signal *or* the other signal causes the device to operate.

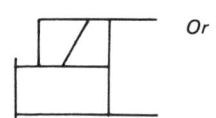

Or

The solenoid *and* and pilot or the manual override alone causes the device to operate.

And/Or

The solenoid *and* the pilot *or* the manual override *and* and pilot.

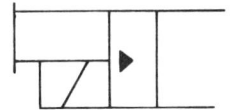

[The solenoid *and* and the pilot] *or* [a manual override *and* and pilot] *or* [a manual override alone].

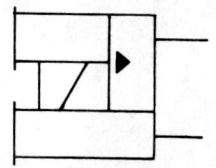

Rotary devices

Basic symbol

With ports

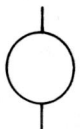

With rotating shaft, with control, and with drain

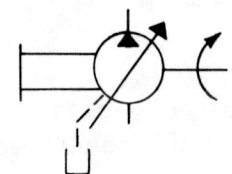

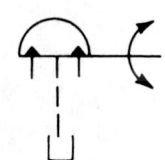

Hydraulic pump

Fixed displacement *Undirectional*

Bidirectional

Variable displacement, noncompensated

Unidirectional

Simplified

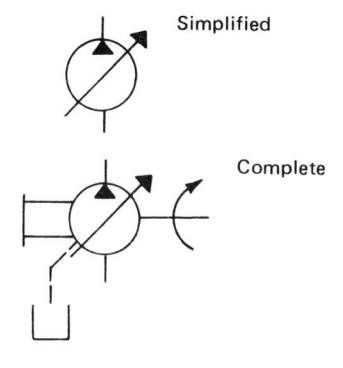

Complete

Bidirectional

Simplified

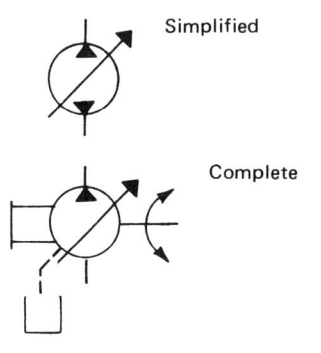

Complete

Variable displacement, pressure compensated

Unidirectional

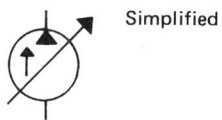

Simplified

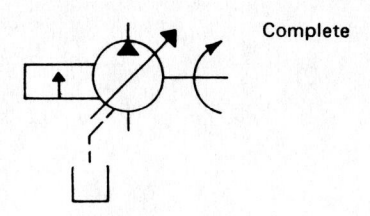

Complete

Bidirectional

Simplified

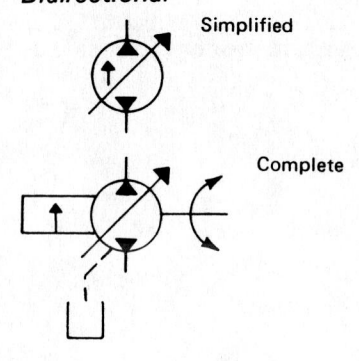

Complete

Hydraulic motor

Fixed displacement *Unidirectional*

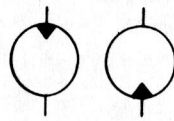

Bidirectional

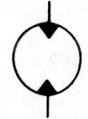

Variable displacement *Unidirectional*

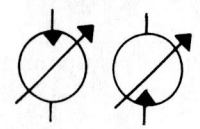

Bidirectional

373 Abridged graphic symbols for fluid power diagramming

Pump-motor, hydraulic Operating in one direction as a pump. Operating in the other direction as a motor.

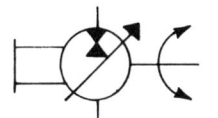

Complete symbol

Simplified symbol

Operating one direction of flow as either a pump or as a motor.

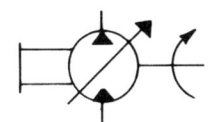

Complete symbol

Simplified symbol

Operating in both directions of flow either as a pump or as a motor. (Variable displacement, pressure compensated shown.)

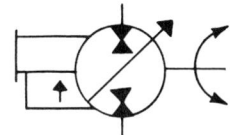

Complete symbol

Simplified symbol

Pump, pneumatic

Compressor, fixed displacement

374 Abridged graphic symbols for fluid power diagramming

Vacuum pump, fixed displacement

Motor, pneumatic

Unidirectional

Bidirectional

Oscillator

Motors, engines

Electric motor

Heat engine Internal combustion engine

375 Abridged graphic symbols for fluid power diagramming

Instruments and accessories

Indicating and recording

Pressure

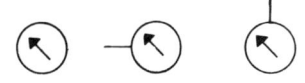

Temperature

Flow meter *Flow rate*

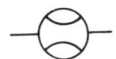

Totalizing

Sensing

Venturi

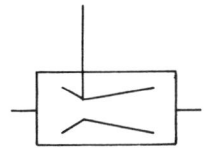

Orifice plate

Pilot tube

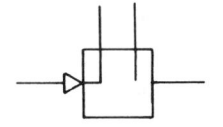

Nozzle

→ Hydraulic

→ Pneumatic

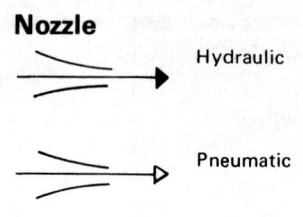

Accessories

Pressure switch

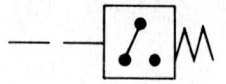

Muffler

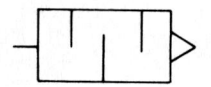

Valves

A basic valve symbol is composed of one or more envelopes with lines inside the envelope to represent flow paths and flow conditions between ports. Three symbol systems are used to represent valve types: single envelope, both finite and infinite position; multiple envelope, finite position; and multiple envelope, infinite position.

In infinite-position, single-envelope valves, the envelope is imagined to move to illustrate how pressure or flow conditions are controlled as the valve is actuated.

Multiple envelopes symbolize valves providing more than one finite flow path option for the fluid. The multiple envelope moves to represent how flow paths change when the valving element within the component is shifted to its finite positions.

Multiple-envelope valves capable of infinite positioning between certain limits are symbolized with the addition of horizontal bars which are drawn parallel to the envelope. The horizontal bars are the clues to the infinite-positioning function possessed by the valve represented.

377 Abridged graphic symbols for fluid power diagramming

Envelopes

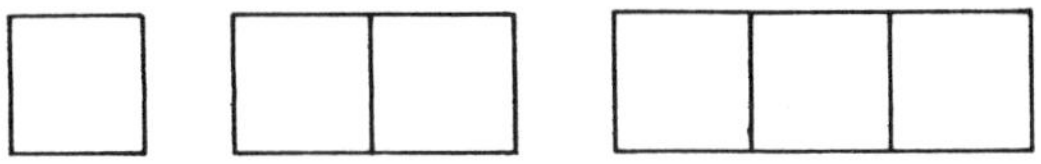

Ports

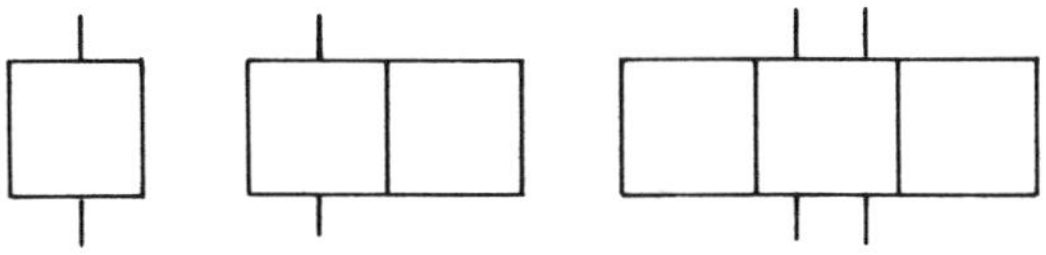

Ports, internally blocked

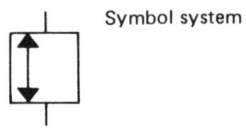

Symbol system

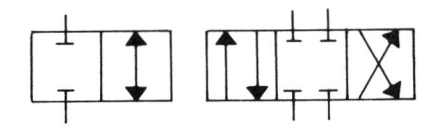

Flow paths, internally open (symbol systems)

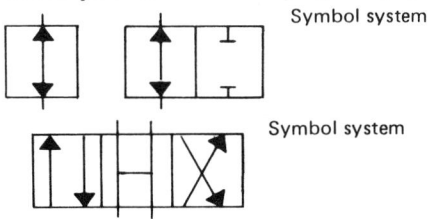

Symbol system

Symbol system

Flow paths, internally open (symbol system)

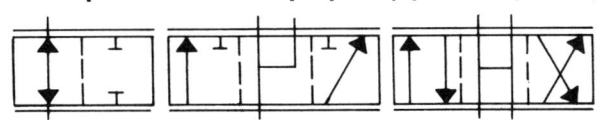

Two-way valves (two-ported valves)

On-off (manual shut-off)

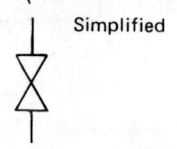

Simplified

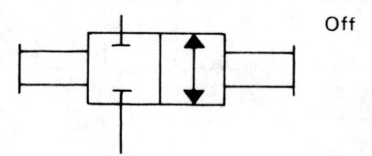

Off

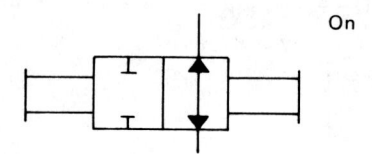

On

Check

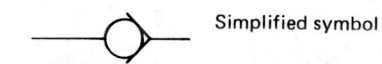

Simplified symbol

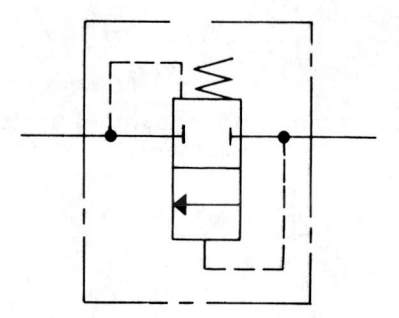

Check, pilot-operated to open

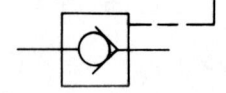

Check, pilot-operated to close

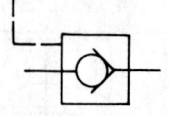

Two-way valves *Two-position*

Normally closed

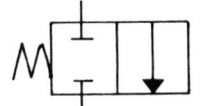

Normally open

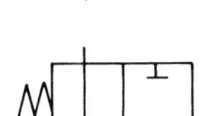

*Normally open
Infinite position*

Normally closed

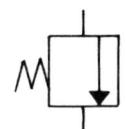

Normally open

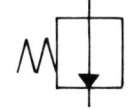

Three-way valves

Two-position *Normally open*

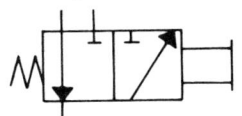

Normally closed

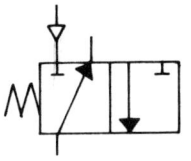

Distributor (Pressure is distributed first to one port, then the other.)

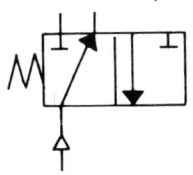

Two-pressure

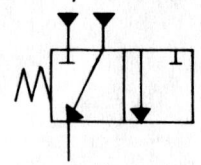

Double check valve Double check valves can be built with and without "cross bleed." Such valves with two poppets do not usually allow pressure to momentarily "cross bleed" to return during transition. Valves with one poppet may allow "cross bleed" as these symbols illustrate.

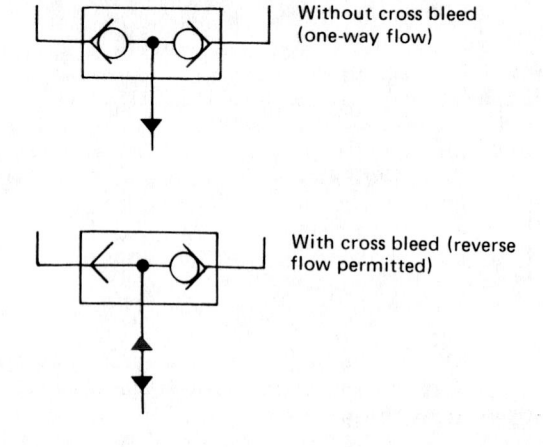

Without cross bleed (one-way flow)

With cross bleed (reverse flow permitted)

Four-way valves

Two position

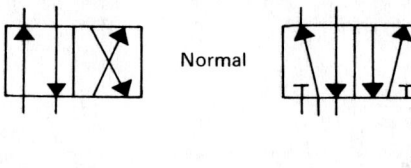

Normal

Actuated

Three position

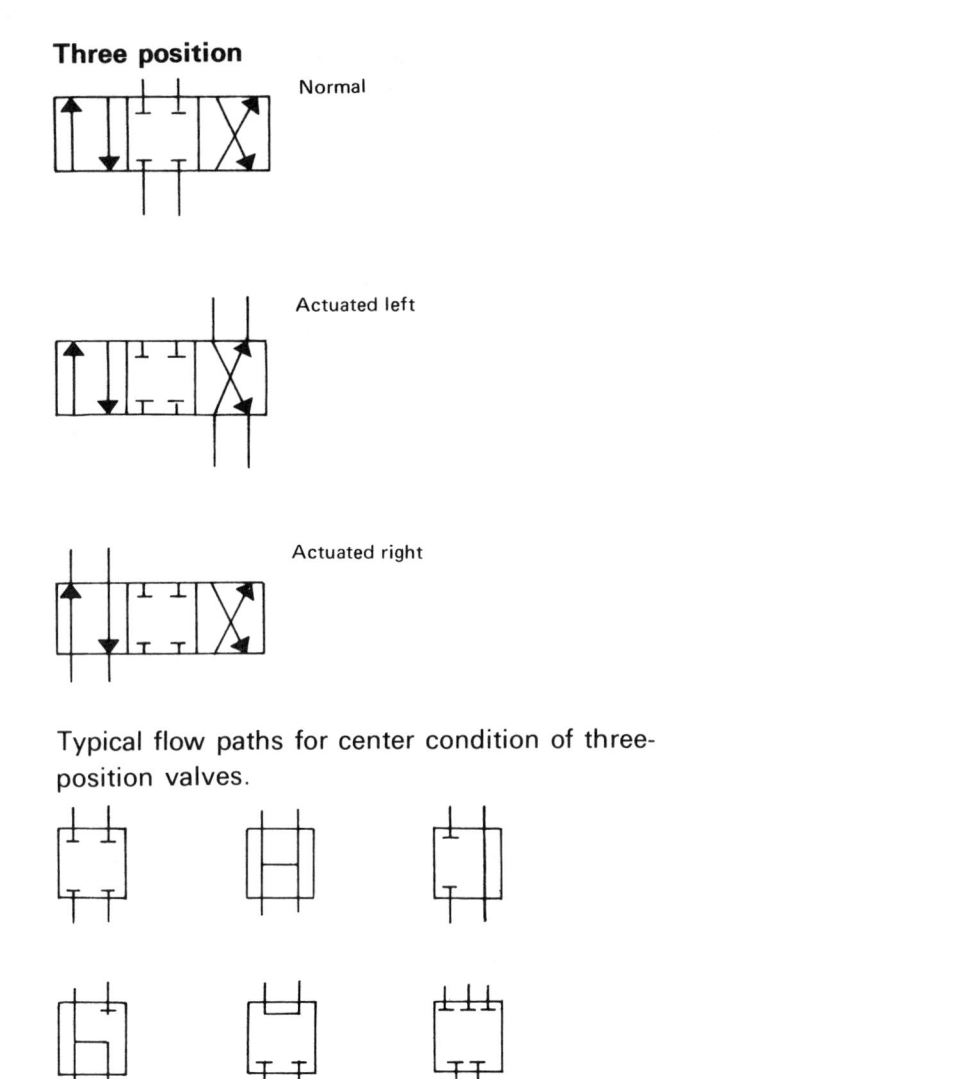

Normal

Actuated left

Actuated right

Typical flow paths for center condition of three-position valves.

Two-position, snap action with transition

As the valve element shifts from one position to the other, it passes through an intermediate position. If it is essential to circuit function to symbolize this "in transit" condition, it can be shown in the center position, enclosed by dashed lines.

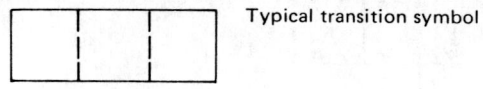

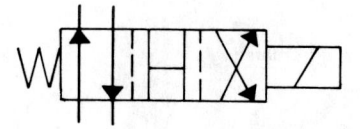

Typical transition symbol

Infinite positioning (between open and closed)

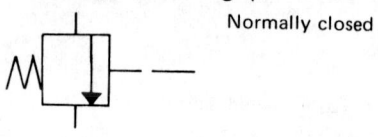

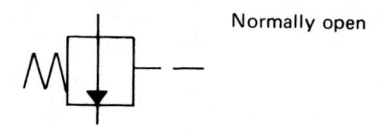

Normally closed

Normally open

Pressure control valves *Pressure relief*

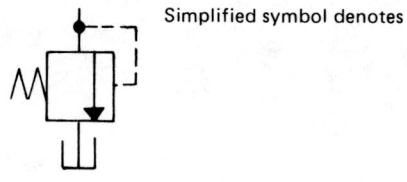

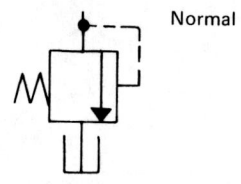

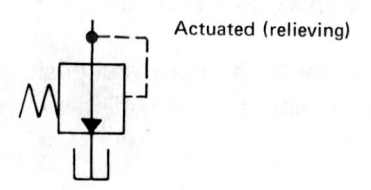

Simplified symbol denotes

Normal

Actuated (relieving)

383 Abridged graphic symbols for fluid power diagramming

Sequence

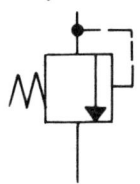

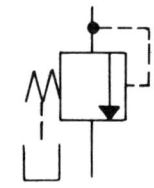

Pressure reducing

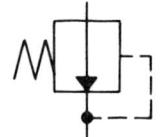

ISO

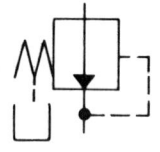

Pressure reducing and relieving

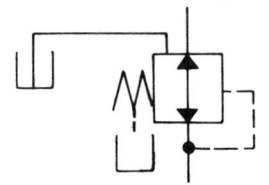

Airline pressure regulator (adjustable, relieving)

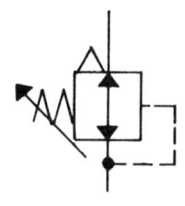

Infinite-positioning three-way valves

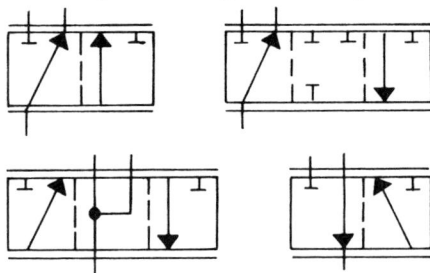

Infinite positioning four-way valves

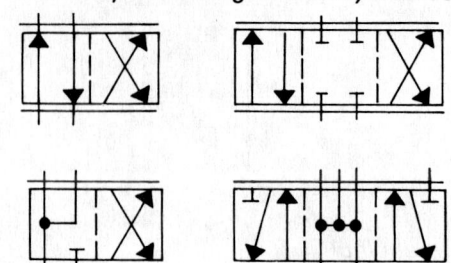

Flow-control valves Adjustable, noncompensated (flow control in each direction)

Adjustable with bypass Flow is controlled to the right. Flow to the left bypasses control.

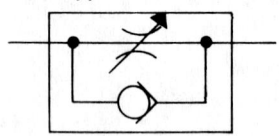

Adjustable and pressure compensated with bypass

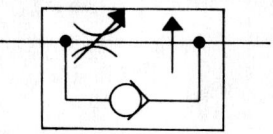

Adjustable, temperature and pressure compensated

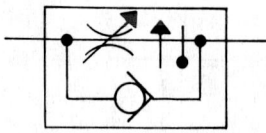

Appendix E
Symbols and abbreviations used in this text

Symbols

Symbol	Definition	Units	Example of Units
A (A_p, A_o)	area (subscript used to denote a specific area)	length^2	in^2, ft^2, cm^2, m^2
a	acceleration	$\dfrac{\text{length}}{\text{time}^2}$	$\dfrac{\text{in.}}{\text{sec}^2}$, $\dfrac{\text{ft}}{\text{sec}^2}$, $\dfrac{\text{cm}}{\text{sec}^2}$, $\dfrac{\text{m}}{\text{sec}^2}$
C	constants	usually dimensionless	
D (D_o, D_i)	diameter (subscripts used to denote outside and inside diameter)	length	in., ft, cm, m
d	distance as applied to work or torque	length	in., ft, cm, m
e (e_m, e_v, e_o)	efficiency (subscript denotes a specific type of efficiency)	dimensionless	A number less than 1 or a percentage less than 100%
F (F_1, F_2, F_p)	force (subscript used to denote force at a location)	force	lb, dynes, N
f	friction factor	dimensionless	
G (G_o, G_i)	work (subscript used to denote output or input to a device)	length $\times$ force	in-lb, ft-lb, dyne-cm, J(N · m)
g	acceleration due to gravity	$\dfrac{\text{length}}{\text{time}^2}$	$386\dfrac{\text{in.}}{\text{sec}^2}$, $32.2\dfrac{\text{ft}}{\text{sec}^2}$, $981\dfrac{\text{cm}}{\text{sec}^2}$, $9.81\dfrac{\text{m}}{\text{sec}^2}$
h	height	length	in., ft, cm, m
H	pressure head		

Symbols and abbreviations used in this text

Symbol	Definition	Units	Example of Units
K, K'	constants usually associated with flow loss	dimensionless	
KE	kinetic energy	length × force	in-lb, ft-lb, J(N · m)
L	linear distance	length	in., ft, cm, m
M	mass flow rate	$\dfrac{\text{mass}}{\text{time}}$	$\dfrac{\text{lb-sec}}{\text{ft}}, \dfrac{\text{slug}}{\text{sec}}, \dfrac{\text{g}}{\text{sec}}, \dfrac{\text{kg}}{\text{sec}}$
M_c, M_p, M_r	momentum as applied to fluid flow	$\dfrac{\text{mass} \times \text{length}}{\text{time}}$	lb-sec, $\dfrac{\text{kg} \cdot \text{m}}{\text{sec}}$
m	mass	mass	$\dfrac{\text{lb-sec}^2}{\text{ft}}$, slug, g, kg
N	rotational speed	$\dfrac{\text{revolutions}}{\text{time}}$	rpm, rev/min
N_R	Reynolds number	dimensionless	
$P\ (P_o, P_i)$	power (subscripts used to denote input and output to a device)	$\dfrac{\text{length} \times \text{force}}{\text{time}}$	$\dfrac{\text{in-lb}}{\text{sec}}, \dfrac{\text{ft-lb}}{\text{sec}}, \dfrac{\text{ft-lb}}{\text{min}}, \dfrac{\text{dyne-cm}}{\text{sec}}, W(J/S)$
p (p_1, p_2, p_a)	pressure (subscript used to denote pressure at a location)	$\dfrac{\text{force}}{\text{length}^2}$	$\dfrac{\text{lb}}{\text{in}^2}(\text{psi}), \dfrac{\text{lb}}{\text{ft}^2}, Pa(N/m^2)$
$Q\ (Q_p, Q_v)$	volume flow rate (subscript used to denote flow from a source or at a location)	$\dfrac{\text{length}^3}{\text{time}}$	$\dfrac{\text{in}^3}{\text{sec}}, \dfrac{\text{gal}}{\text{min}}, \dfrac{\text{cm}^3}{\text{sec}}, \dfrac{\text{liter}}{\text{sec}}, \dfrac{\text{m}^3}{\text{sec}}$
q	volume flow rate (a small quantity)	$\dfrac{\text{length}^3}{\text{time}}$	$\dfrac{\text{in}^3}{\text{sec}}, \dfrac{\text{gal}}{\text{min}}, \dfrac{\text{cm}^3}{\text{sec}}, \dfrac{\text{liter}}{\text{sec}}, \dfrac{\text{m}^3}{\text{sec}}$
r	radius of circle	length	in., ft, cm, m
s	distance (usually stroke length of piston)	length	in., ft, cm, m
S_g	specific gravity (ratio of weight of a substance to weight of water)	dimensionless	
T	torque	force × length	lb-in, lb-ft, N · m

387 Symbols and abbreviations used in this text

Symbol	Definition	Units	Example of Units
T_F, T_C, T_R, T_K	temperature (other subscripts used to denote various states)	degrees temperature	F, C, R, K
t (t_1, t_2, t_3)	time (subscripts used to denote specific times)	time	sec, min, hr
V (V_a, V_t, V_o)	volume (subscripts used to denote actual and theoretical output from a device)	length^3	in^3, ft^3, cm^3, m^3
V_D	displacement volume per minute	$\dfrac{\text{length}^3}{\text{time}}$	$\dfrac{\text{in}^3}{\text{min}}, \dfrac{\text{ft}^3}{\text{min}}, \dfrac{\text{cm}^3}{\text{min}}, \dfrac{\text{m}^3}{\text{min}}$
v	velocity	$\dfrac{\text{length}}{\text{time}}$	$\dfrac{\text{in.}}{\text{min}}, \dfrac{\text{ft}}{\text{min}}, \dfrac{\text{ft}}{\text{sec}}, \dfrac{\text{cm}}{\text{sec}}, \dfrac{\text{m}}{\text{sec}}$
W	weight (subscripts used to denote type of substance)	force	lb, dyne, N
w	weight density or specific weight	$\dfrac{\text{force}}{\text{length}^3}$	$\dfrac{\text{lb}}{\text{in}^3}, \dfrac{\text{lb}}{\text{ft}^3}, \dfrac{\text{dyne}}{\text{cm}^3}, \dfrac{\text{N}}{\text{m}^3}$
x	small distance associated with motion in instruments	length	in., ft, cm, mm
Y	small distance used to determine viscosity	length	in., ft, cm, mm
α	angle	degrees or dimensionless	°, deg, rad, 2π rad = 360 deg
θ	angle	degrees or dimensionless	°, deg, rad 2π rad = 360 deg
Δ	change in quantity or small quantity	dimensionless	
μ	absolute viscosity	$\dfrac{\text{force} \times \text{distance}}{\text{area} \times \text{velocity}}$	$\dfrac{\text{lb-sec}}{\text{in}^2}, \dfrac{\text{lb-sec}}{\text{ft}^2},$ poise, Pa·s $\dfrac{(\text{N}\cdot\text{s})}{\text{m}^2}$
ν	kinematic viscosity	$\dfrac{\dfrac{\text{force} \times \text{distance}}{\text{area} \times \text{velocity}}}{\text{mass density}}$	$\dfrac{\text{in}^2}{\text{sec}}, \dfrac{\text{ft}^2}{\text{sec}},$ stroke, $\dfrac{\text{m}^2}{\text{s}}$
π	constant ratio of circumference of circle to diameter	dimensionless	3.1416

Symbols and abbreviations used in this text

Symbol	Definition	Units	Example of Units
ρ	mass density	$\dfrac{\text{mass}}{\text{volume}}$	$\dfrac{\text{lb-sec}^2}{\text{in}^4}, \dfrac{\text{lb-sec}^2}{\text{ft}^4},$ $\dfrac{\text{slug}}{\text{ft}^3}, \dfrac{\text{g}}{\text{cm}^3}, \dfrac{\text{kg}}{\text{m}^3}$
ω	angular velocity	$\dfrac{\text{angular displacement}}{\text{time}}$	$\dfrac{\text{rad}}{\text{sec}}, \dfrac{1}{\text{s}}$

Abbreviations

AB.	Definition	Units	Example
cfm	volume flow rate	$\dfrac{\text{length}^3}{\text{time}}$	cubic feet per minute
cu ft	volume	length^3	cubic feet
hp	horsepower	$\dfrac{\text{length} \times \text{force}}{\text{time}}$	1 hp = 33,000 $\dfrac{\text{ft-lb}}{\text{min}}$
bhp	brake horsepower	$\dfrac{\text{ft-lb}}{\text{sec}}$	1 hp = 550 $\dfrac{\text{ft-lb}}{\text{sec}}$
ihp	indicated horsepower	$\dfrac{\text{ft-lb}}{\text{sec}}$	
ft	feet	length	feet
in.	inch	length	inch
gpm	volume flow rate	$\dfrac{\text{length}^3}{\text{time}}$	gallons per minute
lb	pounds force	force	pounds
1b$_m$	pounds mass	mass	pounds; 1 lb$_m$ = $\dfrac{1}{32.2}$ lb$_f$
lb$_f$	pounds force	force	pounds; 1 lb$_f$ = 32.2 lb$_f$
psi	pressure	$\dfrac{\text{force}}{\text{area}}$	pounds per square inch
psia	absolute pressure	$\dfrac{\text{force}}{\text{area}}$	pounds per square inch absolute
psig	gage pressure	$\dfrac{\text{force}}{\text{area}}$	pounds per square inch gage
cu in.	volume	length^3	cubic inch
lb-in.	torque*	force $\times$ length	pound-inch

389 Symbols and abbreviations used in this text

AB.	Definition	Units	Example
lb-ft	torque*	force × length	pound-foot
in-lb	work or energy*	length × force	inch-pound
ft-lb	work or energy*	length × force	foot-pound
OD or O.D.	outside diameter, usually associated with piping or tubing	length	outside diameter ft, in., cm, m
ID or I.D.	inside diameter, usually associated with piping or tubing	length	inside diameter ft, in., cm, m
rpm	speed in revolutions per minute	$\frac{\text{revolutions}}{\text{time}}$	$\frac{\text{rev}}{\text{min}}$, revolutions per minute
SCFM	standard cubic feet per minute—air flow at 68 °F, 14.7 psi, 36% relative humidity	$\frac{\text{length}^3}{\text{time}}$	standard cubic feet per minute
emf	electromotive force or voltage	volt	v
P.D.	positive displacement or volume	length^3	in^3, ft^3, m^3
Hz	frequency in cycles per second	$\frac{\text{cycle}}{\text{time}}$	hertz, $\frac{\text{cycles}}{\text{sec}}$
sq in.	area in square inches	length^2	square inches
sq ft	area in square feet	length^2	square feet
gal	volume in gallons	length^3	gallons
fpm	velocity in feet per minute	$\frac{\text{length}}{\text{time}}$	feet per minute

*Note difference between torque and work in length-force units.

Index

A

Absolute pressure, 23, 29, 273
Absolute temperature, 23
Absolute viscosity, 23, 40–41
Absorbent (inactive) filters, 74
Acceleration, 23
Measuring, 289
Accumulators, 85
In circuits, 236–238
Accuracy, of instrument readings, 264
Active devices, 298, 314–321
Active filters, 74–75
Actual capacity, of compressors, 130
Actuation, 158–160
Actuator(s)
Defined, 171, 172
Limited-rotation motor, 209–211
Motor synchronization, 244–247
Principles of operation, 172–174
Sizing in circuits, 224
Actuator control circuit, 217
Additives, fluid, 61–62, 65, 68
Defined, 57
Adiabatic compression, of air, 131–132
Adsorbent filters, 74–75
Air
Adiabatic/isothermal compression, 131–132
Density of, 26
Air coolers, 78–79
Air-oil systems, 252, 256–257
Aluminum-alloy seamless conductors, 86
American National Standards Institute (ANSI), 20
American Standards Association, 85
Amplifier(s), 298, 321–328
Applications, 336–337
Cascading, 328
Diaphragm, 325
Double-leg elbow, 327–328
Focused jet, 324–325
Impact modulators, 326–327
Induction, 327
Jet interaction and, 306–307, 309
Momentum exchange and, 307–308
Submerged, 337
Turbulence, 323–324
Vortex, 322–323
Analog devices, 264, 283, 298, 308, 321, 323
AND devices, 298, 316–317
Assembly device, 299
Atmospheric pressure, 23, 29, 273
Axial-piston motor, 202, 203
Pneumatic, 206
Axial-piston pumps
Defined, 97, 117
Design, 117–118

B

Ball-joint mount, cylinders, 189
Ball-type directional-control valve, 153
Bernoulli, Daniel, 6
Bernoulli's theorem, 6, 36–39
Bias port, 298
Bimetallic temperature measurement, 281
Binary arithmetic, 298, 319–320, 331–335
And decimal equivalents, 333–334
Binary counters, 318–320
Bistable devices, 298, 312, 313, 324, 337
Bits, 298, 333–335
Bleed-off circuit, 240
Boole, George, 298
Boolean algebra, 298, 331
Bourdon tube, 274, 276
Boyle, Robert, 30, 31
Boyle's law, 30–31
Bramah, Joseph, 6
Built-in contaminants, 72

C

Cap, of cylinder, 171, 185, 190
Capacitors, 298, 328
Celsius temperature scale, 27, 280
Centerline mounting, 171, 188, 189
Centrifugal pumps, 104–105
Characteristic curve, 264, 269
Charles, J. A. C., 31
Charles' law, 31–32, 33
Check valve, 154–155

Chlorinated-hydrocarbon-base fluids, 67
Circuits, see Fluid power circuits
Clearance, piston cylinders, 131, 135
Clevis mount, cylinders, 189
Closed-center circuits, 214, 217–218, 225–231
Closed-loop systems, 214, 215–216
Feedback in, 215
Coanda, Henry, 303, 304
Coanda (wall-attachment) effect, 299, 304, 308, 313, 327
Component, 299
Compound relief valves, 145–147
Compression efficiency, 131, 136–137
Compressor(s)
Classification, 122–123
Defined, 97
Displacement, 122, 133
Efficiency, 127–130, 131
Maintenance, 137–138
Performance, 130–138
Reciprocating, 125–127
Rotary, 123–125
Compressor capacity, 132–135
Actual, 130
Theoretical, 133
Connectors, 57, 90–93
Constant overrunning load, 219
Constant power, cycle goal, 220
Constant resistive load, 219
Constant speed-variable power, cycle goal, 221
Contaminants, 57, 64
Types, 72
Control
Fluidics applied, 340–341
Fluid power systems generally, 11
Control jet, 299
Control port, 299, 306
Convoluted metal bellows, pressure measurement, 274
Copper seamless conductors, 86
Counter, 299
Counterbalance system, 214, 257
Counterbalance valves, 141, 149–150
Cracking pressure, 144
Critical sector, cycle part, 220
Cushions, 171, 182, 190–191, 192, 211
Cycle
Defined, 97
Time requirements in circuits, 220–222
Cylinders
Air, 190–193
Body style, 185–186
Cap, 171, 185, 190
Construction, 181–185, 191–192
Controlling pressure, 234–236
Defined, 171, 175
Mountings, 186–190, 192

Plunger/piston form, 175
Selection factors, 192–193
Standard identification code, 186
Types, 175–177, 180, 185–186
Velocity, 177–181

D

D'Alembert's principle, 289
Darcy's formula, 41
Darcy-Weisbach equation, 42
Deadband, 264, 271
Dead-head pressure, 228
Deceleration control circuit, 242–243
Delivery, 97, 114, 123
Pump pressure and, 222
Demand system, 229
Density, 25–26. See also Mass density; Specific gravity; weight density
Dependent variable, 264, 269
Design
Conductor elements, 84–85
Cylinders, 181–186
Fail-safe circuits, 248–249
Filtration approach, 75
Fluid power systems generally, 10, 13, 15
Fluid reservoirs, 79–82
Limited-rotation motors, 209–211
Open-loop circuits, 218–225
Synchronization of components, 244–247
Devices, fluidics, 229
Construction of, 309–314
Diamond Ordinance Fuze Laboratories, 301, 304
Diaphragm amplifier, 325
Diaphragm sensing device, 275, 276
Differential (regenerative) circuit, 240–241
Differential cylinder, 175–177
Differential pressure gage, 274
Digital devices, 299, 313, 320–321, 327, 331
Output, 283–284
Diodes, fluidic, 330
Directional-control circuit, 217
Pneumatic, 252
Directional-control (directional-flow) valve(s), 234
Actuation, 158–160
Center, 158
Defined, 141
Internal elements, 152–154
Position, 157–158
Ways, 154–156
Displacement
Compressors, 122, 133
Hydraulic motors, 194–195
Measurement, 282, 284–288

Motors, 198, 203
Pumps, 97, 99, 101
Double-acting cylinder, 171, 175, 181, 190
Double-end rod cylinder, 177
Double-leg elbow amplifier, 327–328
Double pump, 115
"Drag-cup" viscosimeter, 290
Dynamic displacement compressors, 122

E

Efficiency
Compressor, 131, 136–137
Fluid motors, 198
Fluid power systems generally, 11
Instruments, 269
Mechanical, 97, 127, 128–130, 131
Over-all, 97, 130, 131
Transducers, 269. *See also* Volumetric efficiency
Electrically welded conductors, semirigid, 85
Electrohydraulic servo controls, 164
Electromechanical instrumentation, pressure, 175–276
Element, 299
Elevation head, 36
Emulsions, 57, 64, 65, 66–67
Energy, 1
Forms, 46
Kinetic, 1, 36–37, 99
Law of conservation of, 6, 36–42
Potential, 1, 36, 99
Sources, 3–4
Work distinguished from, 46
Entrainment, 305–306
Equation of continuity, steady-rate flow, 34–35
Exclusive OR device, 299
External-gear motors, 200
External-gear pumps, 107–109

F

Fahrenheit temperature scale, 27–28, 280
Fan-in arrangement, fluidics, 299, 330
Fan-out arrangement, fluidics, 299
Feedback
Closed-loop circuits, 215
Defined, 214–215
Servo systems, 162–163, 216
Filters, 57, 73
Location of, 75–76
Materials, 73–75
Mechanical, 74
Selection, 76–77

Filter-regulator-lubricator (FRL) unit, 253
Filtration, 73–77
Design, 75
Fire-resistance, fluids, 66–68
Fittings, 90–93
Fixed-delivery pumps, 101
Vane pump as, 114
Fixed-displacement motor, 195
Fixed-displacement pumps, 97, 222, 225
Inline axial-piston pump as, 119
Fixed mounting, 171, 188
Flared fittings, 91
Flareless fittings, 92
Flexible conductors, 83, 88–90
Flip-Flop devices, 300, 312, 314–315, 320, 336
Flow
Laminar, 24, 40–42, 294, 323–324
Mass measurement, 293–294
Pressure distinguished, 99
Turbulent, 24, 40, 42, 294, 323
Volume measurement, 290–293
Flow control, pressure compensated valve, 141, 166
Pressure-temperature compensated valve, 141, 166
Flow-control circuit, 216, 238–244
Intermittent, 241–242
Pneumatic, 253–254
Flow-control valves, 141, 164
In circuits, 238, 239
Compensated, 166–168
For motor synchronization, 247
Noncompensated, 165
Flow-divider circuit, 243
Flow-divider valves, 247
Flow rates, 33–36, 52
Fluid(s)
Compressibility, 29–33
Conditioning, 71–79
Defined, 1, 25
Effect of heat on, 33
Effect on motor performance, 249–252
In motion, 33–36
Physical properties, 24–29, 58–65
Types, 65–71
Fluid conductors
Fittings, 90–93
Defined, 57
Selection, 83–84
Sizing, 84–85, 86, 89
Types, 83
Fluidics, 298–342
Active devices, 314–321
Advantages/disadvantages, 302–303
Amplifiers, 321–328
Applications, 336–340
Construction of devices, 309–314
Defined, 299, 301

Development of, 301–302
Passive elements, 328–331
Principles of, 303–309
Terms used, 298–300
Fluid mechanics, 23–54
Defined, 24
Fluid power
Applications, 19–20
Defined, 1, 4
Historical development of, 4–7
Technology of, 1–20
Fluid power circuits, 14
Accumulators in, 236–238
Classification, 216–218
Closed-center, 225–230, 231
Design, 218–225
Fail-safe, 248–249, 255–256
Flow-control, 216, 238–244
Fluidic applications, 336–340
Load requirements, 218–220
Open-loop, 216–225, 230–231
Output speed control, 224–225
Pneumatic, 249–258
Pressure control, 216–217, 231–236
Pressure requirements, 222–223
Selection criteria, 218
Synchronization within, 244–247
Fluid Power Society, 20
Fluid power systems
Advantages of, 7–11, 174
Applications, 8–10, 19–20, 50–54
Function of fluid in, 58
Functions, 14
Limitations, 13
Open-/closed-loop, 215–216
Operation, 15–19
Viscosity and, 59–61. *See also* Fluid power circuits; Hydraulic systems; Pneumatic systems
Fluid reservoirs, 79–82
Accessories, 81–82
Features, 79
Sizing, 80–81
Foaming, resistance of fluids to, 63–64, 65
Focused jet amplifier, 324–325
Force(s)
Centrifugal, 105
Defined, 23
Measurement of, 276–278
Multiplication of in liquid, 15–16, 18
Pressure distinguished from, 43
Force pumps, 98
Four-way valves, 156
Free air, 130
Friction factor, laminar/turbulent flow, 41–42
Full-flow filtration, 75
Full-flow pressure, 144–145
Full-scale accuracy, 264

G

Gage pressure, 23, 29, 273
Gas(es)
Compressibility, 18, 29
Conditioning, 71
Defined, 25
Laws regarding, 30–33
Pascal's law and, 18
Gas-bulb thermomenter, 281
Gate, 299, 316, 320
Gay-Lussac, J. L., 32
Gay-Lussac's law, 32–33
Gear motors, 198, 199–200
External/internal, 200
Slippage, 200
Gear pumps
Output capabilities, 111–112
Types, 107–111
Gear rack actuator, 210–211
Generated contaminants, 72

H

Hagen-Poiseuille formula, 41
Hall, Robert M., 9
Head, cylinder, 171
Heat, effect on fluids, 33
Heaters, 79
Heat exchanger, 57, 78–79
Heat transfer, and fluid selection, 77
Helical gage, 274
Helical-screw compressors, 124–125
Helix-type actuator, 209
High-low circuits, 233–234
Horsepower, 23, 49, 197, 205, 207
Hose, *see* Flexible conductors
Hot-wire anemometry, 294
Hydraulic fuse, 232
Hydraulic motors, displacement, 194–195
Hydraulic systems
Components, 18–19
And properties of fluids, 58, 63, 65
Hydraulic torque, 48
Hydraulics, 1, 4, 98
Potential energy and, 36
Hysteresis, 264, 277, 299
Losses, 271

I

Impact modulator amplifiers, 299, 326–327
Inactive (absorbent) filters, 74

Independent variable, 265, 269
Indicator cards, compressors, 135–136
Induction amplifier, 327
Inertia load, 219
Instrument(s), 265, 268
Accuracy, 264, 270–271
Classification, 272
Efficiency, 269
Output, 266–270
Instrumentation
Accuracy, 264, 270–271
Electric, 267–268, 275–276, 284–286
Fluid-effect devices, 268, 277
Measurement standards, 271–272
Mechanical, 267, 268, 274–275, 287–288
Optical, 279, 285–286
Output, 266–270
Thermal, 268
Interface, 299
Intermittent-feed control circuit, 241–242
Internal-gear motors, 200
Internal-gear pumps, 110
Introduced contaminants, 72
Isothermal compression, air, 131–132
Isothermal compression efficiency, 136

J

Jet-impingement reaction principle, 291–292
Jet interaction, 306–307
Proportional, 308. *See also* Momentum-exchange devices

K

Kelvin temperature scale, 27, 280
Kinematic viscosity, 24
Kinetic energy, 1, 36–37, 99

L

Laminar (streamline) flow, 24, 40–42, 294, 323–324
Meter, 294
Linear variable differential transformer (LVDT), 265, 276, 283
Liquid(s)
Force multiplication in, 15–16, 18, 172–174
Incompressibility of, 15–16, 29, 172
Pascal's law and, 16–18
Liquid-piston compressors, 124
Load, circuit
Defined, 219
Types, 219–220

Load-locking circuit, 248
Lobed-rotor compressors, 124
Logical state, 299
Logic devices, 299, 317, 320–321, 324, 331
Pneumatic, 341
Logic functions, 317
Digital, 331–335
Lubrication, fluid selection and, 77
Lubricity, property of fluids, 64
LVDT (linear variable differential transformer), 265, 276, 283

M

Manifold, 57
Manometer, 275
Mass, 24
Mass density, 24, 25
Mechanical efficiency, 97, 127, 128–130, 131
Mechanical filters, 74
Mechanical torque, 48
Memory devices, 300, 314
Merriman, Mansfield, 98
Meter-in circuit, 238–239, 241, 243
Meter-out circuit, 239–240, 241, 243
Micron, 57, 73
Mill-type cylinders, 185
Minimum overall cycle time, 220
Mobile cylinders, 186
Momentum, conservation of, 322
Momentum exchange, 307–308, 326
Devices, 308
Monostable device, 300, 309
Motor(s), fluid
Defined, 171, 193
Efficiency, 198
Fixed-/variable-displacement, 194–195
Limited-rotation, 209–211
Operation of, 193
Pneumatic, 204–209
Speed, 195–196, 207–209
Synchronization with actuator, 244–247
Torque, 196–199
Types, 198–203
Motor torque, 196
Mountings, types for cylinders, 187–190
Multiple-screw pumps, 111
Multiple-stage compressor, 97

N

NAND devices, 300, 316–317
Naphthenic oils, 65

Viscosity of, 61, 65
National Bureau of Standards, 271
National Fluid Power Association, 20, 301
Negative load, 219, 240
Newton's law, 24, 219
Nominal displacement, compressors, 133
Nomograph, 85
Noncenterline mounting, 188
Nonpositive-displacement pumps, 99-100
NOR devices, 300, 315-316, 320, 324
Normally closed valves, 141, 144, 147, 148, 149
Normally open valves, 141, 150
NOT device, 300
Nutating-disk meter, 292

O

Oil-coolers, 78, 82
Oils, types of hydraulic, 65-71
One-way valve, 154-155
Open-center circuit, 215, 217, 222-223, 224, 230-231, 246
Open-loop system, 215-216
Classifications, 216-218
Operator-protection circuit, 248-249
Optical sensing instruments
Torque measurement, 279
Transducers, 285-286
OR devices, 300, 315-316
Orifice, 141
Sharp-edged, 290
Orifice meter, 290-291
Oscillators, fluidic, 330-331
Output port, 300, 319
Overall efficiency, 97, 130, 131
Overrunning load, 219, 240
Oxidation, resistance of fluids to, 62-63

P

Package units, 82
Paraffinic oils, 66
Viscosity of, 61, 66
Pascal, Blaise, 6
Pascal's law, 6, 16-18, 28, 43, 58, 172, 181, 196, 273
Passage, 300
Passive devices, 300, 328-331
Capacitors, 328
Diodes, 330
Oscillators, 330-331
Resistors, 329
Petroleum oils, viscosity of, 61
Phosphate esters, 67
Viscosity of, 62
Piezoelectric effect, 265
And acceleration measurement, 289
And pressure measurement, 276
Pilot-relief valve, 145-146, 235
Piston, 175
Piston-chain actuator, 209-210
Piston cylinders, 131, 135
Piston displacement, 130
Piston motors, 198, 202-203
Pneumatic, 206-207
Types, 202
Piston pumps, 117-122
Advantages, 121-122
Types, 117-121
Pivot mounting, 171, 189
Plunger, 175
Stepped, 180
Tapered, 184
Pneumatics, 1, 4
Pneumatic systems, 19, 30
Compressors, 97, 122-138
Cylinders for, 190-193
Fluidics and, 341-342
Motors, 204-209
Power circuits, 249-258
Point-to-point accuracy, 265
Poppet valve, 153
Port, 97
Position, directional-control valves, 157
Positive-displacement compressors, 122
Positive-displacement (P.D.) meter, 292
Positive-displacement pumps, 99, 101
Piston pumps as, 121
Rotary gear pumps as, 107
Positive load, 219, 239
Potential energy, 1, 36, 99
Potentiometric signal generator, 275
Pour point, 64
Power, 1, 48
Unit, 97
Power transmission
Electric, 11-12
Fluid, 12-15
Mechanical, 12
Pressure
Absolute, 23, 29, 273
Atmospheric, 23, 29, 273
Cracking, 144
Dead-head, 228
Defined, 1, 28, 265, 273
Flow distinguished from, 99
Force distinguished from, 43
Gage, 23, 29, 273
Measuring, 273-276
Pumps and circuit requirements, 222-223
Pressure-compensated pump, 115, 218, 228, 229, 232

Pressure-control circuits, 216-217
High-low, 233-234
Maximum pressure-limiting, 231-232
Unloading, 232-233
Pressure-control valve(s), 234
Relief valve, 141
Selection of, 143
Sequence valve, 141
Types, 143-151
Unloading valve, 142
Pressure differential, 308-309
Pressure gage, 265
Pressure head, 36, 39
Pressure-reducing valves, 142, 150-151, 236
Prime mover, 98
Prony brake, 278
Proportional filtration, 75
Pump control circuit, 217
Pumps, hydraulic
Defined, 97
Design classes, 104-122
Efficiency, 127-130
Performance and rating, 98-104
As rotational velocity transducer, 288
Slippage, 102-103
Types, 99-100
Viscosity and, 60-61
Pumps, pneumatic, *see* Compressors

Q

Quick-disconnect couplings, 92-93

R

Radial-piston motor, 202, 203
Pneumatic, 206-207
Radial-piston pumps, 98, 119-121
Rankine temperature scale, 27-28, 280
Read-outs, 275
Reciprocating compressors, 125-127
Single-/double-acting, 126
Reciprocating-piston meter, 292
Reciprocating pumps, 105-106
Reducing, 215
Regenerative (differential) circuit, 240-241
Relief valves, 143-145
Reservoir, 79. *See also* Fluid reservoirs
Resistive load, 219, 239
Resistor devices, 300, 329, 337
Variable, 329
Resolver, 286
Response time, 265

Reynolds, Osborne, 40
Reynolds number, 40-41
Rigid conductors, 83
Rotary compressors, 123-125
Rotary-plate valve, 153
Rotary potentiometers, 284
Rotary pumps
Gear, 107-112
Piston, 117-122
Vane, 112-117
Rotary-spool valve, 153
Rusting, resistance of systems to, 63, 67

S

Safety
Circuit design, 248-249, 255-256
Fluid power systems, generally, 11
Saybolt Seconds Universal (SSU), viscosity measurement, 59, 290
Schmitt trigger, 317-318
Screw pumps, 110-111
Sealing device, 171, 192
Seamless conductors, semirigid, 85-86
Seating action, 142
Seebeck effect, 281
Semirigid conductors, 83, 85-88
Types, 85-86
Sensitivity, 265
Sensor devices, 300, 329
Sequence valves, 148-149
Sequencing circuit, 215, 243-244
Servo systems, 162
Feedback in, 162-163, 216
Pneumatic, 338
Synchronization with, 246
Valve applications, 162-164
Shear rate, 23
Shearing stress, 23
Signal, 300, 306
Signal-generating instruments, 272, 279
Signal-processing instruments, 272, 279
Silicone-base fluids, 68
Single-acting cylinder, 171, 175, 181, 190
Single-ended cylinder, 190
Sliding-spool valve, 153
Sliding-vane compressors, 123-124
Slippage
Motors, 200, 201
Pumps, 102-103
Specific gravity, 26
Spring condition, valves, 160-162
SSU (Saybolt Seconds Universal), 59, 290
Stainless steel seamless conductors, 86
Steady rate, fluid flow, 33-36

Steel seamless conductors, 86
Stepped-plunger, 180
Strainers
Filters distinguished from, 73
Location of, 75-76
Strain gage(s), 265
Acceleration measurement, 289
Pressure measurement, 276
Torque measurement, 279
Stream-deflection amplifier, 300
Streamline (laminar) flow, 24, 40-42, 294, 323-324
Streamlines, 33, 34
Synchros, 283, 286

T

Tachometer generator, 288
Teflon tape, 91
Telescopic cylinder, 180
Temperature
Absolute, 23
Defined, 281
Measurement, 280-282
And oil oxidation, 62-63
Scales, 26-28, 280-281
And viscosity, 61-62
Theoretical displacement, 101
Thermocouples, 281-282
Thermometer, 281
Threaded construction, cylinders, 185
Threaded fittings, 90-91
Three-position, four-way valves, 158
Three-way valves, 156
Tie-rod construction, cylinders, 185
Time
Circuit cycle factor, 220-222
Measurement, 280
Standard second, 265
Torque
Defined, 48, 196, 278
Fluid motors, 196-199
Hydraulic, 48
Measuring, 278-282
Mechanical, 48
Motor, 196
Torricelli's theorem, 39-40
Transducers, 265, 268, 300
Digital, 285
Efficiency, 269
Electric signals generated by, 282-284
Optical, 285-286
Piezoelectric, 276
Pneumatic, 287-288
Variable inductance, 289
Variable reluctance, 276

Trigger, 300
Trunnion mounts, cylinders, 189
Truth table, 300
Tube fittings, 91
Turbine, 205
Turbulence amplifier, 300, 323-324
Turbulent flow, 24, 40, 42, 294, 323
Two-position valves, 157
Two-way valves, 155

U

Ultrasonic vibrations, and flow velocity, 294
Unloading circuits, 215, 232-233
Unloading valves, 147-148, 232-233
U.S. Army Harry Diamond Laboratory, 301, 304

V

Valve(s)
Defined, 142
Directional-control, 151-164
Flow-control, 164-168
Position, 142
Pressure-control, 143-151
Servo system, 162-164
Two-position, 157
Two-way, 155
Types, 142-143
Variable fluid resistor as, 329
Valve actuator, 142
Valve control circuit, 217
Vane motors, 198, 200-202
Pneumatic, 204-206
Slippage, 201
Vane pumps
Applications, 117
Balanced/unbalanced, 113
In combination, 115-117
Delivery, 114
Operation of, 112-113
Vane actuator, 209
Variable-delivery pumps, 101
Vane pump as, 114
Variable-displacement motor, 195
Variable-displacement pump, 98
Inline axial-piston pump as, 119
Variable overrunning load, 219
Variable resistive load, 219
Variable transformers, 284-285
Velocity
Controlled output, 251
Of a cylinder, 177-181
Defined, 24, 52

Measurement, 288, 294
Tangential, 105, 323
Vent, 300, 309, 323
Venturi meter, 291
Viscosimeter, 59
Viscosity
Absolute, 23, 40–41
Defining, 24, 57, 59
Kinematic, 24
Measuring, 59, 290
Temperature and, 61–62
Viscosity index (VI), 57, 61–62
Volume-control valves, defined, 142
Volumetric efficiency, 98, 103, 127–128
Compressors, 131, 133
Gear pumps, 112
Vortex amplifiers, 300, 322–323

W X Y Z

Wall-attachment amplifier, 300, 309
Wall-attachment effect (Coanda effect), 299, 304, 308, 313, 327
Water
Density of, 26
Separation, 64–65
Temperature scales and, 280–281
Use of for performing work, 4–5, 6, 66
Water-base fluids, 66
Water-glycol fluids, 67
Water-in-oil emulsions, 66–67
Watt, James, 49
Weight density, 25
Welded connectors, 92
Westinghouse, George, 6
Wind, use of for performing work, 4, 6
Work, 1, 45
Energy distinguished from, 46
Young, Thomas, 303